W9-CID-606

SCIENCE READING ROOM

SCIENCE READING ROOM

SCIENCE READING ROOM

SCIENCE READING ROOM

Robust Estimates
of
Location

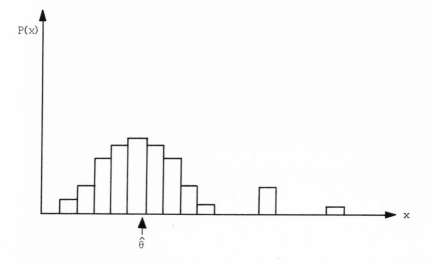

Robust Estimates of Location

of

Location

Survey and Advances

D. F. ANDREWS

P. J. BICKEL

F. R. HAMPEL

P. J. HUBER

W. H. ROGERS

J. W. TUKEY

Princeton University Press

Princeton, New Jersey, 1972

QA
276
.8
.R6

Copyright © 1972, by Princeton University Press

All rights reserved. No part of this book may be reproduced in any form or by any electronic or mechanical means including information storage and retrieval systems without permission in writing from the publisher, except by a reviewer who may quote brief passages in a review.

LC Card: 72-39019

ISBN: 0-691-08113-1 (hard cover edition)
ISBN: 0-691-08116-6 (paperback edition)

AMS 1970: primary, 62G35
secondary, 62C05, 62F10, 62G05

Printed in the United States of America
by Princeton University Press, Princeton, New Jersey

PREFACE

Since John Tukey is one of the originators of robust estimation, I invited other workers in the field to visit Princeton during the 1970-71 academic year. It was hoped that the year-long seminar would culminate in a conference dedicated to the memory of Samuel S. Wilks. The nature of the work being done suggested that a written summary would be more appropriate. This book follows Harald Cramér's as our second tribute to Wilks.

The visitors were Peter J. Huber, Swiss Federal Institute of Technology (Eidg. Technische Hochschule), Zurich, Peter Bickel, University of California, Berkeley, and Frank Hampel, University of Zurich. Not only Princeton staff and students, but also colleagues from other universities and Bell Telephone Laboratories, contributed to the seminar. It began by the participants outlining past, present, and future research. It was soon apparent that there were quite different opinions held concerning what were the important and relevant problems. The term robust clearly had many interpretations, and the intersection of these seemed void. Much vigorous discussion ensued but little unanimity was reached. In order to clarify some issues and to take advantage of the enthusiasm and diversity of the participants, a comparative survey was started after Christmas at the suggestion of David Andrews.

The organization of this project is outlined in the next paragraphs. This outline may be of some interest for understanding the structure of the following report and our experience may be useful in the planning of similar projects.

Any group endeavor requires the participation of all the members and for this, some agreement on goals and objectives is required. The Fall seminar had shown conclusively that no consensus of all the members could be obtained on central issues. Thus in the initial proposed survey for assessing robust estimates of location, each seminar member wrote down one estimate, one property deemed relevant and one set of conditions under which this property would be assessed. As will be evident from the chapters that follow, this plan was but a catalyst for a project which grew vigorously in the next months. The important result of the plan was immediate submission of contributions or entries and an immediate start on the work.

The seminar continued to meet to discuss methods for evaluating properties and in the latter stages to discuss the results and the methods for summarizing these.

v

The project had no official organizer and no methods of encouraging participation other than the interest in the results themselves.

Toward the end of May, results were collected and the task of writing these up was partitioned among the group. Drafts of each section were circulated for comments or additions, and a draft appeared in July. Andrews, Rogers and Tukey sustained most of this preparation.

A large number of people contributed to this survey. The central group was Andrews, Bickel, Hampel, Huber, Rogers and Tukey. William Rogers, a Princeton undergraduate, experimented with and planned the Monte Carlo calculations. He was also in charge of the extensive computing in which he was assisted by William Eddy, another undergraduate and by two graduate students, Dennis Ash and Gunnar Gruvaeus. A host of other statisticians helped, through discussions, to shape the project -- to name a few, Louis Jaeckel, Vernon Johns, Colin Mallows, Richard Olshen, L. J. Savage, and Leonard Steinberg. Linda Andrews helped in the preparation of several drafts. Finally the production of the book owes much to the efforts of Diana Webster, Linda Pelech and Mary Bittrich.

The research in this book was supported in many ways, for all of which we are very grateful. The extensive computing was done at the Princeton University Computer Center which has been assisted by grants from the National Science Foundation, NSF-GJ-34, NSF-GW-3157. Computing, programming charges, and salary components were paid in part by contracts to Princeton University from Army Research Office (Durham) DA-31-124-ARO-(D)-215 and from the Office of Naval Research N00014-67A-0151-0017. Other salary sources were Princeton University, University of California, Guggenheim Foundation, Princeton University Higgins Fund, National Science Foundation, Bell Telephone Laboratories and the University of Zurich.

By individuals, the support was:

Andrews (BTL, A.R.O.(D))	Hampel (ONR, Higgins, Zurich)
Ash (ONR)	Huber (N.S.F., ONR, Higgins)
Bickel (U.C., Guggenheim, ONR)	Rogers (ONR, A.R.O.(D))
	Tukey (A.R.O.(D), BTL)
Eddy (ONR)	
Gruvaeus (A.R.O.(D))	

G. S. Watson, Princeton August 1971

TABLE OF CONTENTS

1. INTRODUCTION

Estimation is the art of inferring information about some unknown quantity on the basis of available data. Typically an estimator of some sort is used. The estimator is chosen to perform well under the conditions that are assumed to underly the data. Since these conditions are never known exactly, estimators must be chosen which are robust, which perform well under a variety of underlying conditions.

The theory of robust estimation is based on the specified properties of specified estimators under specified conditions. This book is the result of a large study designed to investigate the interaction of these three components over as wide a range as possible. The results are comprehensive, not exhaustive.

The literature on robust estimation is a varied mixture of theoretical and practical proposals. Some comparative studies are available concentrating primarily on asymptotic theory, including the papers of Huber (1964), Bickel (1965), Gastwirth (1966), Crow and Siddiqui (1967), and Birnbaum and Laska (1967). In addition, various useful small sample studies have been made, based either on Monte Carlo methods or exact moments of order statistics. Among the former we include Leone, Jayachandran, and Eisenstat (1967) and Takeuchi (1969) while the latter include Gastwirth and Cohen (1970) and Filliben (1969).

As do these authors, we address ourselves to the problem of point estimation of location. The significant advantages of this study are size and scope.

The aims of this survey are, by bringing together a wide experience with a variety of concepts, to
 (i)   augment the known properties of robust estimators
(ii)   develop and study new estimates.

A great variety of some 68 estimates were studied, some well known, others developed during this study; some hand-computable, others requiring a great deal of computer time; some suitable for sample size 5, others best suited to sample sizes over 100. In particular, trimmed means, adaptive estimates, Huber estimates, skipped procedures and the Hodges-Lehmann estimate are among the estimates considered. The estimates are outlined in Chapter 2. Computer programs used to implement these procedures are given in Appendix 11.

The diversity of the survey extends to the variety of properties of both large and small samples which were investigated. Variances of the estimates were calculated

1

for samples of 5, 10, 20 and 40. Percentage points were calculated. Breakdown bounds were studied. Hampel's (1968) influence, or Tukey's (1970) sensitivity, curves were calculated to study the dependence of estimators on isolated parts of the data. The ease of computation was investigated. A wide variety of sampling situations were considered, some designed deliberately to test weak points of the procedures considered.

Several new asymptotic results were established to evaluate variances and influence curves. While these were not found for all the estimates in this study, a large number are recorded in Chapter 3. Of particular interest are the formulae for skipped procedures and the unpleasant features of the "shorth".

In the course of this study, W. H. Rogers developed Monte Carlo techniques for sampling from a family of distributions proposed by J. W. Tukey (scale mixtures of zero mean Gaussians) which yielded very great accuracy with a moderate number of samples. The theory used in this study is reported in Chapter 4.

In Chapter 5, numerical results are given for a variety of characteristics. In Chapter 6, these results are analyzed in detail, introducing new methods for the summarization of such a large body of data. J. W. Tukey shouldered this responsibility. Brief additional summaries reflecting the varied experience of some of the contributors are included in Chapter 7.

In the appendices, computer programs for the estimates are given and some material related to the study of robust procedures is included. Appendix 11, as mentioned above, contains the estimate programs used in this study. Appendix 12 contains programs and details for the random number generation of the study. New methods for displaying and assessing variances were developed. The general theory and details relevant to particular distributions are given in Appendix 13. Various numerical integration and Monte Carlo sampling schemes were considered for use in this project. Some considerations and resulting developments are included in Appendix 14. Finally in Appendix 15 are outlined some possible extensions for more efficient Monte Carlo methods.

The results of this study are too diverse to summarize at all completely. We can, however, offer the following guidance for a quick exploration, whether or not this is to be followed by careful study: read Chapter 2 as far as you are strongly interested, then do the same for Chapter 3. Read sections 5E and 5F, then sections 6A, 6B, 6C, continuing as far in six as strongly interested. Begin Chapter 7, continuing as long as interested.

2. ESTIMATES

2. ESTIMATES

2A INTRODUCTION AND INDEX OF ESTIMATES

A great variety of some 68 estimates have been
included in this study. All of these have the property
that they estimate the center of a symmetric distribution
(assured by constraint (ii) below). Indeed some were
designed to optimize some aspect of this estimation procedure. Other estimates were designed under different conditions or with different objectives in mind. The only
constraints on the estimates have been:
  (i)   all estimates must be represented by computable
        algorithms,
  (ii) all estimates must be location and scale invariant
        in the sense that if each datum is transformed by
        $x \rightarrow ax + b$ then the estimate $T = T(x)$ is simi-
        larly transformed by $T \rightarrow aT + b$.
     In this chapter we give brief descriptions of the
estimates $T(\underset{\sim}{x})$ in terms of $x_{(1)} \leq x_{(2)} \leq \cdots \leq x_{(n)}$,
the order statistics of a sample $\underset{\sim}{x} = (x_1, \ldots, x_n)$. The
estimates are grouped according to type, e.g., trimmed
means, M-estimates etc. with cross references being given
at the end of each section.
     The descriptions given are not to be taken too liter-
ally for the more complicated procedures. The reader who
desires certainty rather than a very close approximation
should refer to the program as given in Appendix 11.
     In those cases in which the originator of an estimate
is mentioned without a reference it is to be assumed that
the procedures were proposed in the course of this study.
     To each estimate there corresponds a number and a mne-
monic. For example, the mean, estimate 1, is denoted by
M, the median, estimate 6, is denoted by 50%. The number
corresponds roughly to the order of appearance in this
chapter and in the tables. The mnemonics will be introduced as we go along. Exhibit 2-1 gives an index of
the estimates indicating where descriptions may be found.
The index is ordered by the mnemonic of the estimate using
an extension in which the alphabet is prefaced by the digits
and some other symbols.
     Some estimates form families: the "trims" are 5% to
50%, the "hubers" are H07 to H20, the "hampels" are 12A to
25A and ADA, the "sitsteps" are A,B,C.

4

Code	Number	Brief description	Principal reference
5%	2	5% symmetrically trimmed mean	2B1
10%	3	10% symmetrically trimmed mean	2B1
12A	33	M-estimate, ψ bends at 1.2,3.5,8.0	2C3
15%	4	15% symmetrically trimmed mean	2B1
17A	32	M-estimate, ψ bends at 1.7,3.4,8.5	2C3
21A	31	M-estimate, ψ bends at 2.1,4.0,8.2	2C3
22A	30	M-estimate, ψ bends at 2.2,3.7,5.9	2C3
25%	5	25% symmetrically trimmed mean	2B1
25A	29	M-estimate, ψ bends at 2.5,4.5,9.5	2C3
2RM	59	2-times folded median	
33T	43	Multiply-skipped trimean	2D2
3R1	62	3-times folded median trimming 1	2E2
3R2	63	3-times folded median trimming 2	2E2
3RM	60	3-times folded median	2E2
3T0	44	Multiply-skipped mean, 3k deleted	2D2
3T1	45	Multiply-skipped mean, max(3k,2) deleted	2D2
4RM	61	4-times folded median	2E2
50%	6	Median (or 50% symmetr. trimmed mean)	2B1
5T1	46	Multiply-skipped mean, max(5k,2) deleted	2D2
5T4	47	Multiply-skipped mean, ", $\leq$ .6N deleted	2D2
A15	26	Huber M-estimate, k = 1.5, robust scaling	2C1
A20	25	Huber M-estimate, k = 2.0, robust scaling	2C1
ADA	34	Adaptive M-estimate, ψ bends at ADA, 4.5,8.0	2C3
AMT	35	M-estimate, ψ is sin function	2C3
BH	58	Bickel-Hodges estimate	2E2
BIC	10	Bickel modified adaptive trimmed mean	2B3
CML	39	Cauchy maximum likelihood	2C3
CPL	64	Cauchy-Pitman (location only)	2E3
CST	42	Iteratively C-skipped trimean	2D1
CTS	48	CTS-skipped trimean	2D1
D07	24	One-step Huber, k = 0.7, start = median	2C2
D10	23	One-step Huber, k = 1.0, start = median	2C2
D15	22	One-step Huber, k = 1.5, start = median	2C2
D20	21	One-step Huber, k = 2.0, start = median	2C2
DFA	50	Maximum estimated likelihood	2E1

5

Code	Number	Brief description	Principal reference
GAS	7	Gastwirth's estimate	2B2
H07	19	Huber proposal 2, k = 0.7	2C1
H10	18	Huber proposal 2, k = 1.0	2C1
H12	17	Huber proposal 2, k = 1.2	2C1
H15	16	Huber proposal 2, k = 1.5	2C1
H17	15	Huber proposal 2, k = 1.7	2C1
H20	14	Huber proposal 2, k = 2.0	2C1
H/L	57	Hodges-Lehmann estimate	2E2
HGL	53	Hogg 69, based on kurtosis	2E1
HGP	52	Hogg 67, based on kurtosis	2E1
HMD	28	M-estimate, ψ bends at 2.0,2.0,5.5	2C3
JAE	9	Adaptive trimmed mean (Jaeckel)	2B3
JBT	12	Restricted adaptive trimmed mean	2B3
JLJ	13	Adaptive linear combination of trimmed means	2B3
JØH	51	Johns' adaptive estimate	2E1
JWT	56	Adaptive form based on skipping	2E3
LJS	40	Least favorable distribution	2C3
M	1	Mean	2B1
M15	20	One-step Huber, k = 1.5, start = mean	2C2
MEL	38	Mean likelihood	2C3
OLS	36	"Olshen's estimate"	2C3
P15	27	One-step Huber, k = 1.5, start = median, robust scale	2C2
SHO	65	Shorth (for shortest half)	2E3
SJA	11	Symmetrized adaptive trimmed mean	2B3
SST	41	Iteratively s-skipped trimean	2D1
TAK	49	Takeuchi's adaptive estimate	2E1
THL	55	T-skipped Hogg 69	2E1
THP	54	T-skipped Hogg 67	2E2
TOL	37	T-skipped Olshen	2C3
TRI	8	Trimean	2B2
A	66	.75(5T4) + .25(D20)	2E3
B	67	.50(5T4) + .50(D20)	2E3
C	68	.25(5T4) + .75(D20)	2E3

NOTE: For codes arranged by number, see, for example, Exhibit 5-4A.

Many common robust estimates are formed by taking linear combinations of order statistics:

$$T = a_1 x_{(1)} + \cdots + a_n x_{(n)}.$$

The condition of location invariance constrains the a_i to satisfy

$$\Sigma\, a_i = 1 \; .$$

In general these estimates are relatively easy to compute by hand from ordered data. (The properties of such estimates are easily assessed analytically.) The asymptotic theory of these procedures is well known (cf. Chernoff, Gastwirth, Johns (1967), for example) and some small sample properties such as moments are also fairly readily computable (cf. Sarhan and Greenberg (1962)).

2B1   Trimmed Means, Means, Midmeans and Medians

The arithmetic mean is a simple, well understood estimate of location. However it is highly non-robust being very sensitive to extreme outliers. One simple way to make the arithmetic mean insensitive to extreme points is first to delete or 'trim' a proportion of the data from each end and then to calculate the arithmetic mean of the remaining numbers. These trimmed means form a family indexed by α, the proportion of the sample size removed from each end.

If α is a multiple of $1/n$, an integral number of points are deleted from each end and the trimmed mean is a simple average of the remaining points. If α is not a multiple of $1/n$, $[\alpha n]$ points are removed at each end, the largest and smallest remaining points are given a weight $p = 1 + [\alpha n] - \alpha n$. A weighted mean is then taken. Thus,

$$T(\underset{\sim}{x}) = \frac{p x_{([\alpha n+1])} + x_{([\alpha n+2])} + \cdots + p x_{(n-[\alpha n])}}{n(1 - 2\alpha)} \; .$$

In this study trimming proportions $\alpha = 0, (.05), .50$ were used although results for all of these are not reported. The 0 trimmed mean is the usual sample mean, the 0.25 trimmed mean is the midmean. The 0.50 trimmed mean is the median. With the exception of the mean these estimates are defined by their trimming proportion as below.

7

The estimates of this type included in the study are:

1	M	Mean
2	5%	5% symmetrically trimmed mean
3	10%	10% symmetrically trimmed mean
4	15%	15% symmetrically trimmed mean
5	25%	25% symmetrically trimmed mean
6	50%	50% median

Related estimates described elsewhere are:

9	JAE	Adaptive trimmed mean (Jaeckel)
10	BIC	Bickel modified adaptive trimmed mean
11	SJA	Symmetrized adaptive trimmed mean
12	JBT	Restricted adaptive trimmed mean
13	JLJ	Adaptive linear combination of trimmed means
51	JØH	Johns/ adaptive estimate

2B2   Linear Combinations of Selected Order Statistics

Mosteller (1947) introduced a class of estimates which are simple linear combinations of a small number of order statistics. Of this type and in addition to the median mentioned above, we include an estimate proposed by Gastwirth (1966) involving only the median and the upper and lower tertiles defined by,

$$T(\underset{\sim}{x}) = 0.3x_{[\frac{n}{3}+1]} + 0.4(50\%) + 0.3x_{(n-[\frac{n}{3}])}$$

and denoted GAS.

The trimean is another estimate of this form defined in terms of the median and the hinges h_1 and h_2 which are approximately sample quartiles (see 2D1 and Appendix 11). This estimate is defined by

$$T(\underset{\sim}{x}) = (h_1 + 2(50\%) + h_2)/4$$

and is denoted by TRI.

The estimates of this form included in the study are:

7	GAS	Gastwirth's estimate
8	TRI	Trimean

Related estimates discussed in other sections are:

6	50%	Median
41	SST	Iteratively s-skipped trimean
42	CST	Iteratively c-skipped trimean
43	33T	Multiple skipped trimean
48	CTS	CTS skipped trimean

2B3 Adaptive Trimmed Means

Jaeckel (1971) proposed a trimmed mean when the trimming proportion, α, was chosen to minimize an estimate of the asymptotic variance of the estimate, T. If the underlying distribution F is symmetric, this asymptotic variance may be estimated by

$$\hat{A}(\alpha) = \frac{1}{(1-2\alpha)^2} \left\{ \sum_{j=\alpha n+1}^{n-\alpha n} (x_{(j)} - T_\alpha)^2 + \alpha(x_{(\alpha n+1)} - T_\alpha)^2 + \alpha(x_{(n-\alpha n)} - T_\alpha)^2 \right\}$$

for αn an integer (see Chapter 3).

The estimation procedure involves choosing the trimming proportion in the range $0 \leq \alpha \leq .25$, to minimize this quantity for αn integral. This estimate is denoted by JAE.

A number of modifications of Jaeckel's adaptive trimmed mean are considered.

(a) Bickel modified the original proposal in two ways. The first modification is that the estimate of the variance of the trimmed mean is based on a pseudo sample of size 2n at each stage. This pseudo sample is obtained by forming residuals from the median and augmenting the sample by the twenty negatives of these numbers. The estimated variance is now proportional to the sample Winsorized variance of the pseudo residuals. The range $0 \leq \alpha \leq .25$ is still considered but now all α such that $2\alpha n$ is an integer are tested. The second modification is to bias the trimming proportion down when it is suspected that the choice of optimal trimming proportion has been dictated by a wild fluctuation of the Winsorized variance.

If only the first modification is carried out, the resulting estimate is referred to as the symmetrized adaptive trimmed mean and is denoted SJA. When both modifications are carried out, the resulting procedure is identified as BIC.

9

(b) A modification of JAE was proposed which would restrict to two the trimmed means considered. This esti-mate is obtained by calculating $\hat{A}([\frac{n}{12}]/n)$ and $\hat{A}([\frac{n}{4}]/n)$ and then using the trimmed mean having the smaller of these two (estimated) variances.
    This estimate is denoted JBT.

(c) Another modification of the adaptive trimmed mean was proposed by Jaeckel in which a linear combination of two trimmed means is chosen. The estimate is given by

$$T(\underset{\sim}{x}) = c(\underset{\sim}{x})(5\%) + (1 - c(\underset{\sim}{x}))(25\%)$$

with

$$c(\underset{\sim}{x}) = \frac{(V_{12} - V_{22})}{(V_{11} - 2V_{12} + V_{22})}$$

where

$$V_{11} = \hat{\hat{A}}\left(\frac{[.05n]}{n}\right)$$

$$V_{12} = \hat{B}\left(\frac{[.05n]}{n}, \frac{[.25n]}{n}\right)$$

$$V_{22} = \hat{\hat{A}}\left(\frac{[.25n]}{n}\right)$$

and for $n = 2k$,

$$\hat{\hat{A}}(\alpha) = \frac{1}{(1-2\alpha)^2}\left\{\sum_{j=k+1}^{n(1-\alpha)}(x_{(j)}-x_{(n-j+1)})^2 + \alpha(x_{(n-\alpha n)}-x_{(\alpha n+1)})^2\right\}$$

$$\hat{B}(\alpha_1,\alpha_2) = \frac{1}{(1-2\alpha_1)(1-2\alpha_2)}\left\{\sum_{j=k+1}^{n(1-\alpha_2)}(x_{(j)}-x_{(n-j+1)})^2\right.$$

$$+ \sum_{j=n(1-\alpha_2)}^{n(1-\alpha_1)}(x_{(j)}-x_{(n-j+1)})(x_{(n(1-\alpha_2))}-x_{(\alpha_2 n+1)})^2$$

$$\left. + \alpha_1(x_{(n(1-\alpha_2))}-x_{(\alpha_2 n+1)})(x_{(n(1-\alpha_1))}-x_{(\alpha_1 n+1)})\right\}$$

with $\alpha_1 < \alpha_2$, estimating the optimum choice where optimum is defined in terms of asymptotic variances and covariances.

This estimate is denoted by JLJ. Unfortunately only the version for $n = 20$ was programmed.

The estimates dealt with in this section are:

9	JAE	Adaptive trimmed mean (Jaeckel)
10	BIC	Bickel modified adaptive trimmed mean
11	SJA	Symmetrized adaptive trimmed mean
12	JBT	Restricted adaptive trimmed mean
13	JLJ	Adaptive linear combination of trimmed means

Related estimates discussed in other sections are:

51	JØH	Johns' adaptive estimate

2C M-ESTIMATES

In 'regular cases', likelihood estimation of the location and scale parameters θ, σ of a sample from a population with known shape leads to equations of the form

$$\sum_{j=1}^{n} [-f'(z_j)/f(z_j)] = 0 \;,$$

and

$$\sum_{j=1}^{n} [z_j f'(z_j)/f(z_j) - 1] = 0 \;,$$

where f is the density function and $z_j = (x_{(j)}-\theta)/\sigma$.

More generally M-estimates of location are solutions, T, of an equation of the form

$$\sum_{j=1}^{n} \psi\left(\frac{x_{(j)}-T}{s}\right) = 0$$

where ψ is an odd function and s is either estimated independently or simultaneously from an equation of the

11

form

$$\sum_{j=1}^{n} \chi\left(\frac{x_{(j)} - T}{s}\right) = 0 .$$

2C1 Huber Proposal 2

Huber ((1964), page 6) proposed a family of estimates characterized by a function ψ of the form

(a)
$$\psi(x;k) = \begin{cases} -k & x < k \\ x & -k < x < k \\ k & k < x \end{cases}$$

and the function

(b)
$$\chi(x) = \psi^2(x;k) - \beta(k)$$

where

$$\beta(k) = \int \psi(x;k)^2 \, \Phi(dx)$$

These equations (a) and (b) were then solved simultaneously for s and T. The equations were solved iteratively starting with the median and the interquartile range/1.35 as initial values for T and s. The mnemonic for these procedures is Huv. Hampel proposed using ψ as above but with s estimated by $(\text{med}|x_{(i)} - 50\%|)/.6754$. The mnemonic used is Auv. The digits following A and H indicate the value of k.

The estimates of this form included in this study are:

14	H20	Huber proposal 2,k = 2.0
15	H17	Huber proposal 2,k = 1.7
16	H15	Huber proposal 2,k = 1.5
17	H12	Huber proposal 2,k = 1.2
18	H10	Huber proposal 2,k = 1.0
19	H07	Huber proposal 2,k = 0.7
25	A20	Huber M-estimate,k = 2.0,robust scale
26	A15	Huber M-estimate,k = 1.5,robust scale

Related estimates are found in sections

2C2 and 2C3.

2C2 One Step Estimates

These are simple procedures proposed in this context by Bickel (1971) but dating back to LeCam (1956), Neyman (1949) and Fisher which are asymptotically equivalent to estimates of the type proposed by Huber. They are in fact first Gauss-Newton approximations to Huber estimates for fixed scale. To specify one of these estimates we need a parameter k, a preliminary estimate $\hat{\theta}$ (the mean or the median) and a robust scale estimate s (the interquartile range divided by its expected value under normality). The residuals from $\hat{\theta}$ are calculated. All observations (bad) whose residuals exceed ks in absolute value are replaced by ks times the sign of the residual; all other observations (good) are left alone. The estimate is then the sum of the numbers so obtained divided by the total number of good observations. If all observations are bad, the median is used.

These estimates are capable of hand calculation from sorted data.

The estimates of this form included in this study are,

20	M15	One-step Huber, k = 1.5, start = mean
21	D20	One-step Huber, k = 2.0, start = median
22	D15	One-step Huber, k = 1.5, start = median
23	D10	One-step Huber, k = 1.0, start = median
24	D07	One-step Huber, k = 0.7, start = median
27	P15	One-step Huber, k = 1.5, start = median but using a multiple of the median absolute deviation from the median as the estimate of scale, s.

Related estimates are found in sections

2C1 and 2C3.

2C3 Other M-estimates

(a) Independent scale, piecewise linear M-estimates

Hampel introduced a 3 parameter family of M-esti-mates involving ψ functions whose graphs are line seg-ments.

$$\psi(x;abc) = \text{sgn } x \cdot \begin{cases} |x| & 0 \leq |x| < a \\ a & a \leq |x| < b \\ \dfrac{c-|x|}{c-b} a & b \leq |x| < c \\ 0 & |x| \geq c \end{cases}$$

The estimate T is found by solving

$$\Sigma \; \psi\left(\frac{x_{(i)}-T}{s_1}\right) = 0$$

where s_1 is the median of the absolute deviations from the median. The equation is solved iteratively, starting at the median. The estimates, indexed by 3 parameters, included in this study are:

		a	b	c
28	HMD	2.0	2.0	5.5
29	25A	2.5	4.5	9.5
30	22A	2.2	3.7	5.9
31	21A	2.1	4.0	8.2
32	17A	1.7	3.4	8.5
33	12A	1.2	3.5	8.0
34	ADA	ADA	4.5	8.0

The parameter ADA of this last estimate is a func-tion of the estimated tail length of the distribution. This measure of tail length is defined by

$$L = \underset{|x_i-(50\%)|>s_1}{\text{ave}} \{s/|x_i-(50\%)|\}$$

where s is the robust scale estimate defined previously. The constant ADA is defined by

14

$$ADA = \begin{cases} 1.0 & L \leq 0.44 \\ (75L - 25)/8 & 0.44 < L \leq 0.6 \\ 2.5 & 0.6 \leq L \end{cases}$$

(b) A sin function M-estimate

Andrews implemented suggestions by Jaeckel, Hampel and others in an M-estimate for which the function used is

$$\psi(x) = \begin{cases} \sin(x/2.1) & |x| < 2.1\pi \\ 0 & \text{otherwise} \end{cases} .$$

The equation

$$\Sigma \; \psi\left(\frac{x-T}{s_1}\right) \; = \; 0$$

may then be solved. (A closed form results from well known trigonometric identities if the 'end' effects are ignored.) The version implemented here used the median of the absolute deviations about the estimate T as the estimate of scale, s_1. This estimate was revised with every third iteration.
The estimate is denoted by AMT.

(c) "Olshen's" estimate

Tukey developed a suggestion of R. Olshen into the following estimate. The basic idea is that the unbiased minimum variance linear combination of 'independent' variables is proportional to $\Sigma \; x_i/\sigma_i^2$. Here, the variance σ_i^2 is estimated to be proportional to

$$\frac{R^2}{2} + (x_i - T)^2$$

where R is the interquartile range. The equation

$$T = \Sigma \; x_i \; \left/ \; \left(\frac{R^2}{2} + (x_i - T)^2\right)\right.$$

15

is solved iteratively starting at the median. This estimate is approximately a maximum likelihood estimate for
a t distribution with 2.5 degrees of freedom. Another
version of this estimate involves first 't-skipping'
(see section 2D1) the data and then applying this estimate.

The two estimates of this type are

OLS   Tukey's development of Olshen's suggestion
TOL   T-skipped version of the above.

(d) Mean likelihood estimate

Following an assumption implicit in Barnard (1962)
a mean likelihood estimate was developed to maximize
the mean likelihood

$$\Sigma \; \frac{1}{s} \; f\left(\frac{x-\mu}{s}\right)$$

where f is the standard normal density function and s
is the unbiased estimate of the variance

$$s^2 = \Sigma \; (x_i-\bar{x})^2/n-1 \; .$$

The estimate is an M-estimate with

$$\psi\left(\frac{x-\mu}{s}\right) \; = \; \frac{x-\mu}{\sqrt{2\pi} \, s} \; \exp\left\{-\frac{1}{2} \; \frac{(x-\mu)^2}{s^2}\right\} \; .$$

This estimate is denoted by MEL in this study.

(e) Maximum likelihood for Cauchy distribution

The two parameter Cauchy density function

$$f(x:\mu\sigma) = \frac{1}{\pi} \; \frac{\sigma}{\sigma^2 + (x-\mu)^2}$$

yields a likelihood function for μ and σ which may be
differentiated to yield two equations. These equations
were solved iteratively as follows

$$\mu_{j+1} = \frac{\sum\limits_{i=1}^{n} (x_i / [\sigma_j^2 + (x_i - \mu_j)^2])}{\sum\limits_{i=1}^{n} (1 / [\sigma_j^2 + (x_i - \mu_j)^2])}$$

and

$$\sigma_{j+1}^{1/2} = n / [2\sigma_j^{3/2} \Sigma(\sigma_j^2 + (x_i - \mu_j)^2)^{-1}] .$$

This estimate is denoted by CML.

(d) Maximum Likelihood for Least Favorable Distribution

Huber estimates are maximum likelihood estimates for a gross error model with density

$$f_k(x) = \begin{array}{ll} C_1 \exp\{-1/2(x-\theta)^2\} & |x-\theta| \le k \\ C_2 \exp\{-k|x-\theta|\} & |x-\theta| > k \end{array}$$

where C_1 and C_2 are chosen so that the density is continuous at $|x| = k$ and is normalized.

The parameters k, θ may be estimated simultaneously. The location estimate T is a Huber estimate for a particular k. This estimate was included after discussion with L. J. Savage. It is denoted by LJS.

Estimates of this form included in the study are:

28	HMD	M-estimate, ψ bends at 2.0,2.0,5.5
29	25A	M-estimate, ψ bends at 2.5,4.5,9.5
30	22A	M-estimate, ψ bends at 2.2,3.7,5.9
31	21A	M-estimate, ψ bends at 2.1,4.0,8.2
32	17A	M-estimate, ψ bends at 1.7,3.4,8.5
33	12A	M-estimate, ψ bends at 1.2,3.5,8.0
34	ADA	Adaptive M-estimate, ψ bends at ADA,4.5,8.0
35	AMT	M-estimate ψ is sin function
36	OLS	Olshen's estimate
37	TOL	T-skipped Olshen's estimate
38	MEL	Mean likelihood
39	CML	Cauchy maximum likelihood
40	LJS	Least favourable distribution

Related estimates are found in sections 2C1 and 2C2.

A class of estimates has been proposed by J. W. Tukey (1970) which involves simple iterative rejecting based on a simple scale estimate based on the central portion of the data. The estimates are amenable to hand calculation.

The scale estimate is based on the interquartile range, the difference between the value

$$
h_1 = \begin{cases} x_{\left(\left[\frac{n}{4}\right]\right)} & \text{n not a multiple of 4} \\[2ex] \frac{1}{2}\left(x_{\left(\frac{n}{4}\right)} + x_{\left(\frac{n}{4}+1\right)}\right) & \text{n a multiple of 4} \end{cases}
$$

and the value

$$
h_2 = \begin{cases} x_{\left(n+1-\left[\frac{n+3}{4}\right]\right)} & \text{n not a multiple of 4} \\[2ex] \frac{1}{2}\left(x_{\left(n+1-\frac{n}{4}\right)} + x_{\left(n-\frac{n}{4}\right)}\right) & \text{n a multiple of 4} \end{cases}
$$

These points are referred to as 'hinges' in Tukey (1970).

They may be used to define points further from the center of the data:

$$c_1 = h_1 - 2.0(h_2 - h_1)$$

$$c_2 = h_2 + 2.0(h_2 - h_1)$$

$$t_1 = h_1 - 1.5(h_2 - h_1)$$

$$t_2 = h_2 + 1.5(h_2 - h_1)$$

$$s_1 = h_1 - 1.0(h_2 - h_1)$$

$$s_2 = h_2 + 1.0(h_2 - h_1) \ .$$

A skipping procedure involves deleting or skipping those observations lying "outside" c_i, t_i, or s_i. The hinges are then recomputed and the procedure repeated until no further skipping is performed. A single estimate of the remaining points is then calculated. For singly-skipped estimates the skipping procedure is not iterated.

These calculations are relatively simple to perform once the data is ordered.

(a) Iteratively s-skipped trimean SST

The trimean of the sample remaining after iterative s-skipping is calculated.

(b) Iteratively c-skipped trimean CST

The trimean of the sample remaining after iterative c-skipping is calculated.

(c) CTS skipped trimean CTS

Single skipping first at the c-level, then at the t and s levels, respectively is performed. The trimean of the skipped sample is then calculated.

Estimates of this form included in this study are:

41   SST   Iteratively s-skipped trimean
42   CST   Iteratively c-skipped trimean
48   CTS   CTS skipped trimean

Related estimates discussed in other sections are:

37   TOL   T-skipped Olshen
54   THP   T-skipped Hogg 69
55   THL   T-skipped Hogg 67

and section 2D2.

2D2 Multiple Skipping

The skipping procedure is used to identify and delete extreme points. Many more points may be also deleted. The estimates considered here involve single

skipping at the t level and counting the number of
points deleted. A small multiple of this number may then
be further deleted from each end and the mean of the
remaining points calculated. These procedures also were
proposed by J. W. Tukey.

(a) Multiply-skipped mean, 3k deleted

   If single skipping at the t level deletes a total
of k points, a further k points are deleted from each
end and the mean of the remaining sample calculated is the
estimate (3T0).

(b) Multiply-skipped trimean, 3k deleted

   Only the final procedure in the above estimate is
changed. Rather than the mean we calculate the average
of the median and the extremes of the remaining sample
(33T).

(c) Multiply-skipped mean, max(3k,2) deleted

   The same procedure is followed as is outlined in (a)
except that at least 1 point is deleted from each end
after skipping (3T1).

(d) Multiply-skipped mean, max(5k,2) deleted

   If single t-skipping deletes a total of $k \geq 1$ points
a further max(2k,1) are deleted from each end and the
mean of the remaining points is calculated as the estimate
(5T1).

(e) Multiply-skipped mean, max(5k,2) but $\leq$ .6n deleted

   If single t-skipping deletes a total of k points
a further ℓ points are deleted from each end where

$$\ell = \min(\max(1,2k),.6n-k) \ .$$

The mean of the remaining sample is used as the estimate
(5T4).

The estimates of this form included in this study are:
43	33T	Multiply-skipped trimean
44	3T0	Multiply-skipped mean, 3k deleted
45	3T1	Multiply-skipped mean, max(3k,2) deleted
46	5T1	Multiply-skipped mean, max(5k,2) deleted
47	5T4	Multiply-skipped mean, ", $\leq$ .6n deleted

Related estimates are discussed in section 2D1.

2E1   Other Adaptive Estimates

Some of the estimates such as trimmed means are relatively inflexible; the form of the estimate is not affected by the data. Other estimates, such as the adaptive trimmed means are more fundamentally affected by the sample, though only through a small number of parameters.

Some estimates were designed explicitly to estimate many aspects of the underlying distribution and then to incorporate this information in the estimate of location.

(a)   Takeuchi's adaptive estimate

Takeuchi (1970) proposed an estimate which involved estimating the minimum variance unbiased linear combination of order statistics from a subsample of size k. The expected value of this linear combination under all permutations was then calculated as the estimate.

More precisely, let T_α be the expected value of the αth order statistic from a subsample of size k given $x_{(1)}, \ldots, x_{(n)}$. Then

$$T_\alpha = \sum_{i=1}^{n} P_\alpha^i x_{(i)}$$

where

$$P_\alpha^i = \begin{cases} {}_{i-1}C_{\alpha-1} \cdot {}_{n-i}C_{k-\alpha} / {}_nC_k & \alpha < i < n-k+\alpha \\ 0 & \text{otherwise} \end{cases}.$$

Then the covariance of the α and β order statistics may be estimated by

$$s_{\alpha\beta} = \sum_{i<j} P_{\alpha\beta}^{ij} x_{(i)} x_{(j)} - T_\alpha T_\beta \qquad \alpha < \beta$$

$$s_{\alpha\alpha} = \sum P_\alpha^i x_{(i)}^2 - T_\alpha^2$$

where

$$p_{\alpha\beta}^{ij} = \begin{cases} {}_{i-1}C_{\alpha-1}\ {}_{j-i-1}C_{\beta-\alpha-1}\ {}_{n-j}C_{k-\beta}\big/{}_{n}C_{k} & \alpha < i,\ \beta \le n-j+k \\ 0 & \text{otherwise} \end{cases}.$$

These estimated covariances may be symmetrized

$$\tilde{s}_{\alpha\beta} = (s_{\alpha\beta} + s_{\beta\alpha})/2 \ ,$$

and the inverse matrix calculated

$$\tilde{s}^{\alpha\beta} = \{\tilde{s}_{\alpha\beta}\}^{-1} \ .$$

The estimate is then

$$T = \Sigma\ \Sigma\ \tilde{s}^{\alpha\beta}T_{\alpha} \Big/ \Sigma\ \Sigma\ \tilde{s}^{\alpha\beta} \ .$$
$$\quad \alpha\ \beta \qquad\qquad \alpha\ \beta$$

The size of the subsample, k, may increase with n. In this study k was chosen by

$$k = [(37n/20)^{\frac{1}{2}}] \ .$$

The estimate is adaptive and as n increases is sensitive to more and more aspects of the distribution. The estimate is denoted TAK.

(b) A maximum likelihood estimate

A maximum likelihood estimate was developed by D. F. Andrews to (i) estimate the density function $f(x:\theta)$ and (ii) choose T to maximize the likelihood for this estimated density function. More precisely, for trial values of θ the sample is transformed to $|x-\theta|$ and r, the number of positive values of $(x-\theta)$. Using the symmetric assumption of $f(x-\theta)$, the binomial probability $P(r)$ of r out of n positive values is calculated. A simple algorithm was developed to find the density maximizing $\Pi\ f|x-\theta| = Q(\theta)$ among the class of densities monotone in $|x-\theta|$. The value of θ maximizing $L(\theta) = Q(\theta)P(r)$ is found and the estimate T is the mid-range of those values satisfying $L(\theta) > 0.1L(\hat{\theta})$. The estimate is denoted by DFA.

(c) An adaptive estimate due to Vernon Johns (1971)

This estimate can be written as a linear combination of two trimmed means or more readily in the form

$$T(\underset{\sim}{x}) = c_1^*(\underset{\sim}{x})T_1(\underset{\sim}{x}) + c_2^*(\underset{\sim}{x})T_2(\underset{\sim}{x})$$

where

$$T_1(\underset{\sim}{x}) = \frac{1}{2s} \sum_{j=r+1}^{r+s} [x_{(j)} + x_{(n-j+1)}]$$

$$T_2(\underset{\sim}{x}) = \frac{1}{n-2r-2s} \sum_{j=r+s+1}^{n/2} [x_{(j)} + x_{(n-j+1)}]$$

with c_1^*, c_2^* determined from the sample by,

$$c_1^*(\underset{\sim}{x}) + c_2^*(\underset{\sim}{x}) = 1$$

and

$$c_1^*(\underset{\sim}{x}) \propto \frac{1}{D_1^2} \left(\frac{2s(n+r+2s)}{(2r+s)(n-2r)} \right) - \frac{2}{D_1 D_2} \frac{n-2r-2s}{n-2r}$$

$$c_2^*(\underset{\sim}{x}) \propto \frac{1}{D_2^2} \left(\frac{2(n-2r-2s)}{n-2r} \right) - \frac{4}{D_1 D_2} \left(\frac{s}{n-2r} \right).$$

the D_i being proportional to the lengths of the segments of the sample entering in the T_i. Values of r and s are prescribed as $r = 1$, $s \doteq n/2 - r - s$, $n \le 20$. However the same rule was used in our program for all n. The full motivation behind this estimate was not available at the time this was written. However it was indicated that as the sample size increased the integers $1,\ldots,n/2$ would be partitioned into a slowly increasing number of successive intervals $r_1 < r_2 < \cdots < r_k < n/2$ $= r_{k+1}$, the statistics $S_1,\ldots,S_k$ where

$$S_j = \sum_{\ell=r_j}^{r_{j+1}} (x_{(\ell)} + x_{(n-\ell+1)})$$ would be found and an estimate

of the type $\sum_{j=1}^{k} c_j S_j$ would be used with c_j suitably

estimated from the sample. The (crucial) details will appear in a forthcoming paper by Johns. The estimate is denoted by JOH.

(d) Estimates based on sample kurtosis

Hogg (1967) suggested the choice among a set of 3 estimates, the choice to be made on the basis of the sample kurtosis

$$k = n\Sigma(x_i-\bar{x})^4/[\Sigma(x_i-\bar{x})^2]^2 \; .$$

A subsequent modification is also recorded below

Hogg 67 (HGP)	
k	estimate
k < 2	outer mean
2 < k < 4	mean
4 < k < 5.5	.25 trimmed mean
5.5 < k	median

Hogg 69 (HGL)	
k	estimate
k < 1.9	outer mean
1.9 < k < 3.1	mean
3.1 < k < 4.5	.25 trimmed mean
4.5 < k	Gastwirth

The outer mean referred to in the table is the mean of the upper and lower quarters of the data. These estimates may be applied to a skipped sample. In this study t-skipping was applied.

The two estimates of this form in this study are HGP and HGL. The t-skipped versions of these are THP and THL respectively.

Estimates of this form included in this study are:

49	TAK	Takeuchi's adaptive estimate
50	DFA	Maximum estimated likelihood
51	JOH	Johns' adaptive estimate
52	HGP	Hogg 67 based on kurtosis
53	HGL	Hogg 69 based on kurtosis
54	THP	T-skipped Hogg 67
55	THL	T-skipped Hogg 69

Related estimates discussed in other sections are:

discussed in section 2B3.

2E2   Folded Medians

The Hodges-Lehmann estimate is the median of all
pairwise means of the sample. For its rationale, see
Hodges and Lehmann(1963). The number of such pairwise
means, $_{\sim n}C_2$ can be prohibitively large for large samples.
The following estimates take means of only symmetrically
placed observations. There are only $\sim n/2$ of these
means.
A set of estimates was proposed by Tukey, designed
'for those who do not like to take means of more than two
numbers'. The calculation procedure is iterative but simple. At each stage the set of numbers is 'folded' by
taking means of symmetrically placed observations (in
an odd count sample the middle observation is folded with
itself):

$$x_{(1)} \cdots x_{(n)} \rightarrow \frac{x_{(1)}+x_{(n)}}{2}, \frac{x_{(2)}+x_{(n-1)}}{2}, \cdots .$$

The folded sample of $\left[\frac{n+1}{2}\right]$ points may then be sorted and
the process repeated.
After a fixed number of iterations the median of the
remaining numbers is taken. At each iteration of the
above procedure a fixed number of points may be trimmed
from the data before it is folded. The number of points
trimmed is given in the brief description.
One of this family had been studied. The Bickel-
Hodges estimate involves folding once without trimming
and taking the median of these numbers. For more details

see Bickel and Hodges (1967) and Hodges (1967).

The estimates of this form included in the study are:

57	H/L	Hodges-Lehmann
58	BH	Bickel-Hodges
59	2RM	2-times folded median
60	3RM	3-times folded median
61	4RM	4-times folded median
62	3R1	3-times folded median,trimming 1
63	3R2	3-times folded median,trimming 2

2E3 Other Estimates

(a) Pitman location estimate for the Cauchy distribution

The Pitman estimator of only the location of the Cauchy distribution was included for reference purposes. This estimate may be conveniently described as the Bayes estimate for the invariant prior: $d\theta$:

$$T = \frac{\int_{-\infty}^{\infty} \theta \prod_{i=1}^{n} f(x_{(i)} - \theta) \, d\theta}{\int_{-\infty}^{\infty} \prod_{i=1}^{n} f(x_{(i)} - \theta) \, d\theta}$$

where f is the standard Cauchy density function. It may be shown to be the minimum variance location invariant estimate (see Ferguson (1967)). The variance of this estimate is a (unsharp) lower bound to the variances of the estimates considered in this study (like others this bound may be violated (slightly) in a particular realization of a Monte Carlo simulation). The estimate is denoted by CPL.

(b) Shorth or Shortest half

This simple estimate is the mean of the shortest half of the data, i.e., $(x_{([\frac{n}{2}]+k)} - x_{(k)})$ is minimized as a function of k and the mean of $x_{(k^*)}, \ldots, x_{(k^*+[\frac{n}{2}])}$ is taken where k^* is the minimizing value. The estimate is denoted by SHO.

(c) A simple adaptive estimate with form depending on
     skipping.

     The amount of skipping gives an indication of what
type of estimate is best suited to the data. Analogous
to the Hogg estimates (HGP, HGL), Tukey proposed an estimate whose form depends on whether points fall outside the
c or s points (for definition, see section 2D1). The
details are summarized below.

Condition	Estimate
If points fall outside s	$\frac{1}{10}$(median × 4 + quantiles × 2 + "eights")
outside s but not outside c	trimean (TRI)
outside c	single c-skipped median

This estimate is denoted by JWT.

Estimates of this form included in the study are:

56   JWT   Adaptive form based on skipping
64   CPL   Cauchy Pitman location
65   SHO   Shorth (or shortest half)

2E4 Linear Combinations

     In addition to the properties of the individual
estimates, covariances were calculated between each pair
of estimates studies, as were values reflecting efficiencies for linear combinations of each pair of estimates
of the forms

$$\frac{5}{4}L - \frac{1}{4}M$$

$$L$$

$$\frac{3}{4}L + \frac{1}{4}M$$

$$\frac{1}{2}L + \frac{1}{2}M$$

$$\frac{1}{4}L + \frac{3}{4}M$$

$$M$$

$$-\frac{1}{4}L + \frac{5}{4}M$$

Thus 65 estimates generate some information for $65 + 5(65)(64)/2 = 10,465$ estimates.

Three of these are given special attention in the subsequent chapters. These are:

66	A	.75(5T4) + .25(D20)
67	B	.50(5T4) + .50(D20)
68	C	.25(5T4) + .75(D20)

3. ASYMPTOTIC CHARACTERISTICS OF THE ESTIMATES

3A   GENERAL REMARKS

In this section various asymptotic characteristics
of some of the procedures introduced in Chapter 2 will
be stated. Where the formulae are novel, a brief semi-
rigorous derivation will be given. In all other cases
references to the literature are provided.
The following notation will be used throughout. The
sample $x_1, \ldots, x_n$ is a set of independent and identically
distributed observations from a parent distribution F
which is assumed continuous. If F is absolutely continuous, its density is denoted by f. The *empirical
distribution* of the sample will be denoted by F_n and as
usual defined by

$$F_n(t) = \frac{1}{n} \sum_{i=1}^{n} I_{[x_i \leq t]} \; .$$

The *order statistics*, denoted by $x_{(1)} < \cdots < x_{(n)}$, are
the sample values in ascending order of magnitude.
If F is any distribution function, its inverse is
defined as usual by

$$F^{-1}(t) = \inf\{x: F(x) \geq t\}, \; 0 < t < 1 \; .$$

The *quantile function* F_n^{-1} is the inverse of the empirical distribution function.
All estimates for which asymptotic approximations
are derived can be thought of as functionals $T_n(F_n)$ of
the empirical distribution, where $T_n(\cdot)$ is defined on a
domain which includes the empirical distributions and the
true distribution F. In essentially all the cases, T_n
either is independent of n or has an asymptotically
equivalent version T which is independent of n. A
discussion of this point of view may be found in Hampel
(1971).
In all cases under discussion but one (the shorth)
under mild regularity conditions

$$\mathcal{L}(\sqrt{n}(T(F_n) - T(F))) \rightarrow N(0, A(F)) \; .$$

29

where A(F) is the *asymptotic variance* of $T(F_n)$. Since
all estimates under consideration are translation invariant we may without loss of generality suppose that our
target value is 0, so that T(F) measures the *asymptotic
bias*.

A useful concept discussed in Hampel (1968) is the
influence curve IC(x;F,T), or short IC(x), defined as
the von Mises derivative (cf. von Mises (1947)) of T
when it exists. Thus

$$\int IC(x;F,T) \ G(dx) = \lim_{\varepsilon \to 0} \frac{T((1-\varepsilon)F + \varepsilon G) - T(F)}{\varepsilon} \ .$$

In particular

$$IC(x;F,T) = \lim_{\varepsilon \to 0} \frac{T((1-\varepsilon)F + \varepsilon \delta_x) - T(F)}{\varepsilon}$$

where δ_x is the distribution function of a point mass
1 at x.

In many cases a "Taylor expansion" is valid; that
is

$$\sqrt{n} \, [T(F_n) - T(F) - \int IC(x;F,T)F_n(dx)] \to 0$$

in probability (see Filippova (1962)). Thus

(0) $$\qquad A(F) = \int [IC(x;F,T)]^2 F(dx) \ .$$

where F(dx) symbolizes integration according to the measure
corresponding to F(x). (The notations dF(x) and f(x)dx are
more classical equivalents.)

A caveat: The estimation procedures described in this
chapter are often only asymptotically equivalent to those
actually used, for which see Chapter 2, or, in critical
cases, the subroutines in Appendix 11. Using an asymptotically equivalent T, independent of n, was often
much more convenient.

The asymptotic results given in this chapter for more
than half of the 65 estimates are:

1) The form of T(·) asymptotically equivalent to
the description in Chapter 2.

2) An expression for the asymptotic variance $A(F)$.
3) An expression for the influence curve $IC(\cdot)$.
Exhibit 3-1 gives equation numbers for these expressions.

3B   LINEAR COMBINATIONS OF ORDER STATISTICS

3B1   Trimmed Means, Means, Midmeans and Medians

(a)   The trimmed mean (100α %)

The α-trimmed mean $(0 \le \alpha < 1/2)$ is defined by the functional

(1)
$$T_\alpha(F) = \frac{1}{1-2\alpha} \int_\alpha^{1-\alpha} F^{-1}(t)\ dt$$

$$= \frac{1}{1-2\alpha} \int_{F^{-1}(\alpha)}^{F^{-1}(1-\alpha)} x\ F(dx)\ .$$

Thus the estimate is given by

(2)
$$T(F_n) = \frac{1}{n(1-2\alpha)} \left\{ \sum_{[n\alpha]+2}^{[n(1-\alpha)]} X_{(k)} \right.$$
$$\left. + (1-n\alpha+[n\alpha])(X_{([n\alpha]+1)} + X_{([n(1-\alpha)]+1)}) \right\}\ .$$

The 0-trimmed mean is the *sample mean* (M).
As $\alpha \to 1/2$,

$$T(F) \to X_{(k+1)} \qquad\qquad \text{if}\ \ n = 2k + 1$$

$$T(F) \to \frac{1}{2}\left(X_{(k)} + X_{(k+1)}\right) \qquad \text{if}\ \ n = 2k,$$

i.e., the estimate we know as the *sample median* (50%).
For our purposes it is more convenient to work with the (asymptotically) equivalent definition of the median as

(3)
$$T(F) = F^{-1}\left(\frac{1}{2}\right)\ .$$

The influence curve of the α-trimmed mean is

Tag	No	Formula[1] for IC	Formula[2] for A(F)	Supplementary formulae	Discussed in section
GAS	7	(18)	(19)		3B2
HO7	19	(29),(31)*	←, (38)*	(34) to (37)	3C, 3C1
H10	18	(29),(31)*	←, (38)*	(34) to (37)	3C, 3C1
H12	17	(29),(31)*	←, (38)*	(34) to (37)	3C, 3C1
H15	16	(29),(31)*	←, (38)*	(34) to (37)	3C, 3C1
H17	15	(39),(31)*	←, (38)*	(34) to (37)	3C, 3C1
H20	14	(29),(31)*	←, (38)*	(34) to (37)	3C, 3C1
H/L	57	(65),(67)*	←, (68)*		3E1
HGL	53	(82)	←	(4),(18),(81),(83)	3E3
HGP	52	(82)	←	(4),(81),(83)	3E3
HMD	28	(31)*	←	(40),(41)	3C, 3C3(b)
JAE	9	(4)*	(7)*	(22)	3B3
JRT	12	(4)	(24a)	(24a)	3B3
JLJ	13	none	(24e)*		3B3
JOH	51	none			
JWT	56	none			
LJS	40				
M	1	(4)	(7)		3B1
M15	20	(31)*	←	(39)	3C, 3C2
MEL	38	(31)*	←	2C(d)	2C(d)
OLS	36	(31)*	←	(45),(45a)	3C, 3C3(e)
P15	77	(31)*	←	(40)	3C, 3C2
SHO	65	none			3E2
SJA	11	(4)*	(7)*	(23)	3B3
SST	41	(53),(59)*	←, (60)*	(46),(47), with p=1	
TAK	49	none			3E4
THL	55				
THP	54				
TOL	37				
TRI	8	(18)	(20)		3B2

FOOTNOTES:
1) *: this formula holds for symmetric F only
   none: not well defined (estimate usually not asymptotically normal)
   (blank): not given here.

2) Where IC is given, $A(F) = \int IC(x)^2 F(dx)$, but sometimes this can be implied.
   ←: compute A(F) from IC.

3) Values of parameters etc. for institution in IC, A(F).

$$IC(x) = \frac{1}{1-2\alpha} (F^{-1}(\alpha) - c(\alpha)) \qquad \text{for} \quad x < F^{-1}(\alpha)$$

$$(4) \qquad = \frac{1}{1-2\alpha} (x - c(\alpha)) \qquad \text{for} \quad F^{-1}(\alpha) \le x \le F^{-1}(1-\alpha)$$

$$= \frac{1}{1-2\alpha} (F^{-1}(1-\alpha) - c(\alpha)) \quad \text{for} \quad x > F^{-1}(1-\alpha),$$

where

$$(5) \qquad c(\alpha) = \int_{\alpha}^{1-\alpha} F^{-1}(t)\, dt + \alpha(F^{-1}(\alpha) + F^{-1}(1-\alpha))$$

$$= T(F) + \alpha(F^{-1}(\alpha) + F^{-1}(1-\alpha)) \ .$$

Note that $c(\alpha) = 0$ if F is symmetric:

$$(6) \qquad\qquad F(x) = \overline{F}(x) = 1 - F(-x) \ .$$

The asymptotic variance is

$$(7) \quad A(F) = \frac{1}{(1-2\alpha)^2} \left\{ \int_{F^{-1}(\alpha)}^{F^{-1}(1-\alpha)} (x-c(\alpha))^2 \, F(dx) \right.$$

$$\left. + \alpha((F^{-1}(\alpha)-c(\alpha))^2 + (F^{-1}(1-\alpha)-c(\alpha))^2) \right\} \ .$$

For the sample median,

$$(8) \qquad\qquad IC(x) = \frac{\operatorname{sign}\left(x - F^{-1}\left(\frac{1}{2}\right)\right)}{2f\left(F^{-1}\left(\frac{1}{2}\right)\right)}$$

and

$$(9) \qquad\qquad A(F) = \frac{1}{\left(2f\left(F^{-1}\left(\frac{1}{2}\right)\right)\right)^2} \ .$$

The variances are well known (cf. Bickel (1965) for example). The influence curves are derived in Hampel (1968). A simple approach is the following. The influence function $q(x;t)$ of the statistic $F^{-1}(t)$ (the t-quantile) is obtained by differentiating

$$F^{-1}_{x,\varepsilon}(t)$$

with respect to ε, at $\varepsilon = 0$, where

$$F_{x,\varepsilon} = (1-\varepsilon)F + \varepsilon\delta_x \ .$$

This gives

(10) $$q(x,t) = \frac{t - \delta_x(F^{-1}(t))}{f(F^{-1}(t))} \ .$$

Evidently, if M assigns mass 1 to $(0,1)$, and

(11) $$T(F) = \int_0^1 F^{-1}(t) \ M(dt) \ ,$$

then

(12) $$IC(x;F,T) = \int_0^1 q(x,t) \ M(dt) \ ,$$

which yields (4) and (8) in a straightforward way.

3B2 Linear Combinations Of Selected Order Statistics (GAS, TRI)

Define an estimate by the functional

(14) $$T(F) = \sum_1^m \beta_i F^{-1}(t_i)$$

with $0 < t_1 < t_2 < \cdots < t_m < 1$ and $\Sigma \beta_i = 1$.
    It follows at once from (10), (11), (12) that the influence curve has jumps of the size

(15)
$$\frac{\beta_i}{f(F^{-1}(t_i))}$$

at the points $F^{-1}(t_i)$ and is otherwise constant, thus

(16) $IC(x;F,T) = \underset{\{i:\ F^{-1}(t_i)<x\}}{\Sigma} \dfrac{\beta_i}{f(F^{-1}(t_i))} - c(F)$

with $c(F)$ such that $\int IC(x)\, F(dx) = 0$. The asymptotic variance then is

(17) $A(F) = \int IC(x)^2 F(dx)$.

Only two of this class of procedures, GAS, an estimate of Gastwirth's (1966), and TRI, Tukey's trimean, were used in the study. The Gastwirth estimate corresponds to $m = 3$, $(t_1,t_2,t_3) = (1/3,1/2,2/3)$, $(\beta_1,\beta_2,\beta_3) = (0.3,0.4,0.3)$, while the trimean corresponds to $m = 3$, $(t_1,t_2,t_3) = (1/4,1/2,3/4)$, $(\beta_1,\beta_2,\beta_3) = (1/4,1/2,1/4)$. The trimean also appears in skipped form, see below.

In these two cases, for symmetric F, the influence curves are

$IC(x;F,T) = \dfrac{\beta_2}{2f(0)}$ for $0 < x < F^{-1}(t_3)$

(18) $= \dfrac{\beta_2}{2f(0)} + \dfrac{\beta_3}{f(F^{-1}(t_3))}$ for $x > F^{-1}(t_3)$

          $= -IC(-x;F,T)$ for $x < 0$.

The variances are, for the Gastwirth estimate

(19) $A(F) = \dfrac{1}{3}\left(\dfrac{0.2}{f(0)}\right)^2 + \dfrac{2}{3}\left(\dfrac{0.2}{f(0)} + \dfrac{0.3}{f\left(F^{-1}\left(\frac{2}{3}\right)\right)}\right)^2$

and for the trimean

$$(20) \qquad A(F) = \frac{1}{32} \left\{ \left(\frac{1}{f(0)} \right)^2 + \left(\frac{1}{f(0)} + \frac{1}{f\left(F^{-1}\left(\frac{3}{4}\right)\right)} \right)^2 \right\} .$$

3B3 Adaptive Trimmed Means (JAE, BIC, SJA)

The procedure JAE is due to Jaeckel; the modification BIC proposed by Bickel is asymptotically equivalent. Here, for JAE,

$$(21) \qquad T(F) = \frac{1}{1-2\alpha(F)} \int_{\alpha(F)}^{1-\alpha(F)} F^{-1}(t) \; dt$$

where $\alpha(F)$ is that value of α which minimizes

$$
\begin{aligned}
A^*(F,\alpha) = \frac{1}{(1-2\alpha)^2} & \left\{ \int_{F^{-1}(\alpha)}^{F^{-1}(1-\alpha)} (x-T\alpha(F))^2 \; F(dx) \right. \\
(22) & \\
& \left. + \alpha[F^{-1}(\alpha)-T_\alpha(F)]^2 + \alpha[F^{-1}(1-\alpha)-T_\alpha(F)]^2 \right\} ,
\end{aligned}
$$

in the range $0 \leq \alpha \leq 1/4$, with $T_\alpha(F)$ given by (1).

Then, the limiting behavior of $T(F_n)$ coincides with that of $T_{\alpha(F)}(F_n)$, as given in (1) to (7). In particular, if F is symmetric, $A(F) = A^*(F,\alpha(F))$.

If the centering in (22) had been $c(\alpha)$ instead of $T_\alpha(F)$, the analogy with (7) and $A(F) = A^*(F,\alpha(F))$ would hold for the asymmetric case also. (In the symmetric case, $c(\alpha) = T_\alpha(F) = 0$.)

Bickel's modification (BIC) uses an $\tilde{\alpha}(F)$-trimmed mean where $\tilde{\alpha}(F)$ minimizes

$$(23) \quad B(F,\alpha) = \frac{1}{(1-2\alpha)^2} \left\{ \int_\alpha^{1-\alpha} [\tilde{F}^{-1}(t)]^2 dt + 2\alpha[\tilde{F}^{-1}(\alpha)]^2 \right\}$$

with $\tilde{F}$ being a symmetrized version of F:

$$(24) \qquad \tilde{F}(x) = \frac{1}{2}\left\{ F\left(x - F^{-1}\left(\tfrac{1}{2}\right)\right) + \bar{F}\left(x - F^{-1}\left(\tfrac{1}{2}\right)\right) \right\} .$$

The limiting behavior of $T(F_n)$ coincides with that of $T_{\tilde{\alpha}(F)}(F_n)$, as given in (1) to (7). For symmetric F, we have $\tilde{\alpha}(F) = \alpha(F)$. The estimate (SJA) differs from (BIC) in not using the rejection rule. It too is asymptotically equivalent to (JAE).

Note. The idea behind this modification was to smooth the sample cumulative by symmetrizing. In the actual calculation a further attempt was made to eliminate a poor choice of $\tilde{\alpha}(F)$ when the minimum fell into a "rough" part of $B(F_n, \cdot)$. The effect of this vanishes asymptotically.

The simplified adaptive trimmed means (JBT, JLJ)

A simplification (JBT) of Jaeckel's estimate uses the "best" of $T_{.08}$ and $T_{.25}$. Evidently

$$(24(a)) \qquad A(F) = \min\{A(F,.08), A(F,.25)\} .$$

Another modification due to Jaeckel (JLJ) uses the best linear combination of $T_{.05}$ and $T_{.25}$ with estimated weights. This estimate cannot be written in the form $T(F_n)$ and is more closely akin to the adaptive estimates discussed in section 2E1. If F is symmetric, it may be shown that this procedure behaves asymptotically like the statistic $T(F_n)$ corresponding to,

$$(24(b)) \qquad T(F) = c(F)T_{.05}(F) + (1 - c(F))T_{.25}(F)$$

where

$$(24(c)) \qquad c(F) = \frac{A(F,.25) - B(F,.05,.25)}{A(F,.05) - 2B(F,.05,.25) + A(F,.25)}$$

and where for $\alpha_1 < \alpha_2$, and symmetric F,

(24(d))

$$B(F,\alpha_1,\alpha_2) = \frac{2}{(1-2\alpha_1)(1-2\alpha_2)} \left[\int_0^{F^{-1}(1-\alpha_2)} t^2 dF(t) \right.$$

$$\left. + F^{-1}(1-\alpha_2) \int_{F^{-1}(1-\alpha_2)}^{F^{-1}(1-\alpha_1)} t \, dF(t) + \alpha_1 F^{-1}(1-\alpha_2)F^{-1}(1-\alpha_1) \right]$$

is the asymptotic covariance of T_{α_1}, T_{α_2}. This covariance is readily calculable from the influence curves. The variance of the estimate is given by

(24(e))

$$A(F) = c^2(F)A(F,.05) + 2c(F)(1-c(F))B(F,.05,.25)$$

$$+ (1-c(F))^2 A(F,.25) .$$

3C M-ESTIMATES

If the assumed underlying density is $\frac{1}{\sigma} f_0\left(\frac{(x-\theta)}{\sigma}\right)$, then the simultaneous maximum likelihood estimates of location, θ, and scale, σ, can be defined by the functionals $T(F)$, $S(F)$ satisfying

(25)
$$\int \psi \left(\frac{x-T(F)}{S(F)}\right) F(dx) = 0$$

(26)
$$\int \chi \left(\frac{x-T(F)}{S(F)}\right) F(dx) = 0$$

with

(27)
$$\psi(x) = -(\log f_0(x))'$$

(28)
$$\chi(x) = \psi(x)x - 1 .$$

An estimate defined by formulae of the type (25), (26) where ψ, χ need not derive from any probability density f_0, will be called an M-estimate. For the purposes of this section, we shall distinguish this as a *type A* estimate.

If we insert $F_{\varepsilon,x} = (1-\varepsilon)F + \varepsilon\delta_x$ for F and take the derivative of (25) and (26) with respect to ε, we find that the influence curves satisfy

$$IC(x;F,T) \cdot \frac{1}{S(F)} \int \psi'\left(\frac{y-T(F)}{S(F)}\right) F(dy)$$

(29)
$$+ IC(x;F,S) \cdot \frac{1}{S(F)} \int \psi'\left(\frac{y-T(F)}{S(F)}\right)\frac{y-T(F)}{S(F)} F(dy) = \psi\left(\frac{x-T(F)}{S(F)}\right)$$

$$IC(x;F,T) \cdot \frac{1}{S(F)} \int \chi'\left(\frac{y-T(F)}{S(F)}\right) F(dy)$$

(30)
$$+ IC(x;F,S) \cdot \frac{1}{S(F)} \int \chi'\left(\frac{y-T(F)}{S(F)}\right)\frac{y-T(F)}{S(F)} F(dy) = \chi\left(\frac{x-T(F)}{S(F)}\right).$$

If F is symmetric, ψ odd and χ even, this reduces to

$$(31) \qquad IC(x;F,T) = \frac{S(F)}{\int \psi'\left(\frac{y}{S(F)}\right) F(dy)} \psi\left(\frac{x}{S(F)}\right)$$

$$(32) \qquad IC(x;F,S) = \frac{S(F)}{\int \chi'\left(\frac{y}{S(F)}\right) \frac{y}{S(F)} F(dy)} \chi\left(\frac{x}{S(F)}\right).$$

The asymptotic variance of T(F) then is the expected value of the square of IC(x;F,T), as usual.

(*Type B*) Similar results hold if any estimate of scale S(F) is combined with (25) to define an estimate T(F) of location. The general asymmetric case leads to complicated formulae for the influence curve and the variance. But in the symmetric case, $T(F_n)$ and $S(F_n)$ are asymptotically independent, and the influence curve of T is given by the same formula (31), with the appropriate S(F).

(*Type C*) A further variant are the *one-step procedures*: one starts with an estimate of scale S(F) and an initial estimate of location $\widetilde{T}(F)$; then one applies Newton's rule once to (25) to find the *one-step estimate* $T(F)$. The Taylor expansion of (25) starts with

$$0 = \int \psi\left(\frac{x-T(F)}{S(F)}\right) F(dx)$$

$$= \int \psi\left(\frac{x-\widetilde{T}(F)}{S(F)}\right) F(dx) - \frac{(T(F)-\widetilde{T}(F))}{S(F)} \int \psi'\left(\frac{x-\widetilde{T}(F)}{S(F)}\right) F(dx) + \cdots$$

hence one puts

$$(33) \qquad T(F) = \widetilde{T}(F) + \frac{\int \psi\left(\frac{x-\widetilde{T}(F)}{S(F)}\right) F(dx)}{\int \psi'\left(\frac{x-\widetilde{T}(F)}{S(F)}\right) F(dx)} S(F)$$

Although the influence curve and the variance are straightforward to compute in the general asymmetric case, they lead to excessively complicated expressions which depend on the influence curves of $\widetilde{T}$ and S as well as on ψ. However, if $F = \overline{F}$, and ψ and $\widetilde{T}$ are odd (i.e., $\widetilde{T}(\overline{F}) = -\widetilde{T}(F)$ for all F) most terms vanish and one obtains the same form of the influence curve and the same asymptotic variance as type B. The theory of these procedures is discussed in Bickel (1971).

3C1 Huber Proposal 2 [(1969), p. 96] (H07-H20)

These estimates (type A) are defined by (25), (26) with

$$(34) \qquad \psi(x) = \psi(x;k) = \max(-k,\min(k,x))$$

$$(35) \qquad \chi(x) = \psi(x;k)^2 - \beta(k)$$

where

$$(36) \qquad \beta(k) = \int \psi(x;k)^2 \, \Phi(dx)$$

41

(Φ is the standard normal measure). In our Monte Carlo computations, the values, 0.7, 1.0, 1.2, 1.5, 1.7 and 2.0 were used for the parameter k.

For symmetric F, the influence curve is given by (31). Note that the influence curve coincides with that of the α-trimmed mean (4), with α depending on F through

$$(37) \qquad\qquad \alpha = F(-kS(F))$$

($c(\alpha) = 0$ because of symmetry).

The variance can be computed according to (7), or also, in view of (26) and (35), as

$$(38) \qquad\qquad A(F) = \frac{\beta(k)}{(1-2\alpha)^2} S(F)^2 .$$

3C2 One-Step Estimates (type C) (D07-D20, M15, P15)

In our study, the initial estimate $\widetilde{T}(F)$ was either the mean $\int x\, dF$ or the median $F^{-1}\left(\frac{1}{2}\right)$, while

$$(39) \qquad\qquad S(F) = \frac{F^{-1}\left(\frac{3}{4}\right) - F^{-1}\left(\frac{1}{4}\right)}{\Phi^{-1}\left(\frac{3}{4}\right) - \Phi^{-1}\left(\frac{1}{4}\right)} ,$$

with ψ as in (34). For symmetric F, the influence curve is given by (31), while the variance is computed from (37) and (7), with S(F) defined by (39). In one case, (P15) the scale estimate (40) was used.

Note. The asymptotic formulae are inapplicable when $\widetilde{T}$ is the mean and observations are drawn from a longtailed (Cauchy-like) distribution.

3C3 Other M-Estimates

(a) Robust scale Huber (type B) (A15,A20)

In this case, (25), (34) were combined with the scale estimate:

$$(40) \qquad S(F) = \frac{\tilde{F}^{-1}\left(\frac{3}{4}\right) - \tilde{F}^{-1}\left(\frac{1}{4}\right)}{\Phi^{-1}\left(\frac{3}{4}\right) - \Phi^{-1}\left(\frac{1}{4}\right)}$$

where $\tilde{F}$ is the symmetrized distribution function (24). The asymptotic properties coincide with that of the 1-step estimate (for $k = 1.5$) in the symmetric case.

(b) M-estimate, ψ bends at $(2.0, 2.0, 5.5)$, type B (HMD)

In this case, (25) is used with

$$(41) \qquad \psi(x) = x \qquad\qquad 0 \le x \le 2$$

$$= -\frac{4}{7}x + \frac{22}{7} \qquad 2 \le x \le 5.5$$

$$= 0 \qquad\qquad x \ge 5.5$$

$$= -\psi(-x) \qquad x < 0$$

and (40) as an estimate of scale. In the symmetric case, the influence curve is again given by (31).

(c) M-estimate, ψ bends at a,b,c (12A-25A, ADA)

In this case, (25) is used with (40) as an estimate of scale and

$$(42) \qquad \psi(x; a, b, c) = x \qquad\qquad |x| \le a$$

$$= a \ \text{sign}(x) \qquad a < |x| \le b$$

$$= \frac{c - |x|}{c - b} a \qquad b < |x| \le c$$

$$= 0 \qquad\qquad |x| > c .$$

In the symmetric case, the influence curve is again given by (31).

An adaptive version of these estimates (ADA) fixes b and c at 4.5, 8.0 respectively but uses a random $a(F_n)$ given by,

$$a(F) = 1.0 \qquad L(F) \leq 0.44$$

$$= \frac{75L(F) - 25}{8} \qquad 0.44 < L(F) \leq 0.6$$

$$= 2.5 \qquad 0.6 < L(F)$$

where

$$L(F) = S(F) \int_{\{|z| > S(F)\}} \frac{1}{z} \, d\tilde{F}(z)$$

with S given by (40). Evidently, if F is symmetric, the influence curve of ADA for $F = F_0$ is that of the M estimate with b, c as above and $a = a(F_0)$.

(d)   The sine curve M estimate (AMT)
      Here (25) is used with,

(44) $$\psi(x) = \sin\left(\frac{x}{2.1}\right) \qquad |x| < 2.1\pi$$

$$= 0 \qquad \text{otherwise}$$

and the scale estimate (40).

(e)   "Olshen's" estimate (OLS) solves (25) with

(45) $$\psi(x) = \frac{x}{\mu + x^2}$$

and

(45a) $$S(F) = F^{-1}\left(\frac{3}{4}\right) - F^{-1}\left(\frac{1}{4}\right) ;$$

for μ, the arbitrary value $\mu = \frac{1}{2}$ was used. The influence curve is given by (31) for symmetric F.

3D SKIPPED PROCEDURES

Skipping removes all observations from the sample whose distance to the nearest quartile exceeds p times the interquartile range (usually, $p = 1$, $3/2$, or 2).

We are mainly concerned here with iterative skipping, where the above procedure is iterated until the sample stays fixed.

Any estimation procedure can then be applied to the (skipped) samples so obtained.

To describe skipping more formally, we introduce population values of the skipped quartiles or hinges $H_1(F)$, $H_2(F)$ and the skipped p-corners $C_1(F)$, $C_2(F)$ defined by

$$(46) \qquad C_2 - H_2 = H_1 - C_1 = p(H_2 - H_1)$$

$$(47) \quad F(C_2) - F(H_2) = \frac{1}{2}\left\{F(H_2) - F(H_1)\right\} = F(H_1) - F(C_1)$$

We assume that these quantities are uniquely determined which, of course, restricts F.

3D1 Influence Curves for Skipped Quantiles

We will first determine the influence curves of the skipped quantiles. Denote the $\left(\frac{1}{2}+u\right)$-quantile of the skipped population by $Q_u = Q(u)$; evidently, $Q\left(\pm\frac{1}{2}\right)$ are the p-corners, and $Q\left(\pm\frac{1}{4}\right)$ are the hinges, and more generally,

$$(48) \qquad Q(u) = F^{-1}(\mu + \gamma\mu) ,$$

where $\gamma = F(C_2) - F(C_1)$ is the fraction of the population retained after skipping, and $\mu = F(m)$, where m is the skipped median.

Insert

$$(49) \qquad F_\varepsilon = (1-\varepsilon)F + \varepsilon\delta_x$$

for F in (48) and take the derivative with respect to

ε. For $\varepsilon = 0$ this gives (with $\displaystyle \cdot = \frac{d}{d\varepsilon}$)

(50)
$$\dot{Q}_u = IC(x;F,Q_u)$$

$$= \frac{1}{f(Q_u')} \left\{ (\mu+\dot{\mu}) + (\gamma+\dot{\gamma})u - \delta_x(Q_u') \right\} .$$

Here, $\dot{\mu}$ and $\dot{\gamma}$ do not depend on u, and can be found as follows.
Rewrite (46) as

(51)
$$Q\left(-\frac{1}{2}\right) - Q\left(-\frac{1}{4}\right) - Q\left(\frac{1}{4}\right) + Q\left(\frac{1}{2}\right) = 0$$

(52)
$$-Q\left(-\frac{1}{2}\right) + (2p+1)Q\left(-\frac{1}{4}\right) - (2p+1)Q\left(\frac{1}{4}\right) + Q\left(\frac{1}{2}\right) = 0$$

and differentiate with respect to ε at $\varepsilon = 0$ (utilizing (50)). This gives two equations from which $\mu+\dot{\mu}$ and $\gamma+\dot{\gamma}$ can be determined. After some minor simplifying transformations one finds that

(53)
$$IC(x;F,Q_u) = \frac{1}{f(Q_u')} (A(x) + B(x) \cdot u)$$

where $A(x)$, $B(x)$ satisfy

(54)
$$a_{11}A + a_{12}B = v(x)$$
$$a_{21}A + a_{22}B = w(x)$$

where

(55)
$$a_{11} = \frac{1}{f(C_1)} - \frac{1}{f(H_1)} - \frac{1}{f(H_2)} + \frac{1}{f(C_2)}$$

$$a_{12} = -\frac{1}{2f(C_1)} + \frac{1}{4f(H_1)} - \frac{1}{4f(H_2)} + \frac{1}{2f(C_2)}$$

$$a_{21} = -\frac{1}{f(C_1)} + \frac{2p+1}{f(H_1)} - \frac{2p+1}{f(H_2)} + \frac{1}{f(C_2)}$$

$$a_{22} = \frac{1}{2f(C_1)} - \frac{2p+1}{4f(H_1)} - \frac{2p+1}{4f(H_1)} + \frac{1}{2f(C_2)} .$$

The functions $v(x)$ and $w(x)$ are piecewise constant and vanish outside of the interval (C_1, C_2); they have jumps at the five points C_1, H_1, H_2, C_2 and Q_u of the respective size

$$
v(x): \quad -\frac{1}{f(C_1)}, \frac{1}{f(H_1)}, \frac{1}{f(H_2)}, -\frac{1}{f(C_2)}, a_{11}
$$

(56)

$$
w(x): \quad \frac{1}{f(C_1)}, -\frac{2p+1}{f(H_1)}, \frac{2p+1}{f(H_1)}, -\frac{1}{f(C_2)}, a_{21} .
$$

Note that this implies that IC vanishes outside of the interval (C_1, C_2) and that it will have jumps at C_1, H_1, H_2 and C_2, caused by skipping, whatever estimate may be used on the skipped sample.

3D2 Symmetric Combinations of Skipped Quantiles

If F is symmetric, and if we consider symmetric combinations of quantiles, there is a great simplification:

(57)
$$
IC\left(x; F, \frac{1}{2}G_u + \frac{1}{2}G_{-u}\right) = \frac{v(x) - v(-x)}{2a_{11}f\ F^{-1}(\gamma u)} .
$$

Evidently, the influence curve of any symmetric linear combination of skipped order statistics can now be found by linear superposition of (57) in the manner of (11), (12).

3D3 Iteratively Skipped Estimates

In particular, consider the (generalized) skipped trimean

(58)
$$
T(F) = \beta H_1(F) + (1-2\beta)M(F) + \beta H_2(F) .
$$

The case $\beta = 0$ gives the skipped median, $\beta = 1/4$ the ordinary skipped trimean, $\beta = 1/2$ the skipped "midhinge". Then, in the symmetric case,

47

$$IC(x;F,T) = \frac{1-2\beta}{2f(m)} \qquad\qquad \text{for}\quad m < x < H_2$$

$$= \left\{ \frac{1-2\beta}{2f(m)} + \frac{\beta}{f(H_2)} \right\} \frac{1}{1 - \frac{f(C_2)}{f(H_2)}} \qquad \text{for}\quad H_2 < x < C_2$$

(59)

$$= 0 \qquad\qquad\qquad\qquad \text{for}\quad x > C_2$$

$$= -IC(x;F,T) \qquad\qquad \text{for}\quad x < 0.$$

The asymptotic variance is

$$(60)\quad A(F) = \frac{\gamma}{2} \left\{ \left(\frac{1-2\beta}{2f(m)} \right)^2 + \left[\frac{1-2\beta}{2f(m)} + \frac{\beta}{f(H_2)} \right]^2 \cdot \frac{1}{\left(1 - \frac{f(C_2)}{f(H_2)} \right)^2} \right\} \cdot$$

In particular, the skipped median $(\beta = 0)$ has

$$(61)\qquad A(F) = \frac{\gamma}{2} \frac{1}{(2f(m))^2} \left\{ 1 + \frac{1}{\left(1 - \frac{f(C_2)}{f(H_2)} \right)^2} \right\} \cdot$$

3E OTHER ESTIMATES

3E1 The Hodges-Lehmann Estimate (H/L)

Rank estimates can be defined as follows. Form a sample of size $2n$ by merging the original sample $\{X_1,\ldots,X_n\}$ with its mirror image around T_n: $\{2T_n-X_1,\ldots,2T_n-X_n\}$. Adjust the estimate T_n such that

$$\sum_1^n J\left(\frac{R_i}{2n+1} \right) = 0 \ ,$$

where R_i is the rank of X_i in the combined sample, and the asymptotic scoring function J satisfies

$\int_0^1 J(t) \, dt = 0.$ The Hodges-Lehmann estimate is the only estimate of this class entering in our study; it corresponds to $J(t) = t - \frac{1}{2}$ and can also be defined to be the median of the pairwise means $(X_i + X_j)/2$.

In terms of functionals, the above formula corresponds to defining $T(F)$ by

$$(62) \qquad \int J\left(\frac{1}{2}\Big[F(x) + 1 - F(2T(F)-x)\Big]\right) F(dx) = 0 \; .$$

The derivative of $T(F_{\varepsilon,x})$ with respect to ε can be computed in a rather straightforward fashion to give the influence curve

$$(63) \qquad IC(x;F,T) = \frac{U(x) - \int U(x)f(x) \, dx}{\int U'(x)f(x) \, dx}$$

where $U(x)$ is an indefinite integral of

$$(64) \qquad U'(x) = J'\left(\frac{F(x) + 1 - F(2T(F)-x)}{2}\right) f(2T(F)-x) \; .$$

In particular, for the Hodges-Lehmann estimate one obtains

$$(65) \qquad IC(x;F,T) = \frac{\frac{1}{2} - F(2T(F)-x)}{\int f(2T(F)-x)f(x) \, dx} \; ,$$

with $T(F)$ defined by

$$(66) \qquad \int F(2T(F)-x) \, F(dx) = \frac{1}{2} \; .$$

In the symmetric case, the influence curve simplifies to

$$(67) \qquad IC(x;F,T) = \frac{F(x) - \frac{1}{2}}{\int f(x)^2 \, dx}$$

and the variance then is

(68)
$$A(F) = \frac{1}{12\left(\int f(x)^2 dx\right)^2} \, .$$

3E2   The Shorth (SHO)

The shorth is the arithmetic mean of the shortest half of the sample; more generally, the α-shorth is the mean of the shortest $(1-2\alpha)$-fraction of the sample.

Somewhat unexpectedly, this estimate has a slower rate of convergence than $\sqrt{n}$; in the symmetric case, we have asymptotically

(69)
$$\mathcal{L}(n^{\frac{1}{3}} T_n) \longrightarrow \mathcal{L}(A \cdot B \cdot \tau)$$

where τ is the random time s for which

(70)
$$s^2 + Z(s)$$

attains its minimum, $Z(s)$ being a two-sided Brownian motion, A is the constant

(71)
$$A = \left\{ \frac{\sqrt{2} f(F^{-1}(1-\alpha))}{-f'(F^{-1}(1-\alpha))} \right\}^{\frac{2}{3}}$$

and

$$B = \frac{2F^{-1}(1-\alpha)}{1 - 2\alpha} \, .$$

The factor A is common to all estimates based on the shortest $(1-2\alpha)$-portion of the sample, while B belongs to the particular estimate -- for instance, B is replaced by $\frac{1}{f(0)}$, if the mean is replaced by the median.

We give a short, heuristic derivation of this result, based on weak convergence.

1. Represent the quantile process by

(73) $X_{(i)} = G(U_{(i)})$

$$= G\left(\frac{i}{n+1}\right) + \frac{1}{\sqrt{n}} g\left(\frac{i}{n+1}\right) Y_n\left(\frac{i}{n+1}\right)$$

$$= G_n\left(\frac{i}{n+1}\right)$$

where $G = F^{-1}$, $g = G'$, and $Y_n(t)$ is weakly converging toward a Brownian bridge $Y(t)$.

2. Assume a symmetric, strongly unimodal density f. Begin to solve the following minimization problem

(74) $G_n(1-\alpha+t) - G_n(\alpha+t) = \min_t$!

or

(75) $G(1-\alpha+t) - G(\alpha+t) + \dfrac{1}{\sqrt{n}} g(1-\alpha+t)Y_n(1-\alpha+t)$

$$- \frac{1}{\sqrt{n}} g(\alpha+t)Y_n(\alpha+t) = \min_t \ ! \ .$$

The leading terms of a Taylor expansion (about 0) of the left hand side of this expression are, because of symmetry of g,

(76) $2G(1-\alpha) + g'(1-\alpha)t^2 + \dfrac{g(1-\alpha)}{\sqrt{n}} \{Y_n(1-\alpha+t) - Y_n(\alpha+t)\}$.

3. For large n, we may for distributional purposes replace $Y_n(t)$ by $Y(t) = X(t) - tX(1)$, where $X(t)$ is a Brownian motion. We may furthermore subtract the following expression, which is independent of t:

(77) $2G(1-\alpha) + \dfrac{g(1-\alpha)}{\sqrt{n}} (Y(1-\alpha) - Y(\alpha))$

and obtain the asymptotically equivalent minimization problem

(78) $$g'(1-\alpha)t^2 + \frac{\sqrt{2}\,g(1-\alpha)}{\sqrt{n}}\ W(t) = \min\ !$$

where

$$W(t) = \frac{1}{\sqrt{2}}\ \{[X(1-\alpha+t) - X(1-\alpha)] - [X(\alpha+t) - X(\alpha)]\}$$

is a standard Brownian motion for $|t| \leq \alpha$. If we substitute $t = n^{-1/3}\,As$, we obtain

(79) $$s^2 + Z(s) = \min\ !$$

where Z is again a standard Brownian motion.

4. If $\hat{t}$ is the minimizing value of t, we find that the main asymptotic variability of the mean of two symmetrically located order statistics of the shortest $(1-2\alpha)$-fraction is caused by $\hat{t}$:

$$\frac{1}{2}\left(F_n^{-1}\left(\frac{1}{2} + \hat{t} + u\right) + F_n^{-1}\left(\frac{1}{2} + \hat{t} - u\right)\right) \approx g\left(\frac{1}{2} + u\right)\cdot\hat{t}\ .$$

In particular, the main variability of the α-shorth T_n is

(80) $$T_n \approx \frac{1}{1-2\alpha}\int_\alpha^{1-\alpha} g(t)\ dt \cdot \hat{t} = \frac{F^{-1}(1-\alpha) - F^{-1}(\alpha)}{1 - 2\alpha}\ \hat{t}\ .$$

Very similar problems occur when one is estimating the mode of a distribution, cf. H. Chernoff (1964), J. H. Venter (1967).

3E3 Simple Adaptive Procedures (HGP,HGL)

Two estimates of this kind suggested by Hogg (1967), (1969) are included in this study. They are both obtained by combining several estimates $T_1(\cdot),\ldots,T_{s+1}(\cdot)$ by means of an estimate of kurtosis $K(\cdot)$ and a partition $a_1 < \cdots < a_s$ of the positive half line.

A Hogg type estimate T is obtained by taking

(81) $T(F) = T_i(F)$ if $a_i \leq K(F) < a_{i+1}$,

where $a_0 = 0$, $a_{s+1} = \infty$.

Evidently,

(82) $IC(x;F,T) = IC(x;F,T_{i(F)})$

where $a_{i(F)} \leq K(F) < a_{i(F)+1}$.

For both estimates considered,

(83) $$K(F) = \int x^4 \, dF \Big/ \left(\int x^2 \, dF \right)^2 .$$

For Hogg (1967) (HGP), $(a_1,a_2,a_3) = (2.0,4.0,5.5)$,

$T_0(\cdot) =$ outer mean

$T_1(\cdot) =$ mean

$T_2(\cdot) = .25$ trimmed mean

$T_3(\cdot) =$ median.

For Hogg (1969) (HGL), $(a_1,a_2,a_3) = (1.9,3.1,4.5)$,

$T_0(\cdot) =$ outer mean

$T_1(\cdot) =$ mean

$T_2(\cdot) = .25$ trimmed mean

$T_3(\cdot) =$ Gastwirth's estimate.

3E4 Other Adaptive Procedures (TAK,JOH)

Two other adaptive procedures one due to K. Takeuchi (1969) and the other to V. Johns (1971) were also included. These cannot readily be described in terms of a functional T independent of n. Under regularity conditions which are not clear to us, if the distribution is symmetric, these estimates are asymptotically fully efficient and thus have "asymptotic" influence curve $-f'/f$ and asymptotic variance which is the reciprocal of the Fisher information for the underlying population. Hampel's ADA procedure is mildly adaptive as is the JAE group. However, these are considered under M estimates and trimmed means respectively.

3E5 Folded Medians (2RM,3RM,BH)

These procedures correspond to functionals of the following form. Let F_k be defined iteratively as follows,

$$F_0 = F$$

$$F_k(s) = \lambda \left[z : \frac{F_{k-1}^{-1}(z) + F_{k-1}^{-1}(1-z)}{2} \leq s \right]$$

where λ is Lebesgue measure on $(0,1)$. Then, kRM is given by

$$T_k(F) = F_k^{-1}\left(\frac{1}{2}\right) .$$

If $k \geq 1$ these procedures do not possess well defined influence curves and in fact are not asymptotically Gaussian. The asymptotic theory of T_1 is dealt with in Bickel and Hodges (1967). The same methods may be applied to T_k.

4. FINITE-SAMPLE CALCULATIONS

The main body of finite-sample results is derived from four basic quantities or properties of the estimators:
1. Average (or bias)
2. Variance (or mean square)
3. Covariance between pairs of estimators
4. Percent-points (or quantiles).

These quantities have been calculated for 65 estimators and 30 different sampling "situations" (parent distributions and sample size) covering the Normal distribution and longer-tailed alternatives. Sample sizes ranged from 5 to 40.

4A SETTING UP THE PROBLEM

One simple but inefficient approach to such a problem is to regard the sampling situation and the estimator as black boxes. The numbers which the boxes produce can be processed through the standard definitions to produce the desired statistics. Percent-points, for example, are simply estimated by the sample quantiles of the estimator's sample distribution. This is quite satisfactory as an initial exploratory procedure, but suppose we require two significant figures on the 1%-point? Hundreds of thousands of samples would be required in almost all cases to achieve such accuracy by this method.

There is much we can do to increase the efficiency (unit accuracy/computation) of our Monte Carlo experiment, provided that the desired quantities are expressed as integrals, expectations, or averages (these are equivalent). Fortunately, our first three quantities are defined that way. However, percent-points need re-expression. If T_n represents the estimator and F the parent sampling distribution, then the percent-point b is defined as the solution of

(1)
$$E_F(I_{[T_n \leq b]}) = \alpha$$

where $I_{[T_n \leq b]}$ is the indicator function defined by

$$I_{[T_n \le b]} = \begin{cases} 1 & T_n \le b \\ 0 & \text{otherwise} \end{cases} .$$

(2)

If G is the distribution function of T_n, (1) is just
another way of writing $b = G^{-1}(\alpha)$. We can re-express
our calculation in terms of α, an expectation, which can
be computed more efficiently. First we select a set of
trial b's, and compute (by Monte Carlo) the corresponding
α's (values of the distribution function). Then we use
these (α, b) pairs to interpolate b as a function of
α for the desired α's.

4B NORMAL/INDEPENDENT DISTRIBUTIONS

The Monte Carlo strategy we used relies heavily on
the Normal/Independent form of the sampling distributions.

Suppose we wish to draw x from a distribution F.
We may always express x in one or more ways as
$x \sim N(\mu, \sigma^2)$ where μ and σ are drawn from some distributions $G(\mu)$ and $H(\sigma)$. We will concentrate on the
symmetric subclass with $\mu \equiv 0$ and $\sigma > 0$. This subclass
is termed "Normal/Independent" because the distribution of
$1/\sigma$ is a standard form for many familiar members of the
subclass, and because an observation x is generated on
the computer as z/y, where z is unit Normal and $y = 1/\sigma$
is generated independently of z. Some prominent members
of this subclass and their independent divisors are given
in Exhibit 4-1.

These distributions generate location-scale families
under the further transformation $x \to ax + b$. The most
tangible benefits of using the Normal/Independent subclass are (1) conditional sufficiency and (2) a known
distribution for the conditionally sufficient statistics.

Conditionally, given σ_i, the variables x_i are
independently and normally distributed. The sufficient
statistics $\hat{x}$ and $\hat{s}_x^2$ have known distributions as given
below

(3)
$$\hat{x} = \frac{\Sigma\, x_i/\sigma_i^2}{\Sigma\, 1/\sigma_i^2} \sim N\left(0, \frac{1}{\Sigma\,(1/\sigma_i^2)}\right)$$

and

Exhibit 4-1

Sampling Distribution	Independent Distribution
$x = z/y$	$F(y)$

Normal — Degenerate $y \equiv 1$

Cauchy — Half-normal

$$F(y) = \begin{cases} 2[\Phi(y) - 0.5] & y > 0 \\ 0 & y \leq 0 \end{cases}$$

Student's t on
n degrees freedom

$$y \sim \sqrt{\chi_n^2/n}$$

Contaminated normal

$F(x) = \alpha\Phi(x/c) + (1-\alpha)\Phi(x)$

$$y = \begin{cases} c & \text{with prob. } \alpha \\ 1 & \text{with prob. } 1-\alpha \end{cases}$$

Laplace or double exponential

$f(x) = \frac{1}{2}\exp\{-|x|\}$

$$f(y) = y^{-3}\exp\{-y^{-2}/2\}$$

— — — — — — — — — — — — — — — —

$$(4) \qquad \hat{s}^2_x = \frac{\Sigma \left(\frac{x_i - \hat{x}}{\sigma_i} \right)^2}{n-1} \sim \chi^2_{n-1}/(n-1) \ .$$

To show how we used $\hat{x}$ and $\hat{s}_x$ profitably, we need to explore invariance.

4C LOCATION AND SCALE INVARIANCE

Consider a statistic $T(\underset{\sim}{x})$. T is said to be *location invariant* if and only if

$$T(\underset{\sim}{x} + m\underset{\sim}{1}) = m + T(\underset{\sim}{x}) \ .$$

T is also *scale invariant* if $T(s\underset{\sim}{x}) = sT(\underset{\sim}{x})$. A *configuration* $\underset{\sim}{c}$ is a standardized (invariant) coding of $\underset{\sim}{x}$, constrained to fewer dimensions than $\underset{\sim}{x}$, but frequently expressed in the same coordinates. In this case, standardization is by location, scale, or both, as for example $\underset{\sim}{c}(\underset{\sim}{x}) = (\underset{\sim}{x} - \hat{x}\underset{\sim}{1})/\hat{s}_x$. Intuitively, a configuration is what is actually invariant.

For any location-scale invariant statistic, T, we have an important result: conditional on the configuration $\underset{\sim}{c}$, the distribution of the statistic is determined only by (i) the value of $T(\underset{\sim}{c})$, and (ii) the distribution of any location and scale equivariants m and $\hat{s}$ (these may depend on $\underset{\sim}{c}$). As a result, conditional expectations involving $T(\underset{\sim}{x})$ given $\underset{\sim}{c}$ may be calculated explicitly.

This technique has been used implicitly on several occasions, notably Dixon and Tukey (1968). The formal structure was exposited by Steinberg in a 1970 seminar based on ideas of Fraser (1968). See also Hodges 1967.

4D MONTE CARLO TECHNIQUES USED IN THIS STUDY

All of the expectations we calculated may be viewed as nested, 2-stage, averaging processes. The inner average, over location and scale was conditioned on both configuration <u>and</u> the vector g of independent scales used to generate the configuration. The outer average (of the two conditional averages) covered both conditions, and was computed by simple (random) sampling.

The inner average -- over location and scale -- turns out in all cases to be a familiar analytic function which may be computed exactly without much work.

4D1 Calculation of Variances and Covariances

For variances and covariances of two statistics T, U, we have

$$Cov[T(\underset{\sim}{x}),U(\underset{\sim}{x})|config.\ \underset{\sim}{c}(\underset{\sim}{x}),g]$$

$$= Ave\left\{\left[\left(\frac{T(x) - \hat{x}}{\hat{s}_x}\right)\hat{s}_x + \hat{x}\right]\cdot\left[\left(\frac{U(\underset{\sim}{x}) - \hat{x}}{\hat{s}_x}\right)\hat{s}_x + \hat{x}\right]\ \middle|\ \underset{\sim}{c}(\underset{\sim}{x}),\underset{\sim}{g}\right\}\ .$$

Note that the averages of T and U are zero for a symmetric parent distribution, since all estimators in the study were *antipodentreu*, i.e., $T(\underset{\sim}{x}) = -T(-\underset{\sim}{x})$. Now the location-scale invariance of T is introduced with $(T(\underset{\sim}{x})-\hat{x})/\hat{s}_x$ being constant depending on $\underset{\sim}{c}$, say $T(\underset{\sim}{c})$. Thus:

$$cov[T,U|\underset{\sim}{c}(\underset{\sim}{x}),\underset{\sim}{g}]$$

$$= Ave\{T(\underset{\sim}{c})\cdot U(\underset{\sim}{c})\hat{s}_x^2 + (T(\underset{\sim}{c})+U(\underset{\sim}{c}))\hat{s}_x s + \hat{x}^2|\underset{\sim}{c}(\underset{\sim}{x}),\underset{\sim}{g}\}\ .$$

Because $\hat{s}_x$ and $\hat{x}$ are conditionally independent with distributions given above, (3), (4), we have

$$Cov(T,U|\underset{\sim}{c}(\underset{\sim}{x}),\underset{\sim}{g}) = T(\underset{\sim}{c})\cdot U(\underset{\sim}{c}) + (1/\textstyle\sum 1/\sigma_i^2)\ .$$

Setting $U = T$ yields variances.

4D2   Calculation of Percent-Points (Quantiles)

For percent-points we note that

$$Pr(T(\underset{\sim}{x}) < b \mid \underset{\sim}{c}(\underset{\sim}{x}), \underset{\sim}{g}) = Pr\left\{ \left(\frac{T(\underset{\sim}{x}) - \hat{x}}{\hat{s}_x} \right) \hat{s}_x + \hat{x} < b \mid \underset{\sim}{c}(\underset{\sim}{x}), \underset{\sim}{g} \right\}$$

$$= Pr\left\{ \frac{\hat{x} - b}{\hat{s}_x} < -T(\underset{\sim}{c}) \mid \underset{\sim}{c}(\underset{\sim}{x}), \underset{\sim}{g} \right\} \quad .$$

This last expression is a non-central Student's t probability, and was numerically calculated with a moderate degree of accuracy by formula 26.7.10 in Abramowitz and Stegun (1964, 1968).

4D3   Calculations for Contaminated Normals

The family of contaminated normal distributions of the form

$$F(x) = \alpha \Phi(x/c) + (1-\alpha) \Phi(x)$$

where α is the 'fraction of contamination' and c^2 is the variance of the contaminant, has been frequently used as a model in studies of robustness. This distribution may be expressed in the form Normal/Independent : x/y where

$$y = \begin{cases} \frac{1}{c} & \text{with probability } \alpha \\ 1 & \text{with probability } 1-\alpha \end{cases} \quad .$$

Alternatively a sample of size n from this distribution may be generated by
  (i)   choosing r, the number of contaminants with binomial probability bin(n,r,α) and
  (ii) drawing r variables from the distribution $N(0,c^2)$ (using y = 1/c) and n-r variables from the distribution N(0,1), using y = 1. The corresponding notation for this procedure would be to express the density function as

$$f(x_1,\ldots,x_n) = \sum_{r=0}^{n} \text{bin}(n,r,\alpha) \prod_{i=r+1}^{n} \phi(x_i) \prod_{i=1}^{r} \frac{1}{c} \phi\left(\frac{x_i}{c}\right)$$

where ϕ is the standard normal density function. The joint density may be expressed as a weighted sum (binomial weights) of joint density functions involving fixed numbers of variables (contamponents), r, from the contaminating distribution. Thus all expectations may be expressed as weighted sums (using the same weights) of the expectations calculated using fixed numbers, r, of contamponents. Once these fundamental expectations have been calculated, marginal expectations for any reasonable contaminating fraction α may be easily calculated. For this reason we tabulate values for the pseudo distribution with fixed number of contamponents.

For small contaminating fractions the binomial weights are large for small numbers of contamponents and small for large numbers of contamponents. Because these latter quantities enter with negligible weight, we chose not to compute them all. As a result the properties of estimators for large contaminating fractions may not be calculated from these tables, although results adequate for moderate fractions are readily attainable.

4E EFFICIENCY

For estimating variances, the use of the above technology resulted in a gain in efficiency (reduction in the number of samples required for a given accuracy) of 2 to more than 1000 over crude methods. Distributions and contamponents near the Normal were handled best, while the Cauchy caused the most difficulty. For a study of variances and covariances, even greater efficiency should be obtainable by exploiting the minimum-attained-variance properties of the Pitman estimate (1939). At this writing, however, adequate Pitman calculation techniques are still being developed. See Hoaglin (1971).

For percent-points, the efficiency gain over crude techniques is several times that of variances. This additional gain accrued in part because (i) percent-points are inherently more stable, and (ii) crude percent-point methods are notoriously inefficient.

4F   ACCURACY OF RESULTS

About 1.8 significant figures are accurate for
variances, about 2.1 are accurate for percent-points.
For comparisons within situations, one additional significant digit may be used. The results are based on 640
to 1000 configurations for each situation.

4G RANDOM NUMBER GENERATION

The need to generate pseudo-random numbers with the appropriate distributions is of first importance in a study of this kind. All of the methods we used rely upon a simple uniform (0,1) program.

Almost all current computer techniques for uniform pseudo random numbers have a multiplicative congruential form. Each random u_{i+1} is determined from its predecessor u_i by

$$u_{i+1} = (au_i + b) \bmod c.$$

(u_i and u_{i+1} are written as integers, with the understanding that they will be scaled into (0,1) before being used.) Marsaglia (1968) has pointed out that congruential generators necessarily put all successor groups $(u_k, \ldots, u_{k+n-1})$ on some number of parallel hyperplanes with dimension n-1. Even taking into account the discrete character of pseudo random numbers, this implies that all consecutively-drawn <u>samples</u> must fall in a set with low intrinsic probability.

We ameliorated Marsaglia's hyperplane problem by permuting the order of the uniform numbers before using them. Specifically, we employed a suggestion of Morven Gentleman: for each batch of 500 numbers needed, we drew 500 uniforms from a well-regarded source (Lewis (1969)), and stored them in a table according to a fixed permutation (Moses and Oakford (1962)), and then drew them out in order.

Normal pseudo-random deviates were produced from uniform numbers by the algorithm of Knuth (1969) derived from Box and Muller (1959).

Computer programs and associated data for reproducing the results of this study are listed in Appendix 12.

5. PROPERTIES

This chapter summarizes the numerical results obtained in the study. The properties of all the estimates studied are recorded; there is no attempt in this chapter to draw conclusions or dwell on the implications of particular results.

The numbers are frequently recorded to precision greater than is warranted for individual results (average 1.8 decimal digits). However since all estimates were evaluated on the same samples, within each situation, relative comparisons may be meaningful when these occur in the third decimal digit.

A brief description of each table is given in Exhibit 5-1.

Exhibit 5-1

INDEX OF EXHIBITS OF CHAPTER 5

5

5A SAMPLING SITUATIONS

Properties of the estimators were assessed under
about 40 different sampling situations. For computational
reasons these were expressed as x/y where x has a
standard normal distribution and y is an independent
variable. The Cauchy, t, double exponential and contaminated normal families are all conveniently expressed in
this way.

Exhibit 5-2 gives a brief description of the sampling
situations used in this study.

The family of contaminated normal distributions may
be expressed as weighted sums (using binomial weights) of
situations in which a fixed number of variables come from
the contaminating normal distribution with large variance
and the rest from the standard normal distribution. Calculations and some results are for these pseudo-distribu-
tions called *contamponents* below as components of the contaminated distributions, involving fixed numbers of values
from the contaminating distribution. These may be combined to yield variances for any proportion of contamination involving that contaminating scale.

All the situations considered were at least as long-
tailed as the normal distribution. Robustness for short-
tailed distributions was thought to be a rather special
case, arising in situations that are usually rather easily
recognized in practice. Those who desire robustness
against both short- and long-tailed situations will have
to look elsewhere for an analysis of their problem.

Exhibit 5-2
INDEX OF SAMPLING SITUATIONS

Tag	Situation	Sample sizes
	DISTRIBUTIONS	
	Normal	5 10 20 40
	Cauchy	5 10 20 40
25% 1/U	.75 Normal + .25 Normal/Uniform(0,1)	5 10 20 40
10% 1/U	.9 Normal + .1 Normal/Uniform(0,1)	20
1/U	Normal/Uniform(0,1)	10 20
.25% 3/U	.75 Normal + .25 Normal/Uniform(0,1/3)	20
10% 3/U	.9 Normal + .1 Normal/Uniform(0,1/3)	20
T3	t distribution 3 degrees freedom	20
D-EX	Laplace distribution (Double Exponential)	20
	CONTAMINATED SITUATIONS	
100r/n% 3N	(20-r) from Normal(0,1) & r from N(0,9) r = 1,2,3,5,10,15	20
100r/n% 3N	(10-r) from Normal(0,1) & r from N(0,9) r = 1,2	10
100r/n% 10N	(20-r) from Normal(0,1) & r from N(0,100) r = 1,2,5	20
	18 from Normal(0,1) & 2 from N(2,1)	20
	18 from Normal(0,1) & 2 from N(4,1)	20

5B VARIANCES

The variances and covariances of the estimates were evaluated for a variety of sampling situations. To facilitate comparisons across sample sizes these variances were standardized by multiplying by the sample size.

Some idea of the precision of these variances may be gained from a comparison in Exhibit 5-3 of exact results calculated for Normal samples of size 20 by Gastwirth and Cohen (1970).

Exhibit 5-3

ACCURACY OF SOME VARIANCES

Variances of Trimmed Means times 20

Normal distribution

Sample size 20

	Exact	Monte Carlo
5%	1.0230	1.0218
10%	1.0552	1.0558
15%	1.0932	1.0978
25%	1.1856	1.1990
50%	1.4688	1.4981

EXHIBIT 5-4A: VARIANCES (MULTIPLIED BY SAMPLE SIZE) OF THE ESTIMATORS.

	CODE	NAME	N O R M A L D I S T R I B U T I O N			
			5	10	20	40
1	M :	MEAN (AVERAGE)	1.000	1.000	1.0C0	1.000
2	5%:	5% SYM TRIM MEAN	1.004	1.009	1.022	1.025
3	10%:	10% SYM TRIM MEAN	1.020	1.048	1.056	1.058
4	15%:	15% SYM TRIM MEAN	1.060	1.069	1.098	1.094
5	25%:	MIDMEAN	1.156	1.148	1.199	1.184
6	50%:	MEDIAN	1.465	1.366	1.498	1.527
7	GAS:	GASTWIRTH	1.156	1.203	1.233	1.248
8	TRI:	TRIMEAN	1.182	1.126	1.153	1.167
9	JAE:	JAECKEL ADAPT TRIM	1.179	1.076	1.105	1.076
10	BIC:	BICKEL ADAPT TRIM	1.085	1.070	1.088	1.071
11	SJA:	SYM JAECKEL AD TRIM	1.083	1.085	1.110	1.089
12	JBT:	2-CHOICE ADAPT TRIM	1.179	1.071	1.110	1.088
13	JLJ:	ADAPT TRIM LINCOM			1.167	
14	H20:	HUBER PROP 2, K=2.0	1.000	1.007	1.009	1.010
15	H17:	HUBER PROP 2, K=1.7	1.015	1.019	1.021	1.024
16	H15:	HUBER PROP 2, K=1.5	1.043	1.031	1.036	1.038
17	H12:	HUBER PROP 2, K=1.2	1.089	1.063	1.074	1.071
18	H10:	HUBER PROP 2, K=1.0	1.112	1.093	1.114	1.106
19	H07:	HUBER PROP 2, K=0.7	1.131	1.165	1.204	1.185
20	M15:	1ST-HU 1.5*IQS MEAN	1.117	1.028	1.046	1.033
21	D20:	1ST-HU 2.0*IQS MED	1.050	1.012	1.017	1.011
22	D15:	1ST-HU 1.5*IQS MED	1.074	1.027	1.045	1.032
23	D10:	1ST-HU 1.0*IQS MED	1.125	1.068	1.110	1.090
24	D07:	1ST-HU 0.7*IQS MED	1.144	1.109	1.174	1.153
25	A20:	HUBER 2.0*ADS	1.074	1.023	1.019	1.015
26	A15:	HUBER 1.5*ADS	1.116	1.049	1.050	1.042
27	P15:	1S-HU 1.5*ADS MED	1.117	1.049	1.051	1.042
28	HMD:	M (2.0,2.0,5.5)*AD	1.375	1.217	1.246	1.193
29	25A:	M (2.5,4.5,9.5)*AD	1.184	1.058	1.046	1.034
30	22A:	M (2.2,3.7,5.9)*AD	1.278	1.121	1.107	1.080
31	21A:	M (2.1,4.0,8.2)*AD	1.231	1.089	1.081	1.061
32	17A:	M (1.7,3.4,8.5)*AD	1.271	1.127	1.130	1.104
33	12A:	M (1.2,3.5,8.0)*AD	1.333	1.187	1.205	1.169
34	ADA:	M (ADA,4.5,8.0)*AD	1.309	1.113	1.102	1.057
35	AMT:	M-TYPE (SINE)	1.232	1.083	1.070	1.056
36	OLS:	"OLSHEN'S"	1.151	1.244	1.347	1.360
37	TOL:	T-SKIP "OLSHEN'S"	1.279	1.294	1.41C	1.413
38	MEL:	MEAN LIKELIHOOD	1.315	1.389	1.518	1.505
39	CML:	CAUCHY MAX LIKELIHD	1.654	1.609	1.720	1.653
40	LJS:	LEAST FAVORABLE	1.073	1.022	1.022	1.012
41	SST:	ITER S-SKIP TRIMEAN	1.388	1.327	1.463	1.482
45	3T1:	MX(3T,2)-SKIP MEAN	1.266	1.138	1.119	1.111
46	5T1:	MX(5T,2)-SKIP MEAN	1.266	1.134	1.125	1.113
47	5T4:	MN(5T,.6)-SKIP MEAN	1.262	1.130	1.123	1.113
48	CTS:	C&T&S-SKIP TRIMEAN	2.062	1.298	1.405	1.400
49	TAK:	TAKEUCHI ADAPTIVE	1.109	1.041	1.050	1.036
50	DFA:	MAXIMUM EST LIKE	1.391	1.206	1.323	1.370
51	JOH:	JOHNS ADAPTIVE	1.465	1.174	1.137	1.101
52	HGP:	HOGG 67 ON KURTOSIS	1.305	1.124	1.055	1.020
53	HGL:	HOGG 69 ON KURTOSIS	1.136	1.074	1.064	1.046
54	THP:	T-SKIP HOGG 67	1.357	1.269	1.152	1.112
55	THL:	T-SKIP HOGG 69	1.303	1.199	1.142	1.112
56	JWT:	ADAPT FROM SKIPPING	1.222	1.141	1.164	1.192
57	H/L:	HODGES-LEHMANN	1.081	1.063	1.063	1.055
58	BH :	BICKEL-HODGES	1.082	1.054	1.047	1.048
59	2RM:	2 TIMES FOLDED MED	1.018	1.013	1.012	1.012
60	3RM:	3 TIMES FOLDED MED	1.018	1.003	1.004	1.004
61	4RM:	4 TIMES FOLDED MED	1.018	1.003	1.001	1.001
62	3R1:	3-FOLDED 1-TRIM MED	1.182	1.048	1.038	1.024
63	3R2:	3-FOLDED 2-TRIM MED	1.182	1.124	1.085	1.044
64	CPL:	CAUCHY PITMAN (LOC)	1.092	1.158	1.256	1.277
65	SHO:	SHORTH	2.453	3.537	4.620	5.378

EXHIBIT 5-4B: VARIANCES (MULTIPLIED BY SAMPLE SIZE) OF THE ESTIMATORS.
25% 1/U A N D 75% N O R M A L

	CODE	NAME	5	10	20	40
1	M :	MEAN (AVERAGE)	77.75	2038.78	*******	2119.31
2	5%:	5% SYM TRIM MEAN	55.35	637.41	3.84	2.10
3	10%:	10% SYM TRIM MEAN	32.91	3.84	1.81	1.61
4	15%:	15% SYM TRIM MEAN	13.14	2.88	1.64	1.56
5	25%:	MIDMEAN	2.59	1.85	1.64	1.62
6	50%:	MEDIAN	2.43	1.87	1.94	2.00
7	GAS:	GASTWIRTH	2.59	1.70	1.67	1.68
8	TRI:	TRIMEAN	2.47	2.52	1.66	1.62
9	JAE:	JAECKEL ADAPT TRIM	2.50	2.10	1.63	1.57
10	BIC:	BICKEL ADAPT TRIM	13.22	16.52	2.34	1.68
11	SJA:	SYM JAECKEL AD TRIM	13.24	16.03	2.03	1.65
12	JBT:	2-CHOICE ADAPT TRIM	2.50	2.09	1.64	1.58
13	JLJ:	ADAPT TRIM LINCOM			1.68	
14	H20:	HUBER PROP 2, K=2.0	77.75	6.28	2.17	1.88
15	H17:	HUBER PROP 2, K=1.7	4.70	4.52	1.87	1.70
16	H15:	HUBER PROP 2, K=1.5	3.88	3.57	1.74	1.62
17	H12:	HUBER PROP 2, K=1.2	3.27	1.94	1.64	1.56
18	H10:	HUBER PROP 2, K=1.0	3.05	1.68	1.62	1.55
19	H07:	HUBER PROP 2, K=0.7	2.88	1.67	1.64	1.62
20	M15:	1ST-HU 1.5*IQS MEAN	6.13	34.20	347.94	2013.36
21	D20:	1ST-HU 2.0*IQS MED	4.17	5.17	1.92	1.81
22	D15:	1ST-HU 1.5*IQS MED	3.88	4.51	1.73	1.64
23	D10:	1ST-HU 1.0*IQS MED	3.61	3.53	1.63	1.55
24	D07:	1ST-HU 0.7*IQS MED	2.87	2.55	1.63	1.59
25	A20:	HUBER 2.0*ADS	3.01	1.97	1.86	1.74
26	A15:	HUBER 1.5*ADS	2.63	1.77	1.69	1.60
27	P15:	1S-HU 1.5*ADS MED	2.65	1.77	1.70	1.60
28	HMD:	M (2.0,2.0,5.5)*AD	2.22	1.63	1.57	1.50
29	25A:	M (2.5,4.5,9.5)*AD	2.51	1.60	1.59	1.50
30	22A:	M (2.2,3.7,5.9)*AD	2.35	1.61	1.55	1.46
31	21A:	M (2.1,4.0,8.2)*AD	2.35	1.57	1.54	1.46
32	17A:	M (1.7,3.4,8.5)*AD	2.23	1.56	1.53	1.47
33	12A:	M (1.2,3.5,8.0)*AD	2.15	1.59	1.56	1.52
34	ADA:	M (ADA,4.5,8.0)*AD	2.29	1.61	1.59	1.50
35	AMT:	M-TYPE (SINE)	2.49	1.58	1.55	1.47
36	OLS:	"OLSHEN'S"	4.31	1.87	1.70	1.73
37	TOL:	T-SKIP "OLSHEN'S"	2.49	1.68	1.74	1.77
38	MEL:	MEAN LIKELIHOOD	3.25	2.92	2.19	2.94
39	CML:	CAUCHY MAX LIKELIHD	2.32	1.98	1.96	1.97
40	LJS:	LEAST FAVORABLE	3.46	4.21	1.78	1.68
41	SST:	ITER S-SKIP TRIMEAN	2.40	1.79	1.86	1.93
42	CST:	ITER C-SKIP TRIMEAN	2.36	1.63	1.61	1.57
43	33T:	MX(3T,2)-SKIP MEXTR	2.94	2.64	1.89	1.82
44	3TO:	3T-SKIP MEAN	3.26	1.96	1.57	1.50
45	3T1:	MX(3T,2)-SKIP MEAN	2.53	1.91	1.56	1.50
46	5T1:	MX(5T,2)-SKIP MEAN	2.53	1.63	1.58	1.53
47	5T4:	MN(5T,.6)-SKIP MEAN	2.53	1.60	1.56	1.53
48	CTS:	C&T&S-SKIP TRIMEAN	12.29	1.77	1.77	1.78
49	TAK:	TAKEUCHI ADAPTIVE	3.70	2.36	1.60	1.52
50	DFA:	MAXIMUM EST LIKE	2.84	1.76	1.89	1.96
51	JOH:	JOHNS ADAPTIVE	2.43	1.78	1.65	1.60
52	HGP:	HOGG 67 ON KURTOSIS	87.07	3.08	1.90	1.90
53	HGL:	HOGG 69 ON KURTOSIS	81.04	2.46	1.68	1.66
54	THP:	T-SKIP HOGG 67	5.43	3.13	1.74	1.53
55	THL:	T-SKIP HOGG 69	4.87	2.96	1.69	1.51
56	JWT:	ADAPT FROM SKIPPING	2.60	1.75	1.84	1.88
57	H/L:	HODGES-LEHMANN	3.54	2.46	1.70	1.62
58	BH :	BICKEL-HODGES	3.54	4.95	1.81	1.69
59	2RM:	2 TIMES FOLDED MED	33.12	7.36	2.97	1.94
60	3RM:	3 TIMES FOLDED MED	33.12	806.64	9.76	4.48
61	4RM:	4 TIMES FOLDED MED	33.12	806.64	4357.23	15.48
62	3R1:	3-FOLDED 1-TRIM MED	2.47	3.84	1.89	1.77
63	3R2:	3-FOLDED 2-TRIM MED	2.47	1.93	1.64	1.64
64	CPL:	CAUCHY PITMAN (LOC)	1.93	1.57	1.65	1.67
65	SHO:	SHORTH	3.34	4.11	4.63	5.63

EXHIBIT 5-4C: VARIANCES (MULTIPLIED BY SAMPLE SIZE) OF THE ESTIMATORS

CAUCHY DISTRIBUTION

	CODE	NAME	5	10A	10B	20	40
1	M :	MEAN (AVERAGE)	4147.3	27314.1	*******	12548.0	*******
2	5%:	5% SYM TRIM MEAN	2886.4	8473.2	*******	24.0	15.79
3	10%:	10% SYM TRIM MEAN	1631.4	27.2	17.2	7.3	5.40
4	15%:	15% SYM TRIM MEAN	542.2	17.1	10.5	4.6	3.64
5	25%:	MIDMEAN	7.1	5.8	4.6	3.1	2.61
6	50%:	MEDIAN	6.3	3.7	3.4	2.9	2.43
7	GAS:	GASTWIRTH	7.1	4.0	3.8	3.1	2.53
8	TRI:	TRIMEAN	6.7	13.3	6.6	3.9	3.16
9	JAE:	JAECKEL ADAPT TRIM	6.8	30.9	6.7	3.5	2.80
10	BIC:	BICKEL ADAPT TRIM	276.9	568.4	12.2	16.6	3.03
11	SJA:	SYM JAECKEL AD TRIM	277.4	11.4	8.8	3.7	2.82
12	JBT:	2-CHOICE ADAPT TRIM	6.8	30.6	6.2	3.3	2.67
13	JLJ:	ADAPT TRIM LINCOM				2.8	
14	H20:	HUBER PROP 2, K=2.0	4147.3	49.8	32.2	9.3	7.39
15	H17:	HUBER PROP 2, K=1.7	17.4	33.5	13.2	6.8	5.48
16	H15:	HUBER PROP 2, K=1.5	10.9	23.7	10.3	5.7	4.55
17	H12:	HUBER PROP 2, K=1.2	8.9	6.8	7.2	4.4	3.56
18	H10:	HUBER PROP 2, K=1.0	8.3	5.3	5.4	3.7	3.06
19	H07:	HUBER PROP 2, K=0.7	7.8	4.2	4.1	3.0	2.58
20	M15:	1ST-HU 1.5*IQS MEAN	17.2	126.1	150.5	719.2	*******
21	D20:	1ST-HU 2.0*IQS MED	11.5	38.0	14.0	6.1	4.90
22	D15:	1ST-HU 1.5*IQS MED	10.4	31.6	11.6	5.0	4.07
23	D10:	1ST-HU 1.0*IQS MED	9.9	23.0	9.1	3.9	3.22
24	D07:	1ST-HU 0.7*IQS MED	7.6	13.2	6.3	3.3	2.77
25	A20:	HUBER 2.0*ADS	9.2	7.4	7.5	5.5	4.53
26	A15:	HUBER 1.5*ADS	7.7	6.0	6.0	4.5	3.73
27	P15:	1S-HU 1.5*ADS MED	7.8	6.0	6.1	4.5	3.74
28	HMD:	M (2.0,2.0,5.5)*AD	6.4	3.9	3.8	2.9	2.82
29	25A:	M (2.5,4.5,9.5)*AD	7.7	5.0	5.0	3.7	3.48
30	22A:	M (2.2,3.7,5.9)*AD	7.1	4.3	4.5	3.3	3.24
31	21A:	M (2.1,4.0,8.2)*AD	7.1	4.4	4.5	3.3	3.16
32	17A:	M (1.7,3.4,8.5)*AD	6.5	4.0	3.9	3.0	2.84
33	12A:	M (1.2,3.5,8.0)*AD	6.1	3.6	3.4	2.7	2.55
34	ADA:	M (ADA,4.5,8.0)*AD	6.3	4.0	3.9	2.9	2.62
35	AMT:	M-TYPE (SINE)	7.8	4.7	4.8	3.5	3.22
36	OLS:	"OLSHEN'S"	92.8	5.9	3.8	2.6	2.21
37	TOL:	T-SKIP "OLSHEN'S"	6.8	5.7	3.4	2.5	2.29
38	MEL:	MEAN LIKELIHOOD	59.8	9.7	22.2	10.0	12.87
39	CML:	CAUCHY MAX LIKELIHD	6.2	3.0	2.7	2.3	2.12
40	LJS:	LEAST FAVORABLE	9.4	30.8	10.5	4.9	3.17
41	SST:	ITER S-SKIP TRIMEAN	6.3	11.8	3.9	2.9	2.72
42	CST:	ITER C-SKIP TRIMEAN	6.6	12.2	4.4	3.0	2.75
43	33T:	MX(3T,2)-SKIP MEXTR	7.4	32.9	7.2	4.0	3.50
44	3TO:	3T-SKIP MEAN	8.2	30.3	6.3	3.3	2.85
45	3T1:	MX(3T,2)-SKIP MEAN	6.7	18.2	5.4	3.3	2.85
46	5T1:	MX(5T,2)-SKIP MEAN	6.7	17.5	4.7	3.0	2.52
47	5T4:	MN(5T,.6)-SKIP MEAN	6.6	17.7	4.7	2.9	2.50
48	CTS:	C&T&S-SKIP TRIMEAN	93.0	11.9	3.9	2.9	2.67
49	TAK:	TAKEUCHI ADAPTIVE	9.1	35.8	11.3	3.5	2.55
50	DFA:	MAXIMUM EST LIKE	8.0	3.9	3.9	2.8	2.53
51	JOH:	JOHNS ADAPTIVE	6.3	6.2	4.4	2.8	2.38
52	HGP:	HOGG 67 ON KURTOSIS	4389.9	45.7	32.2	4.4	2.51
53	HGL:	HOGG 69 ON KURTOSIS	4162.8	41.3	11.9	3.6	2.54
54	THP:	T-SKIP HOGG 67	14.2	36.8	11.1	4.5	3.77
55	THL:	T-SKIP HOGG 69	13.2	35.7	9.8	4.0	3.16
56	JWT:	ADAPT FROM SKIPPING	6.8	20.6	4.6	2.9	2.43
57	H/L:	HODGES-LEHMANN	9.8	12.3	6.9	4.2	3.34
58	BH:	BICKEL-HODGES	9.8	39.3	12.9	5.8	4.37
59	2RM:	2 TIMES FOLDED MED	1633.5	60.0	41.4	21.4	11.67
60	3RM:	3 TIMES FOLDED MED	1633.5	10725.4	*******	67.3	67.17
61	4RM:	4 TIMES FOLDED MED	1633.5	10725.4	*******	5006.6	114.98
62	3R1:	3-FOLDED 1-TRIM MED	6.7	27.2	17.2	7.7	7.89
63	3R2:	3-FOLDED 2-TRIM MED	6.7	6.9	5.2	4.4	5.24
64	CPL:	CAUCHY PITMAN (LOC)	6.2	3.1	2.6	2.3	2.09
65	SHO:	SHORTH	8.0	5.0	4.6	4.1	4.16

EXHIBIT 5-5: VARIANCES (MULTIPLIED BY SAMPLE SIZE) OF THE ESTIMATORS.

	CODE	NAME	N=5 D I S T R I B U T I O N S		
			NORMAL	25% 1/U	CAUCHY
1	M :	MEAN (AVERAGE)	1.000	77.75	4147.3
2	5%:	5% SYM TRIM MEAN	1.004	55.35	2886.4
3	10%:	10% SYM TRIM MEAN	1.020	32.91	1631.4
4	15%:	15% SYM TRIM MEAN	1.060	13.14	542.2
5	25%:	MIDMEAN	1.156	2.59	7.1
6	50%:	MEDIAN	1.465	2.43	6.3
7	GAS:	GASTWIRTH	1.156	2.59	7.1
8	TRI:	TRIMEAN	1.182	2.47	6.7
9	JAE:	JAECKEL ADAPT TRIM	1.179	2.50	6.8
10	BIC:	BICKEL ADAPT TRIM	1.085	13.22	276.9
11	SJA:	SYM JAECKEL AD TRIM	1.083	13.24	277.4
12	JBT:	2-CHOICE ADAPT TRIM	1.179	2.50	6.8
13	JLJ:	ADAPT TRIM LINCOM			
14	H20:	HUBER PROP 2, K=2.0	1.000	77.75	4147.3
15	H17:	HUBER PROP 2, K=1.7	1.015	4.70	17.4
16	H15:	HUBER PROP 2, K=1.5	1.043	3.88	1C.9
17	H12:	HUBER PROP 2, K=1.2	1.089	3.27	8.9
18	H10:	HUBER PROP 2, K=1.0	1.112	3.05	8.3
19	H07:	HUBER PROP 2, K=0.7	1.131	2.88	7.8
20	M15:	1ST-HU 1.5*IQS MEAN	1.117	6.13	17.2
21	D20:	1ST-HU 2.0*IQS MED	1.050	4.17	11.5
22	D15:	1ST-HU 1.5*IQS MED	1.074	3.88	10.4
23	D10:	1ST-HU 1.0*IQS MED	1.125	3.61	9.9
24	D07:	1ST-HU 0.7*IQS MED	1.144	2.87	7.6
25	A20:	HUBER 2.0*ADS	1.074	3.01	9.2
26	A15:	HUBER 1.5*ADS	1.116	2.63	7.7
27	P15:	1S-HU 1.5*ADS MED	1.117	2.65	7.8
28	HMD:	M (2.0,2.0,5.5)*AD	1.375	2.22	6.4
29	25A:	M (2.5,4.5,9.5)*AD	1.184	2.51	7.7
30	22A:	M (2.2,3.7,5.9)*AD	1.278	2.35	7.1
31	21A:	M (2.1,4.0,8.2)*AD	1.231	2.35	7.1
32	17A:	M (1.7,3.4,8.5)*AD	1.271	2.23	6.5
33	12A:	M (1.2,3.5,8.0)*AD	1.333	2.15	6.1
34	ADA:	M (ADA,4.5,8.0)*AD	1.309	2.29	6.3
35	AMT:	M-TYPE (SINE)	1.232	2.49	7.8
36	OLS:	"OLSHEN'S"	1.151	4.31	92.8
37	TOL:	T-SKIP "OLSHEN'S"	1.279	2.49	6.8
38	MEL:	MEAN LIKELIHOOD	1.315	3.25	59.8
39	CML:	CAUCHY MAX LIKELIHD	1.654	2.32	6.2
40	LJS:	LEAST FAVORABLE	1.073	3.46	9.4
41	SST:	ITER S-SKIP TRIMEAN	1.388	2.40	6.3
42	CST:	ITER C-SKIP TRIMEAN	1.272	2.36	6.6
43	33T:	MX(3T,2)-SKIP MEXTR	1.257	2.94	7.4
44	3TO:	3T-SKIP MEAN	1.214	3.26	8.2
45	3T1:	MX(3T,2)-SKIP MEAN	1.266	2.53	6.7
46	5T1:	MX(5T,2)-SKIP MEAN	1.266	2.53	6.7
47	5T4:	MN(5T,.6)-SKIP MEAN	1.262	2.53	6.6
48	CTS:	C&T&S-SKIP TRIMEAN	2.062	12.29	93.0
49	TAK:	TAKEUCHI ADAPTIVE	1.109	3.70	9.1
50	DFA:	MAXIMUM EST LIKE	1.391	2.84	8.0
51	JOH:	JOHNS ADAPTIVE	1.465	2.43	6.3
52	HGP:	HOGG 67 ON KURTOSIS	1.305	87.07	4389.9
53	HGL:	HOGG 69 ON KURTOSIS	1.136	81.04	4162.8
54	THP:	T-SKIP HOGG 67	1.357	5.43	14.2
55	THL:	T-SKIP HOGG 69	1.303	4.87	13.2
56	JWT:	ADAPT FROM SKIPPING	1.222	2.60	6.8
57	H/L:	HODGES-LEHMANN	1.081	3.54	9.8
58	BH :	BICKEL-HODGES	1.082	3.54	9.8
59	2RM:	2 TIMES FOLDED MED	1.018	33.12	1633.5
60	3RM:	3 TIMES FOLDED MED	1.018	33.12	1633.5
61	4RM:	4 TIMES FOLDED MED	1.018	33.12	1633.5
62	3R1:	3-FOLDED 1-TRIM MED	1.182	2.47	6.7
63	3R2:	3-FOLDED 2-TRIM MED	1.182	2.47	6.7
64	CPL:	CAUCHY PITMAN (LOC)	1.092	1.93	6.2
65	SHO:	SHORTH	2.453	3.34	8.0

EXHIBIT 5-6: VARIANCES (MULTIPLIED BY SAMPLE SIZE) OF THE ESTIMATORS.

	CODE	NAME	NORMAL	10% 3N	20% 3N	25% 1/U	ALL 1/U	CAUC
			N=10 D I S T R I B U T I O N S					
1	M :	MEAN (AVERAGE)	1.000	1.803	2.699	2038.78	5255.20	27314.
2	5%:	5% SYM TRIM MEAN	1.009	1.493	2.177	637.41	1665.42	8473.
3	10%:	10% SYM TRIM MEAN	1.048	1.295	1.788	3.84	25.48	27.
4	15%:	15% SYM TRIM MEAN	1.069	1.298	1.690	2.88	15.92	17.
5	25%:	MIDMEAN	1.148	1.358	1.640	1.85	8.41	5.
6	50%:	MEDIAN	1.366	1.598	1.840	1.87	7.33	3.
7	GAS:	GASTWIRTH	1.203	1.416	1.683	1.70	7.53	3.
8	TRI:	TRIMEAN	1.126	1.345	1.652	2.52	10.42	13.
9	JAE:	JAECKEL ADAPT TRIM	1.076	1.327	1.668	2.10	10.69	30.
10	BIC:	BICKEL ADAPT TRIM	1.070	1.372	1.776	16.52	23.22	568.
11	SJA:	SYM JAECKEL AD TRIM	1.085	1.362	1.723	16.03	19.02	11.
12	JBT:	2-CHOICE ADAPT TRIM	1.071	1.353	1.699	2.09	10.27	30.
13	JLJ:	ADAPT TRIM LINCOM						
14	H20:	HUBER PROP 2, K=2.0	1.007	1.410	2.147	6.28	44.77	49.
15	H17:	HUBER PROP 2, K=1.7	1.019	1.345	1.946	4.52	19.51	33.
16	H15:	HUBER PROP 2, K=1.5	1.031	1.320	1.794	3.57	14.58	23.
17	H12:	HUBER PROP 2, K=1.2	1.063	1.308	1.672	1.94	11.57	6.
18	H10:	HUBER PROP 2, K=1.0	1.093	1.319	1.647	1.68	9.75	5.
19	H07:	HUBER PROP 2, K=0.7	1.165	1.380	1.664	1.67	7.96	4.
20	M15:	1ST-HU 1.5*IQS MEAN	1.028	1.369	1.840	34.20	276.61	126.
21	D20:	1ST-HU 2.0*IQS MED	1.012	1.435	2.061	5.17	18.24	38.
22	D15:	1ST-HU 1.5*IQS MED	1.027	1.359	1.846	4.51	14.91	31.
23	D10:	1ST-HU 1.0*IQS MED	1.068	1.316	1.695	3.53	12.33	22.
24	D07:	1ST-HU 0.7*IQS MED	1.109	1.328	1.641	2.55	10.00	13.
25	A20:	HUBER 2.0*ADS	1.023	1.383	1.872	1.97	13.06	7.
26	A15:	HUBER 1.5*ADS	1.049	1.331	1.727	1.77	10.92	6.
27	P15:	1S-HU 1.5*ADS MED	1.049	1.334	1.732	1.77	11.02	6.
28	HMD:	M (2.0,2.0,5.5)*AD	1.217	1.404	1.691	1.63	7.72	3.
29	25A:	M (2.5,4.5,9.5)*AD	1.058	1.345	1.740	1.60	9.63	4.
30	22A:	M (2.2,3.7,5.9)*AD	1.121	1.362	1.700	1.61	8.69	4.
31	21A:	M (2.1,4.0,8.2)*AD	1.089	1.337	1.682	1.57	8.77	4.
32	17A:	M (1.7,3.4,8.5)*AD	1.127	1.342	1.649	1.56	8.04	4.
33	12A:	M (1.2,3.5,8.0)*AD	1.187	1.384	1.657	1.59	7.29	3.
34	ADA:	M (ADA,4.5,8.0)*AD	1.113	1.388	1.745	1.61	7.87	4.
35	AMT:	M-TYPE (SINE)	1.083	1.349	1.714	1.58	9.29	4.
36	OLS:	"OLSHEN'S"	1.244	1.434	1.672	1.87	7.42	5.
37	TOL:	T-SKIP "OLSHEN'S"	1.294	1.468	1.688	1.68	7.15	5.
38	MEL:	MEAN LIKELIHOOD	1.389	1.459	1.666	2.92	17.52	9.
39	CML:	CAUCHY MAX LIKELIHD	1.609	1.775	1.963	1.98	6.56	3.
40	LJS:	LEAST FAVORABLE	1.022	1.346	1.862	4.21	14.63	30.
41	SST:	ITER S-SKIP TRIMEAN	1.327	1.522	1.803	1.79	7.92	11.
42	CST:	ITER C-SKIP TRIMEAN	1.156	1.378	1.668	1.63	8.38	12.
43	33T:	MX(3T,2)-SKIP MEXTR	1.209	1.669	2.131	2.64	10.57	32.
44	3T0:	3T-SKIP MEAN	1.111	1.410	1.840	1.96	10.11	30.
45	3T1:	MX(3T,2)-SKIP MEAN	1.138	1.340	1.693	1.91	9.38	18.
46	5T1:	MX(5T,2)-SKIP MEAN	1.134	1.360	1.711	1.63	8.66	17.
47	5T4:	MN(5T,.6)-SKIP MEAN	1.130	1.354	1.701	1.60	8.59	17.
48	CTS:	C&T&S-SKIP TRIMEAN	1.298	1.494	1.769	1.77	7.90	11.
49	TAK:	TAKEUCHI ADAPTIVE	1.041	1.375	1.914	2.36	13.65	35.
50	DFA:	MAXIMUM EST LIKE	1.206	1.534	1.801	1.76	7.52	3.
51	JOH:	JOHNS ADAPTIVE	1.174	1.463	1.800	1.78	8.97	6.
52	HGP:	HOGG 67 ON KURTOSIS	1.124	1.730	2.660	3.08	27.89	45.
53	HGL:	HOGG 69 ON KURTOSIS	1.074	1.536	2.328	2.46	15.38	41.
54	THP:	T-SKIP HOGG 67	1.269	1.634	2.262	3.13	17.16	36.
55	THL:	T-SKIP HOGG 69	1.199	1.548	2.069	2.96	15.18	35.
56	JWT:	ADAPT FROM SKIPPING	1.141	1.401	1.750	1.75	8.80	20.
57	H/L:	HODGES-LEHMANN	1.063	1.346	1.751	2.46	11.14	12.
58	BH :	BICKEL-HODGES	1.054	1.390	1.866	4.95	15.95	39.
59	2RM:	2 TIMES FOLDED MED	1.013	1.419	2.228	7.36	57.94	59.
60	3RM:	3 TIMES FOLDED MED	1.003	1.600	2.401	806.64	2109.75	10725.4
61	4RM:	4 TIMES FOLDED MED	1.003	1.600	2.401	806.64	2109.75	10725.4
62	3R1:	3-FOLDED 1-TRIM MED	1.048	1.295	1.788	3.84	25.48	27.2
63	3R2:	3-FOLDED 2-TRIM MED	1.124	1.337	1.642	1.93	9.11	6.
64	CPL:	CAUCHY PITMAN (LOC)	1.158	1.355	1.619	1.57	6.31	3.0
65	SHO:	SHORTH	3.537	3.933	3.866	4.11	10.75	4.

EXHIBIT 5-7A: VARIANCES (MULTIPLIED BY SAMPLE SIZE) OF THE ESTIMATORS.

N=20 C O N T A M P O N E N T S

	CODE	NAME	0%	5% 3N	10% 3N	15% 3N	25% 3N
1	M :	MEAN (AVERAGE)	1.000	1.420	1.883	2.259	3.007
2	5%:	5% SYM TRIM MEAN	1.022	1.156	1.389	1.637	2.270
3	10%:	10% SYM TRIM MEAN	1.056	1.166	1.308	1.471	1.926
4	15%:	15% SYM TRIM MEAN	1.098	1.194	1.317	1.443	1.799
5	25%:	MIDMEAN	1.199	1.271	1.406	1.495	1.790
6	50%:	MEDIAN	1.498	1.516	1.704	1.747	2.156
7	GAS:	GASTWIRTH	1.233	1.299	1.451	1.517	1.819
8	TRI:	TRIMEAN	1.153	1.245	1.372	1.468	1.828
9	JAE:	JAECKEL ADAPT TRIM	1.105	1.210	1.367	1.473	1.818
10	BIC:	BICKEL ADAPT TRIM	1.088	1.204	1.390	1.506	1.922
11	SJA:	SYM JAECKEL AD TRIM	1.110	1.224	1.391	1.492	1.860
12	JBT:	2-CHOICE ADAPT TRIM	1.110	1.209	1.370	1.493	1.827
13	JLJ:	ADAPT TRIM LINCOM	1.167	1.267	1.463	1.553	1.906
14	H20:	HUBER PROP 2, K=2.0	1.009	1.171	1.408	1.664	2.305
15	H17:	HUBER PROP 2, K=1.7	1.021	1.156	1.346	1.549	2.087
16	H15:	HUBER PROP 2, K=1.5	1.036	1.158	1.320	1.490	1.956
17	H12:	HUBER PROP 2, K=1.2	1.074	1.180	1.314	1.443	1.824
18	H10:	HUBER PROP 2, K=1.0	1.114	1.208	1.337	1.445	1.781
19	H07:	HUBER PROP 2, K=0.7	1.204	1.270	1.416	1.493	1.793
20	M15:	1ST-HU 1.5*IQS MEAN	1.046	1.170	1.333	1.479	1.927
21	D20:	1ST-HU 2.0*IQS MED	1.017	1.181	1.390	1.593	2.146
22	D15:	1ST-HU 1.5*IQS MED	1.045	1.168	1.333	1.485	1.935
23	D10:	1ST-HU 1.0*IQS MED	1.110	1.205	1.337	1.453	1.790
24	D07:	1ST-HU 0.7*IQS MED	1.174	1.256	1.383	1.485	1.780
25	A20:	HUBER 2.0*ADS	1.019	1.173	1.377	1.578	2.095
26	A15:	HUBER 1.5*ADS	1.050	1.169	1.330	1.474	1.906
27	P15:	1S-HU 1.5*ADS MED	1.051	1.170	1.330	1.475	1.907
28	HMD:	M (2.0,2.0,5.5)*AD	1.246	1.284	1.422	1.476	1.814
29	25A:	M (2.5,4.5,9.5)*AD	1.046	1.156	1.317	1.487	1.936
30	22A:	M (2.2,3.7,5.9)*AD	1.107	1.182	1.339	1.459	1.866
31	21A:	M (2.1,4.0,8.2)*AD	1.081	1.171	1.314	1.447	1.850
32	17A:	M (1.7,3.4,8.5)*AD	1.130	1.204	1.336	1.435	1.784
33	12A:	M (1.2,3.5,8.0)*AD	1.205	1.257	1.398	1.473	1.784
34	ADA:	M (ADA,4.5,8.0)*AD	1.102	1.194	1.380	1.496	1.913
35	AMT:	M-TYPE (SINE)	1.070	1.163	1.314	1.467	1.888
36	OLS:	"OLSHEN'S"	1.347	1.382	1.519	1.570	1.905
37	TOL:	T-SKIP "OLSHEN'S"	1.410	1.422	1.570	1.630	1.963
38	MEL:	MEAN LIKELIHOOD	1.518	1.432	1.478	1.517	1.791
39	CML:	CAUCHY MAX LIKELIHD	1.720	1.660	1.843	1.840	2.142
40	LJS:	LEAST FAVORABLE	1.022	1.175	1.381	1.563	2.066
41	SST:	ITER S-SKIP TRIMEAN	1.463	1.488	1.633	1.749	2.089
42	CST:	ITER C-SKIP TRIMEAN	1.184	1.269	1.386	1.508	1.877
43	33T:	MX(3T,2)-SKIP MEXTR	1.307	1.450	1.784	1.861	2.539
44	3T0:	3T-SKIP MEAN	1.109	1.201	1.379	1.514	1.967
45	3T1:	MX(3T,2)-SKIP MEAN	1.119	1.193	1.340	1.483	1.910
46	5T1:	MX(5T,2)-SKIP MEAN	1.125	1.205	1.356	1.482	1.917
47	5T4:	MN(5T,.6)-SKIP MEAN	1.123	1.202	1.356	1.481	1.892
48	CTS:	C&T&S-SKIP TRIMEAN	1.405	1.443	1.530	1.602	1.984
49	TAK:	TAKEUCHI ADAPTIVE	1.050	1.187	1.383	1.529	2.017
50	DFA:	MAXIMUM EST LIKE	1.323	1.395	1.692	1.704	2.175
51	JOH:	JOHNS ADAPTIVE	1.137	1.259	1.443	1.528	1.904
52	HGP:	HOGG 67 ON KURTOSIS	1.055	1.280	1.557	1.793	2.416
53	HGL:	HOGG 69 ON KURTOSIS	1.064	1.227	1.452	1.606	2.086
54	THP:	T-SKIP HOGG 67	1.152	1.256	1.446	1.613	2.153
55	THL:	T-SKIP HOGG 69	1.142	1.235	1.433	1.580	2.097
56	JWT:	ADAPT FROM SKIPPING	1.164	1.320	1.501	1.634	2.061
57	H/L:	HODGES-LEHMANN	1.063	1.180	1.350	1.497	1.882
58	BH :	BICKEL-HODGES	1.047	1.197	1.388	1.578	1.994
59	2RM:	2 TIMES FOLDED MED	1.012	1.199	1.444	1.746	2.497
60	3RM:	3 TIMES FOLDED MED	1.004	1.217	1.608	1.932	2.749
61	4RM:	4 TIMES FOLDED MED	1.001	1.320	1.724	2.072	2.841
62	3R1:	3-FOLDED 1-TRIM MED	1.038	1.159	1.324	1.518	2.001
63	3R2:	3-FOLDED 2-TRIM MED	1.085	1.187	1.320	1.460	1.820
64	CPL:	CAUCHY PITMAN (LOC)	1.256	1.311	1.450	1.535	1.839
65	SHO:	SHORTH	4.620	4.301	4.412	4.437	4.962

EXHIBIT 5-7B: VARIANCES (MULTIPLIED BY SAMPLE SIZE) OF THE ESTIMATORS.
N=20 C O N T A M P O N E N T S

	CODE	NAME	50% 3N	75% 3N	5% 10N	10% 10N	25%
1	M :	MEAN (AVERAGE)	5.019	6.994	6.485	11.538	26.
2	5%:	5% SYM TRIM MEAN	4.448	6.860	1.232	2.902	14.
3	10%:	10% SYM TRIM MEAN	3.977	6.658	1.205	1.460	6.
4	15%:	15% SYM TRIM MEAN	3.561	6.450	1.215	1.431	3.
5	25%:	MIDMEAN	3.131	6.106	1.288	1.472	2.
6	50%:	MEDIAN	3.369	6.428	1.555	1.804	2.
7	GAS:	GASTWIRTH	3.117	6.228	1.322	1.505	2.
8	TRI:	TRIMEAN	3.457	6.394	1.267	1.478	2.
9	JAE:	JAECKEL ADAPT TRIM	3.545	6.576	1.241	1.448	2.
10	BIC:	BICKEL ADAPT TRIM	3.557	6.391	1.507	1.948	3.
11	SJA:	SYM JAECKEL AD TRIM	3.343	6.253	1.461	1.715	2.
12	JBT:	2-CHOICE ADAPT TRIM	3.529	6.510	1.260	1.464	2.
13	JLJ:	ADAPT TRIM LINCOM	3.692	6.916	1.325	1.475	1.
14	H20:	HUBER PROP 2, K=2.0	4.556	6.921	1.281	1.785	10.
15	H17:	HUBER PROP 2, K=1.7	4.313	6.843	1.232	1.576	5.
16	H15:	HUBER PROP 2, K=1.5	4.089	6.757	1.214	1.496	4.
17	H12:	HUBER PROP 2, K=1.2	3.667	6.538	1.210	1.437	2.
18	H10:	HUBER PROP 2, K=1.0	3.399	6.368	1.224	1.434	2.
19	H07:	HUBER PROP 2, K=0.7	3.126	6.056	1.286	1.470	2.
20	M15:	1ST-HU 1.5*IQS MEAN	3.913	6.664	2.073	13.232	31.
21	D20:	1ST-HU 2.0*IQS MED	4.255	6.780	1.290	1.694	4.
22	D15:	1ST-HU 1.5*IQS MED	3.946	6.691	1.230	1.506	3.
23	D10:	1ST-HU 1.0*IQS MED	3.494	6.464	1.226	1.436	2.
24	D07:	1ST-HU 0.7*IQS MED	3.208	6.169	1.264	1.457	2.
25	A20:	HUBER 2.0*ADS	4.163	6.705	1.273	1.666	3.
26	A15:	HUBER 1.5*ADS	3.785	6.590	1.222	1.491	2.
27	P15:	1S-HU 1.5*ADS MED	3.814	6.632	1.222	1.490	2.
28	HMD:	M (2.0,2.0,5.5)*AD	3.630	7.163	1.231	1.308	1.
29	25A:	M (2.5,4.5,9.5)*AD	3.967	6.727	1.127	1.259	2.
30	22A:	M (2.2,3.7,5.9)*AD	3.918	6.952	1.147	1.249	1.
31	21A:	M (2.1,4.0,8.2)*AD	3.757	6.693	1.136	1.248	1.
32	17A:	M (1.7,3.4,8.5)*AD	3.482	6.573	1.166	1.268	1.
33	12A:	M (1.2,3.5,8.0)*AD	3.241	6.352	1.225	1.315	1.
34	ADA:	M (ADA,4.5,8.0)*AD	3.583	6.567	1.166	1.298	1.
35	AMT:	M-TYPE (SINE)	3.888	6.684	1.131	1.245	1.
36	OLS:	"OLSHEN'S"	3.161	6.140	1.378	1.508	1.
37	TOL:	T-SKIP "OLSHEN'S"	3.382	6.475	1.431	1.527	1.
38	MEL:	MEAN LIKELIHOOD	3.305	6.828	1.204	1.296	2.
39	CML:	CAUCHY MAX LIKELIHD	3.241	6.593	1.644	1.712	1.
40	LJS:	LEAST FAVORABLE	4.288	6.875	1.270	1.565	2.
41	SST:	ITER S-SKIP TRIMEAN	3.870	8.110	1.486	1.616	1.
42	CST:	ITER C-SKIP TRIMEAN	3.693	6.608	1.253	1.390	1.
43	33T:	MX(3T,2)-SKIP MEXTR	5.182	8.009	1.370	1.436	2.
44	3T0:	3T-SKIP MEAN	4.425	7.310	1.178	1.277	1.
45	3T1:	MX(3T,2)-SKIP MEAN	4.334	7.258	1.171	1.270	1.
46	5T1:	MX(5T,2)-SKIP MEAN	4.135	7.189	1.195	1.335	1.
47	5T4:	MN(5T,.6)-SKIP MEAN	4.118	7.175	1.193	1.326	1.
48	CTS:	C&T&S-SKIP TRIMEAN	3.880	7.618	1.391	1.534	1.
49	TAK:	TAKEUCHI ADAPTIVE	4.064	6.926	1.167	1.317	2.
50	DFA:	MAXIMUM EST LIKE	3.623	6.538	1.397	1.584	2.
51	JOH:	JOHNS ADAPTIVE	3.714	6.958	1.318	1.458	1.
52	HGP:	HOGG 67 ON KURTOSIS	4.826	7.440	1.467	1.793	5.
53	HGL:	HOGG 69 ON KURTOSIS	4.206	7.025	1.311	1.515	2.
54	THP:	T-SKIP HOGG 67	5.056	8.073	1.219	1.324	2.
55	THL:	T-SKIP HOGG 69	4.865	7.799	1.200	1.299	2.
56	JWT:	ADAPT FROM SKIPPING	3.817	6.774	1.501	1.681	2.
57	H/L:	HODGES-LEHMANN	3.624	6.320	1.237	1.518	2.
58	BH :	BICKEL-HODGES	4.014	6.786	1.291	1.599	3.
59	2RM:	2 TIMES FOLDED MED	4.764	6.978	1.348	1.895	21.
60	3RM:	3 TIMES FOLDED MED	4.871	7.045	1.410	6.774	23.
61	4RM:	4 TIMES FOLDED MED	4.919	6.986	3.737	8.098	24.
62	3R1:	3-FOLDED 1-TRIM MED	4.097	6.748	1.216	1.525	7.
63	3R2:	3-FOLDED 2-TRIM MED	3.660	6.559	1.216	1.442	2.
64	CPL:	CAUCHY PITMAN (LOC)	3.135	6.778	1.302	1.406	1.
65	SHO:	SHORTH	6.220	14.369	4.480	4.306	3.

EXHIBIT 5-7C: VARIANCES (MULTIPLIED BY SAMPLE SIZE) OF THE ESTIMATORS.

	CODE	NAME	N=20	O T H E R	D I S T R I B U T I O N S		
			10% 1/U	25% 1/U	10% 3/U	25% 3/U	ALL 1/U
1	M :	MEAN (AVERAGE)	2980.848	11128.60	2838.503	77602.06	12951.48
2	5%:	5% SYM TRIM MEAN	1.466	3.84	4.384	35.05	35.94
3	10%:	10% SYM TRIM MEAN	1.264	1.81	1.923	8.30	13.60
4	15%:	15% SYM TRIM MEAN	1.256	1.64	1.519	3.53	9.31
5	25%:	MIDMEAN	1.330	1.64	1.502	2.38	6.62
6	50%:	MEDIAN	1.640	1.94	1.804	2.53	6.60
7	GAS:	GASTWIRTH	1.362	1.67	1.538	2.38	6.59
8	TRI:	TRIMEAN		1.66	1.509	2.94	8.07
9	JAE:	JAECKEL ADAPT TRIM	1.271	1.63	1.479	2.49	7.23
10	BIC:	BICKEL ADAPT TRIM	3.235	2.34	3.183	4.50	9.82
11	SJA:	SYM JAECKEL AD TRIM	833.976	2.03	1617.565	73656.75	7.98
12	JBT:	2-CHOICE ADAPT TRIM	1.275	1.64	1.481	2.42	7.07
13	JLJ:	ADAPT TRIM LINCOM	1.326	1.68	1.468	2.08	6.64
14	H20:	HUBER PROP 2, K=2.0	1.305	2.17	2.202	9.04	18.41
15	H17:	HUBER PROP 2, K=1.7	1.251	1.87	1.786	5.93	14.03
16	H15:	HUBER PROP 2, K=1.5	1.238	1.74	1.613	4.48	11.41
17	H12:	HUBER PROP 2, K=1.2	1.243	1.64	1.498	3.09	8.60
18	H10:	HUBER PROP 2, K=1.0	1.264	1.62	1.477	2.67	7.50
19	H07:	HUBER PROP 2, K=0.7	1.330	1.64	1.509	2.33	6.56
20	M15:	1ST-HU 1.5*IQS MEAN	85.651	347.94	262.421	810.52	2362.78
21	D20:	1ST-HU 2.0*IQS MED	1.271	1.92	1.765	4.70	11.53
22	D15:	1ST-HU 1.5*IQS MED	1.241	1.73	1.571	3.69	9.51
23	D10:	1ST-HU 1.0*IQS MED	1.262	1.63	1.477	2.86	7.82
24	D07:	1ST-HU 0.7*IQS MED	1.310	1.63	1.492	2.51	6.87
25	A20:	HUBER 2.0*ADS	1.261	1.86	1.729	3.90	10.52
26	A15:	HUBER 1.5*ADS	1.238	1.69	1.558	3.08	8.75
27	P15:	1S-HU 1.5*ADS MED	1.239	1.70	1.558	3.09	8.80
28	HMD:	M (2.0,2.0,5.5)*AD	1.600	1.57	1.677	2.19	6.64
29	25A:	M (2.5,4.5,9.5)*AD	1.189	1.59	1.371	2.45	8.01
30	22A:	M (2.2,3.7,5.9)*AD	1.215	1.55	1.337	2.11	7.42
31	21A:	M (2.1,4.0,8.2)*AD	1.202	1.54	1.338	2.19	7.38
32	17A:	M (1.7,3.4,8.5)*AD	1.238	1.53	1.340	2.02	6.76
33	12A:	M (1.2,3.5,8.0)*AD	1.297	1.56	1.380	1.93	6.23
34	ADA:	M (ADA,4.5,8.0)*AD	1.235	1.59	1.388	2.09	6.65
35	AMT:	M-TYPE (SINE)	1.196	1.55	1.339	2.23	7.74
36	OLS:	"OLSHEN'S"	1.450	1.70	1.538	2.12	6.02
37	TOL:	T-SKIP "OLSHEN'S"	1.541	1.74	1.549	2.01	6.10
38	MEL:	MEAN LIKELIHOOD	1.487	2.19	2.204	14.39	13.07
39	CML:	CAUCHY MAX LIKELIHD	1.749	1.96	1.745	2.07	5.78
40	LJS:	LEAST FAVORABLE	1.253	1.78	1.603	3.44	9.83
41	SST:	ITER S-SKIP TRIMEAN	1.686	1.86	1.618	2.08	6.77
42	CST:	ITER C-SKIP TRIMEAN	1.343	1.61	1.427	2.25	6.88
43	33T:	MX(3T,2)-SKIP MEXTR		1.89			9.41
44	3T0:	3T-SKIP MEAN		1.57			7.60
45	3T1:	MX(3T,2)-SKIP MEAN	1.280	1.56	1.351	2.26	7.52
46	5T1:	MX(5T,2)-SKIP MEAN	1.314	1.58	1.389	2.19	6.99
47	5T4:	MN(5T,.6)-SKIP MEAN	1.308	1.56	1.368	2.07	6.86
48	CTS:	C&T&S-SKIP TRIMEAN	1.467	1.77	1.730	4.46	6.73
49	TAK:	TAKEUCHI ADAPTIVE	1.218	1.60	1.425	2.46	7.63
50	DFA:	MAXIMUM EST LIKE	1.492	1.89	1.671	2.24	6.81
51	JOH:	JOHNS ADAPTIVE	1.314	1.65	1.452	2.05	6.22
52	HGP:	HOGG 67 ON KURTOSIS	1.420	1.90	1.842	3.17	9.36
53	HGL:	HOGG 69 ON KURTOSIS	1.295	1.68	1.530	2.54	7.31
54	THP:	T-SKIP HOGG 67	1.301	1.74	1.441	2.90	9.79
55	THL:	T-SKIP HOGG 69	1.291	1.69	1.399	2.46	8.89
56	JWT:	ADAPT FROM SKIPPING	1.481	1.84	1.628	2.33	6.69
57	H/L:	HODGES-LEHMANN	1.257	1.70	1.573	3.21	8.38
58	BH :	BICKEL-HODGES	1.281	1.81	1.651	4.43	11.11
59	2RM:	2 TIMES FOLDED MED	1.569	2.97	5.055	40.49	32.96
60	3RM:	3 TIMES FOLDED MED	2.051	9.76	10.481	98.93	90.80
61	4RM:	4 TIMES FOLDED MED	1166.617	4357.23	1124.892	30456.73	5121.59
62	3R1:	3-FOLDED 1-TRIM MED	1.268	1.89	2.000	8.72	14.26
63	3R2:	3-FOLDED 2-TRIM MED	1.250	1.64	1.510	3.28	8.81
64	CPL:	CAUCHY PITMAN (LOC)	1.366	1.65	1.448	1.87	5.94
65	SHO:	SHORTH	4.708	4.63	4.488	4.45	10.87

	CODE	NAME	CAUCHY	T3	D-EX
1	M :	MEAN (AVERAGE)	12548.0	3.138	2.10
2	5%:	5% SYM TRIM MEAN	24.0	1.883	1.77
3	10%:	10% SYM TRIM MEAN	7.3	1.683	1.60
4	15%:	15% SYM TRIM MEAN	4.6	1.605	1.48
5	25%:	MIDMEAN	3.1	1.591	1.33
6	50%:	MEDIAN	2.9	1.817	1.37
7	GAS:	GASTWIRTH	3.1	1.627	1.35
8	TRI:	TRIMEAN	3.9	1.603	1.43
9	JAE:	JAECKEL ADAPT TRIM	3.5	1.616	1.48
10	BIC:	BICKEL ADAPT TRIM	16.6	1.696	1.45
11	SJA:	SYM JAECKEL AD TRIM	3.7	1.652	1.39
12	JBT:	2-CHOICE ADAPT TRIM	3.3	1.651	1.45
13	JLJ:	ADAPT TRIM LINCOM	2.8	1.713	1.53
14	H20:	HUBER PROP 2, K=2.0	9.3	1.912	1.81
15	H17:	HUBER PROP 2, K=1.7	6.8	1.791	1.71
16	H15:	HUBER PROP 2, K=1.5	5.7	1.713	1.64
17	H12:	HUBER PROP 2, K=1.2	4.4	1.625	1.51
18	H10:	HUBER PROP 2, K=1.0	3.7	1.594	1.43
19	H07:	HUBER PROP 2, K=0.7	3.0	1.600	1.32
20	M15:	1ST-HU 1.5*IQS MEAN	719.2	2.024	1.59
21	D20:	1ST-HU 2.0*IQS MED	6.1	1.825	1.72
22	D15:	1ST-HU 1.5*IQS MED	5.0	1.698	1.61
23	D10:	1ST-HU 1.0*IQS MED	3.9	1.606	1.47
24	D07:	1ST-HU 0.7*IQS MED	3.3	1.588	1.36
25	A20:	HUBER 2.0*ADS	5.5	1.791	1.67
26	A15:	HUBER 1.5*ADS	4.5	1.673	1.55
27	P15:	1S-HU 1.5*ADS MED	4.5	1.680	1.56
28	HMD:	M (2.0,2.0,5.5)*AD	2.9	1.656	1.52
29	25A:	M (2.5,4.5,9.5)*AD	3.7	1.671	1.58
30	22A:	M (2.2,3.7,5.9)*AD	3.3	1.663	1.56
31	21A:	M (2.1,4.0,8.2)*AD	3.3	1.624	1.52
32	17A:	M (1.7,3.4,8.5)*AD	3.0	1.583	1.44
33	12A:	M (1.2,3.5,8.0)*AD	2.7	1.585	1.36
34	ADA:	M (ADA,4.5,8.0)*AD	2.9	1.657	1.48
35	AMT:	M-TYPE (SINE)	3.5	1.643	1.54
36	OLS:	"OLSHEN'S"	2.6	1.650	1.29
37	TOL:	T-SKIP "OLSHEN'S"	2.5	1.715	1.36
38	MEL:	MEAN LIKELIHOOD	10.0	1.630	1.39
39	CML:	CAUCHY MAX LIKELIHD	2.3	1.846	1.35
40	LJS:	LEAST FAVORABLE	4.9	1.804	1.69
41	SST:	ITER S-SKIP TRIMEAN	2.9	1.924	1.66
42	CST:	ITER C-SKIP TRIMEAN	3.0	1.644	1.47
43	33T:	MX(3T,2)-SKIP MEXTR	4.0	2.118	1.92
44	3T0:	3T-SKIP MEAN	3.3	1.737	1.69
45	3T1:	MX(3T,2)-SKIP MEAN	3.3	1.701	1.65
46	5T1:	MX(5T,2)-SKIP MEAN	3.0	1.680	1.61
47	5T4:	MN(5T,.6)-SKIP MEAN	2.9	1.675	1.60
48	CTS:	C&T&S-SKIP TRIMEAN	2.9	1.753	1.58
49	TAK:	TAKEUCHI ADAPTIVE	3.5	1.716	1.62
50	DFA:	MAXIMUM EST LIKE	2.8	1.822	1.45
51	JOH:	JOHNS ADAPTIVE	2.8	1.670	1.51
52	HGP:	HOGG 67 ON KURTOSIS	4.4	1.921	1.87
53	HGL:	HOGG 69 ON KURTOSIS	3.6	1.772	1.67
54	THP:	T-SKIP HOGG 67	4.5	1.889	1.92
55	THL:	T-SKIP HOGG 69	4.0	1.859	1.80
56	JWT:	ADAPT FROM SKIPPING	2.9	1.745	1.47
57	H/L:	HODGES-LEHMANN	4.2	1.658	1.49
58	BH :	BICKEL-HODGES	5.8	1.750	1.67
59	2RM:	2 TIMES FOLDED MED	21.4	2.026	1.90
60	3RM:	3 TIMES FOLDED MED	67.3	2.210	1.98
61	4RM:	4 TIMES FOLDED MED	5006.6	2.622	2.03
62	3R1:	3-FOLDED 1-TRIM MED	7.7	1.734	1.66
63	3R2:	3-FOLDED 2-TRIM MED	4.4	1.632	1.52
64	CPL:	CAUCHY PITMAN (LOC)	2.3	1.616	1.31
65	SHO:	SHORTH	4.1	4.307	2.81

EXHIBIT 5-7E: VARIANCES (MULTIPLIED BY SAMPLE SIZE) OF THE ESTIMATORS.

	CODE	NAME	1% 3N	2.5% 3N	5% 3N	10% 3N
1	M :	MEAN (AVERAGE)	1.085	1.213	1.427	1.838
2	5%:	5% SYM TRIM MEAN	1.050	1.099	1.193	1.411
3	10%:	10% SYM TRIM MEAN	1.078	1.114	1.179	1.326
4	15%:	15% SYM TRIM MEAN	1.117	1.149	1.204	1.325
5	25%:	MIDMEAN	1.215	1.241	1.290	1.394
6	50%:	MEDIAN	1.505	1.522	1.564	1.664
7	GAS:	GASTWIRTH	1.248	1.274	1.322	1.424
8	TRI:	TRIMEAN	1.172	1.202	1.255	1.366
9	JAE:	JAECKEL ADAPT TRIM	1.127	1.162	1.224	1.352
10	BIC:	BICKEL ADAPT TRIM	1.112	1.152	1.223	1.369
11	SJA:	SYM JAECKEL AD TRIM	1.134	1.172	1.237	1.369
12	JBT:	2-CHOICE ADAPT TRIM	1.131	1.165	1.228	1.360
13	JLJ:	ADAPT TRIM LINCOM	1.188	1.225	1.292	1.426
14	H20:	HUBER PROP 2, K=2.0	1.043	1.098	1.200	1.429
15	H17:	HUBER PROP 2, K=1.7	1.049	1.095	1.177	1.362
16	H15:	HUBER PROP 2, K=1.5	1.061	1.101	1.173	1.332
17	H12:	HUBER PROP 2, K=1.2	1.096	1.130	1.190	1.318
18	H10:	HUBER PROP 2, K=1.0	1.133	1.164	1.219	1.334
19	H07:	HUBER PROP 2, K=0.7	1.218	1.244	1.292	1.396
20	M15:	1ST-HU 1.5*IQS MEAN	1.071	1.111	1.182	1.333
21	D20:	1ST-HU 2.0*IQS MED	1.051	1.104	1.197	1.396
22	D15:	1ST-HU 1.5*IQS MED	1.071	1.111	1.182	1.336
23	D10:	1ST-HU 1.0*IQS MED	1.130	1.161	1.217	1.336
24	D07:	1ST-HU 0.7*IQS MED	1.191	1.219	1.270	1.378
25	A20:	HUBER 2.0*ADS	1.051	1.101	1.191	1.384
26	A15:	HUBER 1.5*ADS	1.075	1.114	1.183	1.331
27	P15:	1S-HU 1.5*ADS MED	1.075	1.114	1.183	1.331
28	HMD:	M (2.0,2.0,5.5)*AD	1.255	1.274	1.313	1.400
29	25A:	M (2.5,4.5,9.5)*AD	1.069	1.107	1.176	1.330
30	22A:	M (2.2,3.7,5.9)*AD	1.123	1.152	1.208	1.335
31	21A:	M (2.1,4.0,8.2)*AD	1.100	1.131	1.189	1.319
32	17A:	M (1.7,3.4,8.5)*AD	1.146	1.173	1.222	1.331
33	12A:	M (1.2,3.5,8.0)*AD	1.217	1.239	1.283	1.381
34	ADA:	M (ADA,4.5,8.0)*AD	1.122	1.157	1.222	1.362
35	AMT:	M-TYPE (SINE)	1.090	1.123	1.184	1.325
36	OLS:	"OLSHEN'S"	1.356	1.374	1.411	1.497
37	TOL:	T-SKIP "OLSHEN'S"	1.415	1.429	1.463	1.550
38	MEL:	MEAN LIKELIHOOD	1.503	1.488	1.479	1.498
39	CML:	CAUCHY MAX LIKELIHD	1.712	1.712	1.730	1.790
40	LJS:	LEAST FAVORABLE	1.053	1.103	1.192	1.378
41	SST:	ITER S-SKIP TRIMEAN	1.470	1.487	1.528	1.636
42	CST:	ITER C-SKIP TRIMEAN	1.202	1.230	1.281	1.396
43	33T:	MX(3T,2)-SKIP MEXTR	1.339	1.394	1.496	1.697
44	3T0:	3T-SKIP MEAN	1.129	1.163	1.228	1.373
45	3T1:	MX(3T,2)-SKIP MEAN	1.135	1.164	1.219	1.351
46	5T1:	MX(5T,2)-SKIP MEAN	1.142	1.172	1.229	1.358
47	5T4:	MN(5T,.6)-SKIP MEAN	1.140	1.170	1.226	1.355
48	CTS:	C&T&S-SKIP TRIMEAN	1.413	1.429	1.460	1.538
49	TAK:	TAKEUCHI ADAPTIVE	1.078	1.124	1.204	1.371
50	DFA:	MAXIMUM EST LIKE	1.341	1.378	1.450	1.592
51	JOH:	JOHNS ADAPTIVE	1.162	1.203	1.272	1.408
52	HGP:	HOGG 67 ON KURTOSIS	1.101	1.173	1.295	1.546
53	HGL:	HOGG 69 ON KURTOSIS	1.098	1.151	1.243	1.428
54	THP:	T-SKIP HOGG 67	1.174	1.212	1.285	1.451
55	THL:	T-SKIP HOGG 69	1.163	1.199	1.268	1.427
56	JWT:	ADAPT FROM SKIPPING	1.196	1.244	1.325	1.484
57	H/L:	HODGES-LEHMANN	1.087	1.127	1.197	1.347
58	BH :	BICKEL-HODGES	1.078	1.126	1.212	1.391
59	2RM:	2 TIMES FOLDED MED	1.051	1.112	1.226	1.483
60	3RM:	3 TIMES FOLDED MED	1.050	1.128	1.275	1.598
61	4RM:	4 TIMES FOLDED MED	1.066	1.168	1.345	1.704
62	3R1:	3-FOLDED 1-TRIM MED	1.063	1.104	1.178	1.346
63	3R2:	3-FOLDED 2-TRIM MED	1.106	1.139	1.199	1.329
64	CPL:	CAUCHY PITMAN (LOC)	1.269	1.291	1.336	1.437
65	SHO:	SHORTH	4.564	4.504	4.448	4.440

EXHIBIT 5-8:   VARIANCES (MULTIPLIED BY SAMPLE SIZE) OF THE ESTIMATORS.

	CODE	NAME	N=40 D I S T R I B U T I O N S		
			NORMAL	25% 1/U	CAUCHY
1	M :	MEAN (AVERAGE)	1.000	2119.31	2565124.0
2	5%:	5% SYM TRIM MEAN	1.025	2.10	15.8
3	10%:	10% SYM TRIM MEAN	1.058	1.61	5.4
4	15%:	15% SYM TRIM MEAN	1.094	1.56	3.6
5	25%:	MIDMEAN	1.184	1.62	2.6
6	50%:	MEDIAN	1.527	2.00	2.4
7	GAS:	GASTWIRTH	1.248	1.68	2.5
8	TRI:	TRIMEAN	1.167	1.62	3.2
9	JAE:	JAECKEL ADAPT TRIM	1.076	1.57	2.8
10	BIC:	BICKEL ADAPT TRIM	1.071	1.68	3.0
11	SJA:	SYM JAECKEL AD TRIM	1.089	1.65	2.8
12	JBT:	2-CHOICE ADAPT TRIM	1.088	1.58	2.7
13	JLJ:	ADAPT TRIM LINCOM			
14	H20:	HUBER PROP 2, K=2.0	1.010	1.88	7.4
15	H17:	HUBER PROP 2, K=1.7	1.024	1.70	5.5
16	H15:	HUBER PROP 2, K=1.5	1.038	1.62	4.5
17	H12:	HUBER PROP 2, K=1.2	1.071	1.56	3.6
18	H10:	HUBER PROP 2, K=1.0	1.106	1.55	3.1
19	H07:	HUBER PROP 2, K=0.7	1.185	1.62	2.6
20	M15:	1ST-HU 1.5*IQS MEAN	1.033	2013.36	11033.9
21	D20:	1ST-HU 2.0*IQS MED	1.011	1.81	4.9
22	D15:	1ST-HU 1.5*IQS MED	1.032	1.64	4.1
23	D10:	1ST-HU 1.0*IQS MED	1.090	1.55	3.2
24	D07:	1ST-HU 0.7*IQS MED	1.153	1.59	2.8
25	A20:	HUBER 2.0*ADS	1.015	1.74	4.5
26	A15:	HUBER 1.5*ADS	1.042	1.60	3.7
27	P15:	1S-HU 1.5*ADS MED	1.042	1.60	3.7
28	HMD:	M (2.0,2.0,5.5)*AD	1.193	1.50	2.8
29	25A:	M (2.5,4.5,9.5)*AD	1.034	1.50	3.5
30	22A:	M (2.2,3.7,5.9)*AD	1.080	1.46	3.2
31	21A:	M (2.1,4.0,8.2)*AD	1.061	1.46	3.2
32	17A:	M (1.7,3.4,8.5)*AD	1.104	1.47	2.8
33	12A:	M (1.2,3.5,8.0)*AD	1.169	1.52	2.6
34	ADA:	M (ADA,4.5,8.0)*AD	1.057	1.50	2.6
35	AMT:	M-TYPE (SINE)	1.056	1.47	3.2
36	OLS:	"OLSHEN'S"	1.360	1.73	2.2
37	TOL:	T-SKIP "OLSHEN'S"	1.413	1.77	2.3
38	MEL:	MEAN LIKELIHOOD	1.505	2.94	12.9
39	CML:	CAUCHY MAX LIKELIHD	1.653	1.97	2.1
40	LJS:	LEAST FAVORABLE	1.012	1.68	3.2
41	SST:	ITER S-SKIP TRIMEAN	1.482	1.93	2.7
42	CST:	ITER C-SKIP TRIMEAN	1.191	1.57	2.8
43	33T:	MX(3T,2)-SKIP MEXTR	1.420	1.82	3.5
44	3T0:	3T-SKIP MEAN	1.107	1.50	2.8
45	3T1:	MX(3T,2)-SKIP MEAN	1.111	1.50	2.8
46	5T1:	MX(5T,2)-SKIP MEAN	1.113	1.53	2.5
47	5T4:	MN(5T,.6)-SKIP MEAN	1.113	1.53	2.5
48	CTS:	C&T&S-SKIP TRIMEAN	1.400	1.78	2.7
49	TAK:	TAKEUCHI ADAPTIVE	1.036	1.52	2.6
50	DFA:	MAXIMUM EST LIKE	1.370	1.96	2.5
51	JOH:	JOHNS ADAPTIVE	1.101	1.60	2.4
52	HGP:	HOGG 67 ON KURTOSIS	1.020	1.90	2.5
53	HGL:	HOGG 69 ON KURTOSIS	1.046	1.66	2.5
54	THP:	T-SKIP HOGG 67	1.112	1.53	3.8
55	THL:	T-SKIP HOGG 69	1.112	1.51	3.2
56	JWT:	ADAPT FROM SKIPPING	1.192	1.88	2.4
57	H/L:	HODGES-LEHMANN	1.055	1.62	3.3
58	BH :	BICKEL-HODGES	1.048	1.69	4.4
59	2RM:	2 TIMES FOLDED MED	1.012	1.94	11.7
60	3RM:	3 TIMES FOLDED MED	1.004	4.48	67.2
61	4RM:	4 TIMES FOLDED MED	1.001	15.48	115.0
62	3R1:	3-FOLDED 1-TRIM MED	1.024	1.77	7.9
63	3R2:	3-FOLDED 2-TRIM MED	1.044	1.64	5.2
64	CPL:	CAUCHY PITMAN (LOC)	1.277	1.67	2.1
65	SHO:	SHORTH	5.378	5.63	4.2

5C 2.5% POINTS AS PSEUDO VARIANCES

The variances of 5B are useful measures of accuracy for near normal distributions. However for long tailed distributions the variance, if it exists, is essentially determined by a few observations in the extreme tails of the distribution. Variance is not a robust measure of scale. Percentage points yield more robust measures. Here we use the 2 1/2% point of the distribution of the estimator, square this quantity and divide by the normalizing constant $(1.96)^2$. This pseudo-variance has been further rescaled by multiplying by n to yield quantities comparable across sample sizes.

Pseudo-variances were also calculated for one-sided 50% (central density), 25%, 10%, 2.5%, 1%, .5%, and 0.1% points. Selected values are given in Section 5D.

EXHIBIT 5-9:   PSEUDO-VARIANCES OF THE ESTIMATORS AT THEIR 2.5% POINTS.

	CODE	NAME	NORMAL	25% 1/U	CAUCHY
			N=5 D I S T R I B U T I O N S		
1	M :	MEAN (AVERAGE)	1.000	28.233	177.25
2	5%:	5% SYM TRIM MEAN	1.004	21.293	126.54
3	10%:	10% SYM TRIM MEAN	1.020	14.048	77.63
4	15%:	15% SYM TRIM MEAN	1.059	7.126	34.68
5	25%:	MIDMEAN	1.156	2.121	7.10
6	50%:	MEDIAN	1.469	2.208	5.29
7	GAS:	GASTWIRTH	1.156	2.121	7.10
8	TRI:	TRIMEAN	1.182	2.098	6.61
9	JAE:	JAECKEL ADAPT TRIM	1.180	2.109	6.71
10	BIC:	BICKEL ADAPT TRIM	1.086	5.014	17.44
11	SJA:	SYM JAECKEL AD TRIM	1.084	5.046	18.75
12	JBT:	2-CHOICE ADAPT TRIM	1.180	2.109	6.71
13	JLJ:	ADAPT TRIM LINCOM			
14	H20:	HUBER PROP 2, K=2.0	1.000	28.233	177.25
15	H17:	HUBER PROP 2, K=1.7	1.015	3.781	17.31
16	H15:	HUBER PROP 2, K=1.5	1.043	2.804	11.59
17	H12:	HUBER PROP 2, K=1.2	1.090	2.335	8.97
18	H10:	HUBER PROP 2, K=1.0	1.113	2.230	8.31
19	H07:	HUBER PROP 2, K=0.7	1.132	2.177	7.78
20	M15:	1ST-HU 1.5*IQS MEAN	1.119	3.397	14.81
21	D20:	1ST-HU 2.0*IQS MED	1.051	2.981	11.98
22	D15:	1ST-HU 1.5*IQS MED	1.075	2.554	10.30
23	D10:	1ST-HU 1.0*IQS MED	1.126	2.322	9.57
24	D07:	1ST-HU 0.7*IQS MED	1.144	2.168	7.44
25	A20:	HUBER 2.0*ADS	1.076	2.881	10.13
26	A15:	HUBER 1.5*ADS	1.119	2.427	8.38
27	P15:	1S-HU 1.5*ADS MED	1.120	2.441	8.56
28	HMD:	M (2.0,2.0,5.5)*AD	1.397	1.989	5.89
29	25A:	M (2.5,4.5,5.9)*AD	1.193	2.275	8.02
30	22A:	M (2.2,3.7,5.9)*AD	1.295	2.117	6.84
31	21A:	M (2.1,4.0,8.2)*AD	1.243	2.120	7.06
32	17A:	M (1.7,3.4,8.5)*AD	1.284	2.017	6.24
33	12A:	M (1.2,3.5,5.8)*AD	1.347	1.944	5.56
34	ADA:	M (ADA,4.5,8.0)*AD	1.326	2.055	6.16
35	AMT:	M-TYPE (SINE)	1.247	2.197	8.32
36	OLS:	"OLSHEN'S"	1.151	3.101	12.29
37	TOL:	T-SKIP "OLSHEN'S"	1.284	1.967	6.34
38	MEL:	MEAN LIKELIHOOD	1.310	2.484	10.75
39	CML:	CAUCHY MAX LIKELIHD	1.657	2.109	4.73
40	LJS:	LEAST FAVORABLE	1.074	2.446	9.45
41	SST:	ITER S-SKIP TRIMEAN	1.393	2.040	5.90
42	CST:	ITER C-SKIP TRIMEAN	1.276	1.976	6.17
43	33T:	MX(3T,2)-SKIP MEXTR	1.264	2.139	7.02
44	3TO:	3T-SKIP MEAN	1.224	2.133	7.79
45	3T1:	MX(3T,2)-SKIP MEAN	1.273	1.992	6.45
46	5T1:	MX(5T,2)-SKIP MEAN	1.273	1.992	6.45
47	5T4:	MN(5T,.6)-SKIP MEAN	1.268	1.993	6.32
48	CTS:	C&T&S-SKIP TRIMEAN	1.591	3.268	12.65
49	TAK:	TAKEUCHI ADAPTIVE	1.111	2.292	9.14
50	DFA:	MAXIMUM EST LIKE	1.408	2.455	7.25
51	JOH:	JOHNS ADAPTIVE	1.469	2.208	5.29
52	HGP:	HOGG 67 ON KURTOSIS	1.308	45.443	253.20
53	HGL:	HOGG 69 ON KURTOSIS	1.136	33.071	203.96
54	THP:	T-SKIP HOGG 67	1.361	2.842	12.88
55	THL:	T-SKIP HOGG 69	1.310	2.649	12.03
56	JWT:	ADAPT FROM SKIPPING	1.228	1.998	6.45
57	H/L:	HODGES-LEHMANN	1.082	2.616	10.06
58	BH :	BICKEL-HODGES	1.082	2.615	10.06
59	2RM:	2 TIMES FOLDED MED	1.018	14.299	78.58
60	3RM:	3 TIMES FOLDED MED	1.018	14.299	78.58
61	4RM:	4 TIMES FOLDED MED	1.018	14.299	78.58
62	3R1:	3-FOLDED 1-TRIM MED	1.182	2.098	6.61
63	3R2:	3-FOLDED 2-TRIM MED	1.182	2.098	6.61
64	CPL:	CAUCHY PITMAN (LOC)	1.093	1.720	4.41
65	SHO:	SHORTH	2.359	2.936	5.93

N=10   D I S T R I B U T I O N S

	CODE	NAME	NORMAL	10% 3N	20% 3N	25% 1/U	ALL 1/U	CAUCHY
1	M :	MEAN (AVERAGE)	1.000	1.800	2.696	36.232	769.75	403.87
2	5%:	5% SYM TRIM MEAN	1.009	1.481	2.184	19.921	287.34	133.27
3	10%:	10% SYM TRIM MEAN	1.047	1.292	1.797	2.189	21.60	13.60
4	15%:	15% SYM TRIM MEAN	1.069	1.296	1.689	1.938	15.89	9.50
5	25%:	MIDMEAN	1.147	1.357	1.639	1.681	9.07	4.98
6	50%:	MEDIAN	1.367	1.600	1.843	1.882	7.54	3.89
7	GAS:	GASTWIRTH	1.203	1.417	1.685	1.706	7.97	4.29
8	TRI:	TRIMEAN	1.126	1.343	1.650	1.717	11.53	6.41
9	JAE:	JAECKEL ADAPT TRIM	1.077	1.325	1.669	1.684	11.86	6.81
10	BIC:	BICKEL ADAPT TRIM	1.070	1.369	1.777	2.254	16.43	10.25
11	SJA:	SYM JAECKEL AD TRIM	1.086	1.359	1.723	1.990	13.25	7.50
12	JBT:	2-CHOICE ADAPT TRIM	1.071	1.351	1.703	1.682	11.22	6.43
13	JLJ:	ADAPT TRIM LINCOM						
14	H20:	HUBER PROP 2, K=2.0	1.007	1.400	2.184	2.767	31.45	21.63
15	H17:	HUBER PROP 2, K=1.7	1.018	1.339	1.968	2.191	20.40	12.68
16	H15:	HUBER PROP 2, K=1.5	1.031	1.316	1.800	1.894	15.92	9.35
17	H12:	HUBER PROP 2, K=1.2	1.062	1.305	1.670	1.717	12.85	7.39
18	H10:	HUBER PROP 2, K=1.0	1.093	1.318	1.645	1.680	10.53	5.75
19	H07:	HUBER PROP 2, K=0.7	1.165	1.380	1.664	1.678	8.41	4.52
20	M15:	1ST-HU 1.5*IQS MEAN	1.028	1.364	1.847	4.114	47.00	56.43
21	D20:	1ST-HU 2.0*IQS MED	1.012	1.426	2.078	2.190	19.77	12.36
22	D15:	1ST-HU 1.5*IQS MED	1.027	1.353	1.853	1.905	16.59	10.11
23	D10:	1ST-HU 1.0*IQS MED	1.068	1.313	1.694	1.724	13.78	8.10
24	D07:	1ST-HU 0.7*IQS MED	1.109	1.326	1.639	1.679	10.98	6.14
25	A20:	HUBER 2.0*ADS	1.023	1.377	1.878	1.963	14.01	8.11
26	A15:	HUBER 1.5*ADS	1.049	1.327	1.728	1.776	11.72	6.52
27	P15:	1S-HU 1.5*ADS MED	1.050	1.331	1.733	1.776	11.84	6.56
28	HMD:	M (2.0,2.0,5.5)*AD	1.223	1.409	1.696	1.648	8.04	4.10
29	25A:	M (2.5,4.5,9.5)*AD	1.059	1.342	1.746	1.615	10.38	5.36
30	22A:	M (2.2,3.7,5.9)*AD	1.125	1.363	1.704	1.625	9.15	4.67
31	21A:	M (2.1,4.0,8.2)*AD	1.091	1.336	1.684	1.585	9.35	4.74
32	17A:	M (1.7,3.4,8.5)*AD	1.129	1.343	1.650	1.576	8.49	4.26
33	12A:	M (1.2,3.5,8.0)*AD	1.189	1.386	1.658	1.608	7.62	3.84
34	ADA:	M (ADA,4.5,8.0)*AD	1.116	1.387	1.748	1.628	8.32	4.29
35	AMT:	M-TYPE (SINE)	1.085	1.348	1.719	1.598	9.88	5.02
36	OLS:	"OLSHEN'S"	1.243	1.434	1.674	1.687	7.80	4.18
37	TOL:	T-SKIP "OLSHEN'S"	1.293	1.469	1.690	1.685	7.32	3.71
38	MEL:	MEAN LIKELIHOOD	1.382	1.464	1.673	2.073	14.70	7.25
39	CML:	CAUCHY MAX LIKELIHD	1.612	1.773	1.972	1.980	6.74	3.18
40	LJS:	LEAST FAVORABLE	1.022	1.340	1.875	1.875	15.95	9.43
41	SST:	ITER S-SKIP TRIMEAN	1.337	1.526	1.806	1.810	8.20	4.34
42	CST:	ITER C-SKIP TRIMEAN	1.157	1.377	1.667	1.636	8.95	5.00
43	33T:	MX(3T,2)-SKIP MEXTR	1.212	1.689	2.181	1.883	11.29	6.59
44	3TO:	3T-SKIP MEAN	1.115	1.409	1.861	1.691	1C.61	5.90
45	3T1:	MX(3T,2)-SKIP MEAN	1.142	1.341	1.698	1.637	9.90	5.32
46	5T1:	MX(5T,2)-SKIP MEAN	1.138	1.360	1.714	1.649	9.01	4.67
47	5T4:	MN(5T,.6)-SKIP MEAN	1.133	1.354	1.705	1.620	9.02	4.70
48	CTS:	C&T&S-SKIP TRIMEAN	1.305	1.497	1.772	1.784	8.20	4.36
49	TAK:	TAKEUCHI ADAPTIVE	1.041	1.371	1.937	1.843	14.52	9.62
50	DFA:	MAXIMUM EST LIKE	1.212	1.541	1.809	1.784	7.83	4.29
51	JOH:	JOHNS ADAPTIVE	1.176	1.463	1.807	1.739	8.49	4.46
52	HGP:	HOGG 67 ON KURTOSIS	1.126	1.718	2.720	2.590	22.43	18.39
53	HGL:	HOGG 69 ON KURTOSIS	1.075	1.531	2.410	2.128	16.41	12.45
54	THP:	T-SKIP HOGG 67	1.274	1.638	2.290	2.063	18.11	10.29
55	THL:	T-SKIP HOGG 69	1.203	1.550	2.099	1.884	16.23	8.75
56	JWT:	ADAPT FROM SKIPPING	1.144	1.402	1.749	1.766	9.17	4.93
57	H/L:	HODGES-LEHMANN	1.063	1.343	1.750	1.780	11.98	6.85
58	BH :	BICKEL-HODGES	1.054	1.386	1.871	1.927	17.57	10.71
59	2RM:	2 TIMES FOLDED MED	1.013	1.411	2.273	3.045	38.07	27.40
60	3RM:	3 TIMES FOLDED MED	1.002	1.585	2.406	25.001	362.88	168.76
61	4RM:	4 TIMES FOLDED MED	1.002	1.585	2.406	25.001	362.88	168.76
62	3R1:	3-FOLDED 1-TRIM MED	1.047	1.292	1.797	2.189	21.60	13.60
63	3R2:	3-FOLDED 2-TRIM MED	1.124	1.336	1.640	1.676	9.92	5.47
64	CPL:	CAUCHY PITMAN (LOC)	1.158	1.355	1.621	1.584	6.47	3.03
65	SHO:	SHORTH	3.392	3.718	3.816	3.962	10.78	5.07

EXHIBIT 5-11A: PSEUDO-VARIANCES OF THE ESTIMATORS AT THEIR 2.5% POINTS.

	CODE	NAME	N=20 D I I S T R I B U T I O N S NORMAL	5% 10N	10% 10N	10% 3/U	25% 1/U
1	M :	MEAN (AVERAGE)	1.000	6.484	11.084	139.781	143.717
2	5%:	5% SYM TRIM MEAN	1.021	1.227	3.183	3.593	3.016
3	10%:	10% SYM TRIM MEAN	1.055	1.203	1.455	1.784	1.835
4	15%:	15% SYM TRIM MEAN	1.097	1.214	1.428	1.512	1.652
5	25%:	MIDMEAN	1.198	1.288	1.469	1.506	1.651
6	50%:	MEDIAN	1.497	1.557	1.802	1.809	1.956
7	GAS:	GASTWIRTH	1.233	1.321	1.502	1.542	1.676
8	TRI:	TRIMEAN	1.152	1.267	1.475	1.513	1.674
9	JAE:	JAECKEL ADAPT TRIM	1.106	1.241	1.444	1.483	1.642
10	BIC:	BICKEL ADAPT TRIM	1.089	1.525	1.990	1.923	1.907
11	SJA:	SYM JAECKEL AD TRIM	1.111	1.472	1.724	70.676	1.835
12	JBT:	2-CHOICE ADAPT TRIM	1.111	1.258	1.463	1.487	1.654
13	JLJ:	ADAPT TRIM LINCOM	1.171	1.324	1.477	1.475	1.694
14	H20:	HUBER PROP 2, K=2.0	1.009	1.274	1.772	2.148	2.220
15	H17:	HUBER PROP 2, K=1.7	1.021	1.228	1.568	1.759	1.904
16	H15:	HUBER PROP 2, K=1.5	1.036	1.211	1.490	1.614	1.768
17	H12:	HUBER PROP 2, K=1.2	1.074	1.208	1.433	1.502	1.655
18	H10:	HUBER PROP 2, K=1.0	1.114	1.223	1.431	1.481	1.629
19	H07:	HUBER PROP 2, K=0.7	1.203	1.286	1.468	1.513	1.654
20	M15:	1ST-HU 1.5*IQS MEAN	1.045	1.522	1.901	29.730	12.063
21	D20:	1ST-HU 2.0*IQS MED	1.017	1.282	1.686	1.780	1.955
22	D15:	1ST-HU 1.5*IQS MED	1.045	1.226	1.501	1.579	1.750
23	D10:	1ST-HU 1.0*IQS MED	1.110	1.225	1.432	1.482	1.641
24	D07:	1ST-HU 0.7*IQS MED	1.174	1.264	1.453	1.496	1.639
25	A20:	HUBER 2.0*ADS	1.019	1.266	1.658	1.742	1.898
26	A15:	HUBER 1.5*ADS	1.050	1.219	1.485	1.565	1.714
27	P15:	1S-HU 1.5*ADS MED	1.050	1.219	1.484	1.565	1.717
28	HMD:	M (2.0,2.0,5.5)*AD	1.250	1.233	1.310	1.680	1.579
29	25A:	M (2.5,4.5,9.5)*AD	1.047	1.126	1.259	1.378	1.606
30	22A:	M (2.2,3.7,5.9)*AD	1.108	1.147	1.250	1.342	1.557
31	21A:	M (2.1,4.0,8.2)*AD	1.081	1.136	1.248	1.344	1.555
32	17A:	M (1.7,3.4,8.5)*AD	1.131	1.166	1.268	1.344	1.540
33	12A:	M (1.2,3.5,8.0)*AD	1.206	1.225	1.316	1.384	1.572
34	ADA:	M (ADA,4.5,8.0)*AD	1.104	1.166	1.299	1.395	1.602
35	AMT:	M-TYPE (SINE)	1.071	1.131	1.245	1.345	1.566
36	OLS:	"OLSHEN'S"	1.344	1.379	1.506	1.540	1.707
37	TOL:	T-SKIP "OLSHEN'S"	1.406	1.429	1.526	1.548	1.750
38	MEL:	MEAN LIKELIHOOD	1.508	1.208	1.296	1.936	2.105
39	CML:	CAUCHY MAX LIKELIHD	1.713	1.646	1.712	1.746	1.963
40	LJS:	LEAST FAVORABLE	1.022	1.263	1.556	1.617	1.801
41	SST:	ITER S-SKIP TRIMEAN	1.472	1.496	1.626	1.626	1.877
42	CST:	ITER C-SKIP TRIMEAN	1.185	1.253	1.392	1.431	1.624
43	33T:	MX(3T,2)-SKIP MEXTR	1.308	1.377	1.440		1.906
44	3TO:	3T-SKIP MEAN	1.112	1.180	1.278		1.582
45	3T1:	MX(3T,2)-SKIP MEAN	1.122	1.172	1.271	1.357	1.568
46	5T1:	MX(5T,2)-SKIP MEAN	1.128	1.197	1.335	1.397	1.597
47	5T4:	MN(5T,.6)-SKIP MEAN	1.126	1.195	1.326	1.374	1.564
48	CTS:	C&T&S-SKIP TRIMEAN	1.409	1.394	1.539	1.734	1.777
49	TAK:	TAKEUCHI ADAPTIVE	1.050	1.166	1.317	1.429	1.606
50	DFA:	MAXIMUM EST LIKE	1.331	1.404	1.591	1.688	1.925
51	JOH:	JOHNS ADAPTIVE	1.138	1.319	1.457	1.457	1.657
52	HGP:	HOGG 67 ON KURTOSIS	1.056	1.466	1.792	1.775	1.928
53	HGL:	HOGG 69 ON KURTOSIS	1.065	1.310	1.513	1.536	1.698
54	THP:	T-SKIP HOGG 67	1.154	1.222	1.327	1.452	1.746
55	THL:	T-SKIP HOGG 69	1.145	1.202	1.302	1.409	1.693
56	JWT:	ADAPT FROM SKIPPING	1.166	1.505	1.681	1.637	1.867
57	H/L:	HODGES-LEHMANN	1.063	1.234	1.512	1.582	1.721
58	BH :	BICKEL-HODGES	1.047	1.286	1.591	1.665	1.836
59	2RM:	2 TIMES FOLDED MED	1.012	1.337	1.884	2.778	2.497
60	3RM:	3 TIMES FOLDED MED	1.003	1.393	8.885	7.454	5.137
61	4RM:	4 TIMES FOLDED MED	1.000	3.592	8.278	56.662	58.103
62	3R1:	3-FOLDED 1-TRIM MED	1.038	1.213	1.519	1.864	1.921
63	3R2:	3-FOLDED 2-TRIM MED	1.084	1.215	1.438	1.517	1.657
64	CPL:	CAUCHY PITMAN (LOC)	1.255	1.303	1.407	1.452	1.657
65	SHO:	SHORTH	4.415	4.333	4.185	4.352	4.498

EXHIBIT 5-11B: PSEUDOVARIANCES OF THE ESTIMATORS AT THEIR 2.5% POINTS.

	CODE	NAME	N=20 D I S T R I B U T I O N S T3	D-EX	ALL 1/U	CAUCHY
1	M :	MEAN (AVERAGE)	3.098	2.106	1065.217	922.84
2	5%:	5% SYM TRIM MEAN	1.913	1.795	36.447	17.50
3	10%:	10% SYM TRIM MEAN	1.701	1.635	14.235	7.06
4	15%:	15% SYM TRIM MEAN	1.620	1.521	9.644	4.62
5	25%:	MIDMEAN	1.605	1.364	6.699	3.19
6	50%:	MEDIAN	1.834	1.481	6.620	2.99
7	GAS:	GASTWIRTH	1.641	1.391	6.670	3.18
8	TRI:	TRIMEAN	1.617	1.475	8.381	4.03
9	JAE:	JAECKEL ADAPT TRIM	1.635	1.543	7.377	3.56
10	BIC:	BICKEL ADAPT TRIM	1.717	1.497	9.416	4.57
11	SJA:	SYM JAECKEL AD TRIM	1.673	1.445	8.242	3.85
12	JBT:	2-CHOICE ADAPT TRIM	1.670	1.500	7.159	3.39
13	JLJ:	ADAPT TRIM LINCOM	1.734	1.583	6.761	2.84
14	H20:	HUBER PROP 2, K=2.0	1.946	1.832	18.888	9.22
15	H17:	HUBER PROP 2, K=1.7	1.817	1.746	14.439	6.97
16	H15:	HUBER PROP 2, K=1.5	1.735	1.680	11.934	5.82
17	H12:	HUBER PROP 2, K=1.2	1.642	1.564	8.959	4.47
18	H10:	HUBER PROP 2, K=1.0	1.610	1.483	7.703	3.78
19	H07:	HUBER PROP 2, K=0.7	1.613	1.364	6.641	3.14
20	M15:	1ST-HU 1.5*IQS MEAN	1.749	1.647	2498.084	617.15
21	D20:	1ST-HU 2.0*IQS MED	1.854	1.758	12.217	6.19
22	D15:	1ST-HU 1.5*IQS MED	1.720	1.663	10.029	5.08
23	D10:	1ST-HU 1.0*IQS MED	1.623	1.527	8.130	3.99
24	D07:	1ST-HU 0.7*IQS MED	1.601	1.397	6.998	3.39
25	A20:	HUBER 2.0*ADS	1.819	1.710	10.954	5.60
26	A15:	HUBER 1.5*ADS	1.694	1.601	9.028	4.55
27	P15:	1S-HU 1.5*ADS MED	1.701	1.614	9.091	4.58
28	HMD:	M (2.0,2.0,5.5)*AD	1.679	1.582	6.781	3.06
29	25A:	M (2.5,4.5,9.5)*AD	1.691	1.623	8.484	3.86
30	22A:	M (2.2,3.7,5.9)*AD	1.682	1.611	7.741	3.49
31	21A:	M (2.1,4.0,8.2)*AD	1.642	1.566	7.717	3.46
32	17A:	M (1.7,3.4,8.5)*AD	1.601	1.495	6.998	3.11
33	12A:	M (1.2,3.5,8.0)*AD	1.601	1.406	6.373	2.78
34	ADA:	M (ADA,4.5,8.0)*AD	1.677	1.554	6.849	2.98
35	AMT:	M-TYPE (SINE)	1.660	1.584	8.090	3.59
36	OLS:	"OLSHEN'S"	1.665	1.367	6.053	2.68
37	TOL:	T-SKIP "OLSHEN'S"	1.735	1.429	6.120	2.56
38	MEL:	MEAN LIKELIHOOD	1.645	1.430	13.035	6.92
39	CML:	CAUCHY MAX LIKELIHD	1.857	1.410	5.785	2.40
40	LJS:	LEAST FAVORABLE	1.836	1.731	10.200	5.18
41	SST:	ITER S-SKIP TRIMEAN	1.945	1.713	6.823	2.97
42	CST:	ITER C-SKIP TRIMEAN	1.662	1.507	7.043	3.20
43	33T:	MX(3T,2)-SKIP MEXTR	2.171	2.008	9.969	4.28
44	3T0:	3T-SKIP MEAN	1.761	1.755	7.964	3.51
45	3T1:	MX(3T,2)-SKIP MEAN	1.719	1.712	7.896	3.48
46	5T1:	MX(5T,2)-SKIP MEAN	1.696	1.675	7.177	3.11
47	5T4:	MN(5T,.6)-SKIP MEAN	1.693	1.668	7.101	3.08
48	CTS:	C&T&S-SKIP TRIMEAN	1.774	1.624	6.789	2.96
49	TAK:	TAKEUCHI ADAPTIVE	1.738	1.712	8.006	3.57
50	DFA:	MAXIMUM EST LIKE	1.845	1.529	6.852	2.98
51	JOH:	JOHNS ADAPTIVE	1.689	1.599	6.315	2.87
52	HGP:	HOGG 67 ON KURTOSIS	1.944	1.937	8.762	4.53
53	HGL:	HOGG 69 ON KURTOSIS	1.791	1.753	7.420	3.59
54	THP:	T-SKIP HOGG 67	1.917	1.991	10.571	4.67
55	THL:	T-SKIP HOGG 69	1.877	1.857	9.613	4.20
56	JWT:	ADAPT FROM SKIPPING	1.768	1.560	6.697	3.01
57	H/L:	HODGES-LEHMANN	1.676	1.559	8.631	4.29
58	BH :	BICKEL-HODGES	1.770	1.746	11.851	6.09
59	2RM:	2 TIMES FOLDED MED	2.074	1.944	33.632	15.78
60	3RM:	3 TIMES FOLDED MED	2.260	2.004	88.649	40.20
61	4RM:	4 TIMES FOLDED MED	2.670	2.040	529.281	415.12
62	3R1:	3-FOLDED 1-TRIM MED	1.754	1.704	15.044	7.48
63	3R2:	3-FOLDED 2-TRIM MED	1.647	1.583	9.261	4.55
64	CPL:	CAUCHY PITMAN (LOC)	1.630	1.355	5.931	2.37
65	SHO:	SHORTH	4.358	3.010	11.156	4.23

EXHIBIT 5-12:   PSEUDO-VARIANCES OF THE ESTIMATORS AT THEIR 2.5% POINTS.

	CODE	NAME	NORMAL	25% 1/U	CAUCHY
			N=40 D I S T R I B U T I O N S		
1	M :	MEAN (AVERAGE)	1.000	162.206	21109.17
2	5%:	5% SYM TRIM MEAN	1.024	2.150	14.12
3	10%:	10% SYM TRIM MEAN	1.058	1.614	5.38
4	15%:	15% SYM TRIM MEAN	1.094	1.564	3.76
5	25%:	MIDMEAN	1.183	1.622	2.68
6	50%:	MEDIAN	1.530	2.005	2.49
7	GAS:	GASTWIRTH	1.247	1.681	2.60
8	TRI:	TRIMEAN	1.167	1.620	3.23
9	JAE:	JAECKEL ADAPT TRIM	1.076	1.572	2.88
10	BIC:	BICKEL ADAPT TRIM	1.072	1.687	3.15
11	SJA:	SYM JAECKEL AD TRIM	1.089	1.650	2.90
12	JBT:	2-CHOICE ADAPT TRIM	1.088	1.581	2.75
13	JLJ:	ADAPT TRIM LINCOM			
14	H20:	HUBER PROP 2, K=2.0	1.010	1.900	7.20
15	H17:	HUBER PROP 2, K=1.7	1.024	1.717	5.56
16	H15:	HUBER PROP 2, K=1.5	1.038	1.634	4.67
17	H12:	HUBER PROP 2, K=1.2	1.071	1.566	3.68
18	H10:	HUBER PROP 2, K=1.0	1.105	1.560	3.15
19	H07:	HUBER PROP 2, K=0.7	1.184	1.622	2.65
20	M15:	1ST-HU 1.5*IQS MEAN	1.032	30.841	5659.15
21	D20:	1ST-HU 2.0*IQS MED	1.011	1.824	5.02
22	D15:	1ST-HU 1.5*IQS MED	1.032	1.647	4.19
23	D10:	1ST-HU 1.0*IQS MED	1.090	1.557	3.34
24	D07:	1ST-HU 0.7*IQS MED	1.152	1.593	2.84
25	A20:	HUBER 2.0*ADS	1.015	1.752	4.59
26	A15:	HUBER 1.5*ADS	1.041	1.608	3.83
27	P15:	1S-HU 1.5*ADS MED	1.042	1.609	3.84
28	HMD:	M (2.0,2.0,5.5)*AD	1.195	1.508	2.91
29	25A:	M (2.5,4.5,9.5)*AD	1.034	1.505	3.67
30	22A:	M (2.2,3.7,5.9)*AD	1.080	1.465	3.37
31	21A:	M (2.1,4.0,8.2)*AD	1.061	1.469	3.33
32	17A:	M (1.7,3.4,8.5)*AD	1.104	1.471	2.97
33	12A:	M (1.2,3.5,8.0)*AD	1.169	1.522	2.65
34	ADA:	M (ADA,4.5,8.0)*AD	1.057	1.504	2.73
35	AMT:	M-TYPE (SINE)	1.056	1.471	3.38
36	OLS:	"OLSHEN'S"	1.359	1.728	2.26
37	TOL:	T-SKIP "OLSHEN'S"	1.411	1.776	2.35
38	MEL:	MEAN LIKELIHOOD	1.499	2.302	11.45
39	CML:	CAUCHY MAX LIKELIHD	1.655	1.980	2.17
40	LJS:	LEAST FAVORABLE	1.011	1.687	3.31
41	SST:	ITER S-SKIP TRIMEAN	1.486	1.947	2.78
42	CST:	ITER C-SKIP TRIMEAN	1.190	1.575	2.81
43	33T:	MX(3T,2)-SKIP MEXTR	1.419	1.838	3.64
44	3TO:	3T-SKIP MEAN	1.109	1.505	2.95
45	3T1:	MX(3T,2)-SKIP MEAN	1.112	1.503	2.95
46	5T1:	MX(5T,2)-SKIP MEAN	1.115	1.536	2.59
47	5T4:	MN(5T,.6)-SKIP MEAN	1.115	1.531	2.57
48	CTS:	C&T&S-SKIP TRIMEAN	1.400	1.785	2.75
49	TAK:	TAKEUCHI ADAPTIVE	1.036	1.526	2.65
50	DFA:	MAXIMUM EST LIKE	1.374	1.973	2.62
51	JOH:	JOHNS ADAPTIVE	1.101	1.609	2.45
52	HGP:	HOGG 67 ON KURTOSIS	1.021	1.905	2.58
53	HGL:	HOGG 69 ON KURTOSIS	1.047	1.669	2.60
54	THP:	T-SKIP HOGG 67	1.114	1.533	3.90
55	THL:	T-SKIP HOGG 69	1.113	1.520	3.31
56	JWT:	ADAPT FROM SKIPPING	1.195	1.890	2.50
57	H/L:	HODGES-LEHMANN	1.055	1.625	3.41
58	BH :	BICKEL-HODGES	1.048	1.694	4.54
59	2RM:	2 TIMES FOLDED MED	1.012	1.973	11.42
60	3RM:	3 TIMES FOLDED MED	1.004	4.093	41.63
61	4RM:	4 TIMES FOLDED MED	1.001	11.161	114.54
62	3R1:	3-FOLDED 1-TRIM MED	1.024	1.789	7.65
63	3R2:	3-FOLDED 2-TRIM MED	1.043	1.647	5.32
64	CPL:	CAUCHY PITMAN (LOC)	1.276	1.675	2.14
65	SHO:	SHORTH	5.234	5.471	4.23

5D VARIED % POINTS AS PSEUDO-VARIANCES

Some idea of the accuracy of the percentage points obtained by Monte Carlo can be had from the comparisons of Exhibit 5-13, which, since it is for the Cauchy distribution, is a distinctly unfavorable case. Most combinations of estimate and situation should behave rather better.

For purposes of comparison across %, percentage points are more conveniently given as pseudo-variances. When the estimator has a nearly normal distribution, these values will be almost a constant. Exhibit 5-14 gives pseudovariances for 6 percentages, 17 estimates, and 16 situations, selected to give a good general picture.

Exhibit 5-13

ACCURACY OF SOME PERCENTAGE POINTS

Percentage Points for the median   n = 5

	Normal		Cauchy	
1-α	Exact	Monte Carlo	Exact	Monte Carlo
0.25	0.360	0.3624	0.4727	0.4632
0.10	0.685	0.6934	1.0213	0.9970
0.025	1.051	1.0627	2.0151	2.0155
0.01	1.245	1.2631	2.9017	3.0862
0.005	1.386	1.3997	3.7558	4.2651
0.001	1.690	1.682	6.6441	8.0332

Exhibit 5-14

STANDARDIZED SQUARED NORMALIZED
PERCENTAGE POINTS - $n[F^{-1}(\alpha)/\Phi^{-1}(\alpha)]^2$

TAG		NORMAL DISTRIBUTION			.75N(0,1) + .25N(0,1)/U			
$1-\alpha$	n = 40	n = 20	n = 10	n = 5	n = 40	n = 20	n = 10	n = 5
10%								
.25	1.0593	1.0566	1.0484	1.0212	1.5757	1.657	1.688	2.649
.10	81	54	73	200	5878	1.707	1.796	3.506
.025	80	54	74	199	6143	1.835	2.189	14.048
.010	81	56	76	200	6342	1.962	2.941	39.904
.005	82	57	78	201	6499	2.083	4.471	94.117
.001	84	60	82	202	6876	2.847	13.416	367.2
25%								
.25	1.1869	1.2011	1.1479	1.1560	1.5987	1.606	1.592	1.669
.10	48	1.1992	69	53	1.6054	1.620	1.618	1.768
.025	31	82	75	62	217	1.651	1.681	2.121
.010	23	77	81	71	344	1.675	1.748	2.763
.005	16	74	86	77	447	1.694	1.844	3.933
.001	02	65	97	91	713	1.741	2.493	11.551
H20								
.25	1.0112	1.0098	1.0078	1.0009	1.8135	1.920	1.883	3.422
.10	100	86	66	0.9997	1.8427	2.008	2.055	5.584
.025	099	85	66	6	1.9004	2.226	2.767	28.233
.010	100	87	68	6	1.9409	2.435	4.515	90.990
.005	101	88	69	7	1.9715	2.630	8.211	221.916
.001	102	90	72	8	2.0400	3.102	33.913	954.218
D20								
.25	1.0122	1.0177	1.0126	1.0484	1.7657	1.790	1.834	1.888
.10	11	66	16	485	7857	1.842	1.933	2.124
.025	10	67	18	508	8244	1.955	2.190	2.981
.010	12	70	21	528	8508	2.043	2.472	4.747
.005	13	72	24	543	8702	2.113	2.811	7.905
.001	16	77	30	581	9125	2.270	7.520	22.735
JAE								
.25	1.0745	1.1023	1.0740	1.1800	1.5479	1.582	1.558	1.691
.10	743	029	741	791	5553	1.600	1.594	1.785
.025	761	063	766	796	5720	1.642	1.684	2.109
.010	775	089	785	801	5842	1.675	1.787	2.661
.005	786	109	800	804	5937	1.703	1.904	3.636
.001	811	154	836	812	6162	1.775	3.569	10.668

Exhibit 5-14(cont.)

STANDARDIZED SQUARED NORMALIZED
PERCENTAGE POINTS - $n[F^{-1}(\alpha)/\Phi^{-1}(\alpha)]^2$

TAG	NORMAL DISTRIBUTION				.75N(0,1) + .25N(0,1)/U			
$1-\alpha$	n = 40	n = 20	n = 10	n = 5	n = 40	n = 20	n = 10	n = 5
TAK								
.25	1.0621	1.0496	1.0412	1.1023	1.5066	1.541	1.596	1.644
.10	12	489	05	046	121	560	1.659	1.774
.025	17	496	11	110	262	606	1.844	2.292
.010	23	504	18	159	373	646	2.083	3.496
.005	27	510	23	197	465	682	2.422	5.806
.001	37	523	36	287	702	786	7.501	19.710
HGL								
.25	1.0437	1.0605	1.0726	1.1364	1.6408	1.640	1.783	2.417
.10	440	613	726	56	6492	658	1.869	8.150
.025	467	649	745	63	6692	698	2.128	33.071
.010	488	678	761	70	6845	728	2.503	96.291
.005	506	700	773	75	6968	752	3.099	231.016
.001	548	755	802	85	7267	814	10.746	942.413
SHO								
.25	5.7909	5.1242	3.8876	2.6868	6.0562	4.926	4.421	2.989
.10	5674	4.7989	6683	5443	5.7882	733	4.214	2.911
.025	2341	.4150	3924	3592	5.4709	498	3.962	2.936
.010	4.9999	.2130	2482	2671	5.2835	384	3.852	3.113
.005	8243	0794	1549	2100	5.1498	312	3.792	3.415
.001	4628	3.8207	2.9779	1068	4.8747	172	3.709	6.709
DFA								
.25	1.3558	1.2928	1.1828	1.3535	1.8754	1.782	1.671	1.884
.10	3607	3042	1914	3722	1.9031	1.826	1.707	2.023
.025	3742	3308	2120	4079	1.9732	1.925	1.784	2.455
.010	3854	3522	2286	4312	2.0386	2.006	1.844	3.086
.005	3949	3701	2424	4475	2.0999	2.074	1.895	4.097
.001	4196	4153	2771	4793	2.2833	2.240	2.031	11.868
H/L								
.25	1.0560	1.0630	1.0631	1.0802	1.5887	1.640	1.627	1.791
.10	49	21	25	01	1.6000	665	1.672	1.968
.025	49	26	35	22	1.6252	721	1.780	2.616
.010	51	31	44	39	1.6446	764	1.892	3.889
.005	53	35	51	52	1.6604	799	2.026	6.026
.001	56	44	68	84	1.7016	881	3.130	16.762

Exhibit 5-14(cont.)

STANDARDIZED SQUARED NORMALIZED
PERCENTAGE POINTS - $n[F^{-1}(\alpha)/\Phi^{-1}(\alpha)]^2$

TAG		NORMAL DISTRIBUTION				$.75N(0,1) + .25N(0,1)/U$			
1-α	n = 40	n = 20	n = 10	n = 5	n = 40	n = 20	n = 10	n = 5
JOH									
.25	1.0995	1.1336	1.1697	1.4609	1.5897	1.614	1.642	1.947
.10	0993	342	709	633	5948	627	1.670	2.015
.025	1011	375	756	695	6087	657	1.739	2.208
.010	1026	401	792	734	6200	681	1.808	2.459
.005	1037	421	820	761	6294	700	1.891	2.829
.001	1062	467	890	814	6541	749	2.379	6.844
AMT									
.25	1.0556	1.0666	1.0728	1.1842	1.4497	1.493	1.496	1.659
.10	49	672	759	2031	4559	514	527	1.784
.025	57	706	849	2466	4708	567	598	2.197
.010	65	734	926	2822	4823	613	656	2.850
.005	71	757	992	3116	4914	656	708	3.863
.001	85	816	1176	3822	5137	783	855	9.450
3R2									
.25	1.0441	1.0849	1.1241	1.1828	1.6074	1.586	1.575	1.689
.10	31	40	32	18	6201	608	1.604	1.782
.025	32	43	37	21	6470	657	1.676	2.098
.010	35	47	44	24	6664	697	1.754	2.632
.005	37	50	48	27	6814	729	1.841	3.589
.001	43	57	60	32	7165	808	2.684	10.258
25A									
.25	1.0343	1.0452	1.0528	1.1500	1.4760	1.519	1.511	1.680
.10	33	449	539	1630	4847	545	543	1.817
.025	35	466	585	1934	5053	606	615	2.275
.010	39	481	625	2186	5214	659	675	2.975
.005	42	493	659	2399	5344	706	727	3.960
.001	48	526	753	2937	5667	839	875	8.670
5T4									
.25	1.1073	1.1146	1.1186	1.2470	1.5119	1.517	1.528	1.663
.10	092	177	226	538	173	531	556	1.739
.025	151	259	330	683	307	564	620	1.993
.010	196	320	411	780	408	591	676	2.432
.005	230	367	476	851	488	615	728	3.387
.001	311	477	638	998	684	682	878	12.039

Exhibit 5-14(cont.)

STANDARDIZED SQUARED NORMALIZED
PERCENTAGE POINTS - $n[F^{-1}(\alpha)/\Phi^{-1}(\alpha)]^2$

TAG	NORMAL DISTRIBUTION				.75N(0,1) + .25N(0,1)/U			
$1-\alpha$	n = 40	n = 20	n = 10	n = 5	n = 40	n = 20	n = 10	n = 5
CST
.25	1.1914	1.1812	1.1508	1.2654	1.5639	1.570	1.551	1.669
.10	02	815	521	684	5662	586	1.576	1.741
.025	03	848	572	759	5747	624	1.636	1.976
.010	05	877	614	811	5821	654	1.694	2.357
.005	06	901	647	848	5883	681	1.753	3.100
.001	09	964	734	929	6051	745	2.000	10.132
BH
.25	1.0484	1.0476	1.0542	1.0802	1.6543	1.7376	1.7152	1.7907
.10	74	67	34	01	6667	1.7689	1.7763	1.9677
.025	78	71	40	22	6936	1.8362	1.9273	2.6149
.010	82	76	46	39	7145	1.8886	2.0927	3.8870
.005	86	80	51	52	7321	1.9311	2.3075	6.0211
.001	93	89	62	84	7821	2.0343	8.8400	16.7617

Exhibit 5-14(cont.)

STANDARDIZED SQUARED NORMALIZED
PERCENTAGE POINTS - $n[F^{-1}(\alpha)/\Phi^{-1}(\alpha)]^2$

TAG	15N(0,1) & 5N(0,100) n = 20	.75N + .25N/U(0,1/3) n = 20	Normal/Uniform		Cauchy			
1-α	n = 20	n = 20	n = 20	n = 10	n = 40	n = 20	n = 10	n
10%								
.25	5.44	3.46	9.69	10.13	4.60	5.18	5.44	2
.10	6.12	4.16	11.10	13.11	4.85	5.53	7.52	3
.025	7.41	6.13	14.23	21.60	5.38	7.06	13.60	7
.010	8.03	8.28	17.24	32.76	5.95	9.08	20.63	22
.005	8.30	11.98	19.89	51.54	6.58	11.98	31.46	73
.001	8.50	27.80	28.74	105.74	10.23	20.22	108.89	1462
25%								
.25	2.17	2.15	6.14	6.37	2.39	2.71	2.99	
.10	2.17	2.21	6.30	7.16	2.49	2.84	3.57	
.025	2.19	2.39	6.70	9.07	2.68	3.19	4.98	
.010	2.20	2.60	7.08	10.58	2.83	3.50	6.30	1.
.005	2.21	2.82	7.46	11.71	2.95	3.77	7.46	1
.001	2.22	3.48	8.53	14.32	3.20	4.62	16.09	2
H20								
.25	7.73	4.67	12.14	12.03	6.25	6.87	6.69	6
.10	8.69	5.67	14.19	16.71	6.52	7.32	10.19	7
.025	11.19	8.64	18.89	31.45	7.20	9.22	21.63	17
.010	12.99	11.91	24.39	53.59	8.00	11.37	35.61	55
.005	13.88	15.77	30.23	96.58	8.81	13.69	62.88	185
.001	15.05	39.26	43.02	201.51	15.42	24.66	251.90	3702
D20								
.25	3.66	3.23	9.33	10.06	4.35	4.70	5.19	3
.10	3.90	3.51	10.25	12.50	4.57	5.02	7.24	5
.025	4.36	4.33	12.22	19.77	5.02	6.19	12.36	11
.010	4.71	5.81	14.00	27.63	5.46	7.50	17.47	20
.005	5.01	8.21	15.39	33.74	5.84	8.86	22.43	29
.001	5.51	13.89	17.43	49.29	6.82	13.56	52.12	50
JAE								
.25	2.22	2.19	6.48	7.15	2.53	2.91	3.53	2
.10	2.22	2.26	6.74	8.40	2.66	3.08	4.41	3
.025	2.23	2.47	7.38	11.86	2.88	3.56	6.81	6
.010	2.24	2.73	7.99	14.86	3.06	4.11	9.04	10
.005	2.24	3.06	8.57	17.03	3.21	4.67	11.19	15
.001	2.25	4.22	10.03	20.93	3.50	6.50	34.33	27

Exhibit 5-14(cont.)

STANDARDIZED SQUARED NORMALIZED

PERCENTAGE POINTS $- n[F^{-1}(\alpha)/\Phi^{-1}(\alpha)]^2$

TAG	15N(0,1) & 5N(0,100)	.75N + .25N/U(0,1/3)	Normal/Uniform		Cauchy			
1-α	n = 20	n = 20	n = 20	n = 10	n = 40	n = 20	n = 10	n = 5
TAK								
.25	2.23	2.09	6.47	7.89	2.28	2.89	4.00	2.94
.10	2.34	2.19	6.92	9.46	2.40	3.08	5.27	4.21
.025	2.68	2.49	8.01	14.52	2.65	3.57	9.62	9.14
.010	3.14	2.83	8.88	20.32	2.84	4.08	15.68	14.97
.005	3.77	3.18	9.56	25.64	2.99	4.63	23.07	21.10
.001	6.10	4.19	10.90	37.12	3.27	6.73	99.67	43.18
HGL								
.25	2.21	2.23	6.43	8.29	2.34	2.80	4.13	73.37
.10	2.23	2.31	6.71	10.28	2.42	3.02	5.81	81.20
.025	2.30	2.56	7.42	16.41	2.60	3.59	12.45	203.96
.010	2.55	2.85	8.18	23.18	2.74	4.19	23.26	578.33
.005	2.66	3.17	8.97	29.46	2.85	4.90	37.99	1851.38
.001	11.38	4.06	11.03	44.65	3.05	8.18	218.03	37007.18
SHO								
.25	4.19	4.73	10.05	8.93	4.03	3.40	3.21	3.01
.10	4.05	4.55	10.37	9.46	4.15	3.68	3.74	3.77
.025	3.85	4.33	11.16	10.78	4.23	4.23	5.07	5.93
.010	3.74	4.22	11.68	12.35	4.23	4.66	6.30	9.11
.005	3.67	4.15	12.09	14.24	4.22	5.07	7.36	15.09
.001	3.53	4.02	13.32	19.70	4.16	7.13	10.99	54.22
DFA								
.25	2.05	2.09	5.96	6.11	2.33	2.42	2.55	2.76
.10	2.06	2.14	6.22	6.63	2.43	2.57	3.02	3.76
.025	2.09	2.27	6.85	7.83	2.62	2.98	4.29	7.25
.010	2.11	2.39	7.51	8.91	2.75	3.32	5.42	11.98
.005	2.12	2.51	8.23	9.88	2.83	3.60	6.28	17.55
.001	2.15	2.84	11.14	12.49	2.95	4.16	8.44	40.50
H/L								
.25	2.73	2.56	7.40	7.53	3.04	3.44	3.65	3.30
.10	2.78	2.68	7.79	8.75	3.16	3.65	4.56	4.68
.025	2.90	3.07	8.63	11.98	3.41	4.29	6.85	10.06
.010	3.00	3.60	9.36	14.86	3.64	4.87	9.20	17.20
.005	3.07	4.34	10.01	17.20	3.85	5.43	11.75	24.63
.001	3.19	7.38	11.47	24.03	4.35	7.33	38.45	42.06

Exhibit 5-14(cont.)

STANDARDIZED SQUARED NORMALIZED
PERCENTAGE POINTS - $n[F^{-1}(\alpha)/\Phi^{-1}(\alpha)]^2$

TAG	15N(0,1) & 5N(0,100)	.75N + .25N/U(0,1/3)	Normal/Uniform		Cauchy			
1-α	n = 20	n = 20	n = 20	n = 10	n = 40	n = 20	n = 10	n =
JOH								
.25	1.74	1.92	5.65	6.02	2.19	2.38	2.53	2
.10	1.74	1.96	5.83	6.68	2.28	2.52	3.03	3
.025	1.75	2.05	6.32	8.49	2.45	2.87	4.46	5
.010	1.75	2.14	6.78	10.56	2.58	3.17	5.81	8
.005	1.76	2.25	7.23	13.10	2.68	3.47	7.04	13
.001	1.77	2.63	8.52	31.98	2.87	4.56	29.35	33
AMT								
.25	1.82	1.96	6.65	6.69	2.86	2.93	3.04	3
.10	1.84	2.04	7.11	7.56	3.02	3.15	3.60	4
.025	1.87	2.26	8.09	9.88	3.38	3.59	5.02	8
.010	1.91	2.50	8.87	12.11	3.62	3.98	6.36	12
.005	1.93	2.73	9.50	14.13	3.76	4.33	7.58	17
.001	2.01	3.38	10.75	18.53	3.90	5.30	10.20	29
3R2								
.25	2.61	2.46	7.49	6.63	4.26	3.62	3.22	2
.10	2.67	2.61	8.06	7.56	4.62	3.84	3.88	3
.025	2.88	3.14	9.26	9.92	5.32	4.55	5.47	6
.010	3.15	3.93	10.19	11.84	5.93	5.22	6.98	10
.005	3.41	4.89	10.90	13.25	6.44	5.81	8.36	15
.001	3.89	9.03	12.25	16.28	10.38	7.77	18.54	26
25A								
.25	1.98	2.11	6.89	7.06	3.04	3.25	3.23	2
.10	2.02	2.23	7.42	8.02	3.23	3.42	3.84	4
.025	2.11	2.52	8.48	10.38	3.67	3.86	5.36	8
.010	2.19	2.81	9.20	12.34	3.97	4.26	6.76	12
.005	2.26	3.06	9.66	13.98	4.16	4.61	7.87	17
.001	2.44	3.66	10.37	17.75	4.41	5.52	10.49	30
5T4								
.25	1.72	1.84	5.84	6.17	2.26	2.40	2.79	2
.10	1.73	1.89	6.22	6.90	2.36	2.60	3.32	3
.025	1.75	2.02	7.10	9.02	2.57	3.08	4.70	6
.010	1.76	2.16	7.89	11.38	2.73	3.49	6.02	10
.005	1.78	2.33	8.61	13.62	2.86	3.83	7.49	15
.001	1.86	3.45	10.33	17.65	3.17	4.53	34.38	27

Exhibit 5-14(cont.)
STANDARDIZED SQUARED NORMALIZED
PERCENTAGE POINTS - $n[F^{-1}(\alpha)/\Phi^{-1}(\alpha)]^2$

TAG	15N(0,1) & 5N(0,100)	.75N + .25N/U(0,1/3)	Normal/Uniform		Cauchy			
1-α	n = 20	n = 20	n = 20	n = 10	n = 40	n = 20	n = 10	n = 5
CST								
.25	1.78	1.89	6.11	6.21	2.58	2.57	2.90	2.78
.10	1.78	1.95	6.40	6.94	2.66	2.77	3.49	3.57
.025	1.79	2.11	7.04	8.95	2.81	3.20	5.00	6.17
.010	1.80	2.33	7.60	10.82	2.92	3.54	6.56	10.01
.005	1.81	2.67	8.11	12.39	3.01	3.80	8.31	15.15
.001	1.83	5.60	9.63	15.58	3.21	4.33	26.13	28.70
BH								
.25	3.14	2.85	8.65	8.65	3.49	4.19	4.44	3.30
.10	3.22	3.06	10.67	10.67	3.83	4.59	6.11	4.68
.025	3.69	3.85	17.57	17.57	4.54	6.09	10.71	10.06
.010	4.82	5.54	25.72	25.72	5.24	7.64	15.43	17.19
.005	6.13	7.65	31.34	31.34	5.79	8.98	20.24	24.60
.001	7.49	19.28	40.81	40.81	7.09	11.90	64.93	42.01

5E SENSITIVITY AND INFLUENCE CURVES

Hampel (1968) proposed Influence Curves as a tool for the study of estimates. These curves are limiting forms from which asymptotic properties may be calculated (see Chapter 3). Tukey (1970) uses <u>sensitivity</u> curves to study the finite sample behavior of estimators.

In Exhibit 5-15 we show stylized sensitivity curves for n = 20 which are generated by

(i) starting with a pseudo-sample consisting of the n-1 expected normal order statistics from a sample of n-1.

(ii) adding to this a moving point x

(iii) evaluating this estimate for the comgined sample and denoting this estimate by T(x).

(iv) plotting nT(x) as a function of x.

The following comments may be helpful in learning from such curves:

(i) The mean, a non-robust estimator, is very sensitive to extreme observations: The sensitivity curve is unbounded in the tails.

(ii) The other estimators have curves which are bounded in the extreme tails.

(iii) Trimmed means, Huber estimates, JAE, DFA, HGL H/L, 3R2, and BH estimates all show a nonzero constant value of nT(x) for large x of a given sign. This corresponds to noticing only the sign of extreme outliers.

(iv) Estimates of Hampel type (here 22A and AMT) and those involving skipping (here THL, SST, CST and 5T4) as well as SHO, have nT(x) = 0 for large x. This corresponds to ignoring extreme outliers completely.

(v) Considered as an approximation to the influence curve, the asymptotic variance is approximated by

$$\int (nT(x))^2 F(dx).$$

(vi) These sensitivity curves are rightly said to be stylized for two reasons: First, because the background observations are taken as unusually regularly placed, thus tending to make the sensitivity curves unusually smooth. Second, because the background observation resembles a short-tailed sample, thus compressing the range over which the sensitivity curves change, particularly in the tails.

The program used to produce these plots was based on previous work by L. M. Steinberg.

NOTE: The sensitivity curve for SHO is asymmetric because the program arbitrarily selected the smaller of two equivalent values of the estimate.

Exhibit 5-15a

Stylized Sensitivity Curves for

		M	Mean
10%	10% trimmed mean	50%	Median
TRI	Trimean	GAS	Gastwirth's estimate

To 19 average normal order
statistics a 20-th x was
adjoined. Each estimate, T,
was evaluated and 20T(x)
is plotted against x.

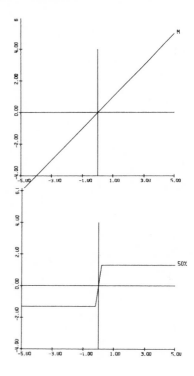

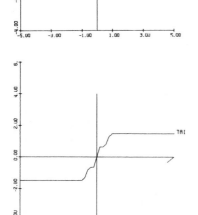

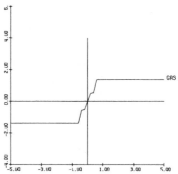

Exhibit 5-15b

Stylized Sensitivity Curves for

H15	Huber k = 1.5	H07	Huber k = 0.7
D15	One step k = 1.5	HMD	M-estimate, ψ bends...
22A	M-estimate, ψ bends...	AMT	Sine M-estimate

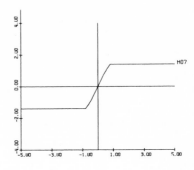

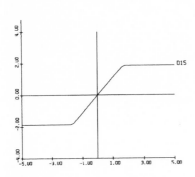

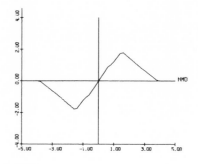

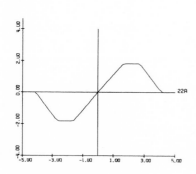

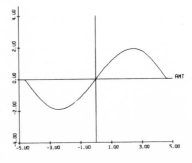

Exhibit 5-15c

Stylized Sensitivity Curves for

JAE	Adaptive trimmed...	JØH	Johns' adaptive estimate
TAK	Takeuchi's adaptive...	DFA	Maximum estimated...
HGL	Hogg 69 based on...	THL	T-skipped Hogg 69

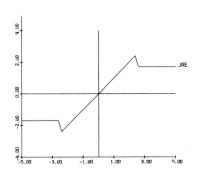

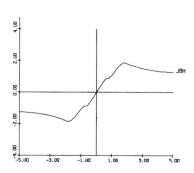

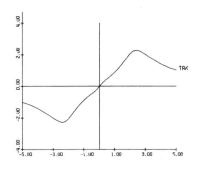

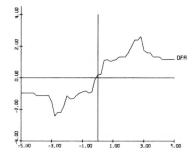

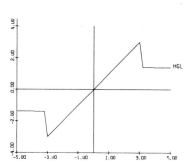

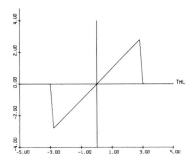

Exhibit 5-15d

Stylized Sensitivity Curves for

SST   Iteratively s-skipped... CST   Iteratively c skipped...
5T4   Multiple skipped...   3R2   3 times folded median...
H/L   Hodges-Lehmann   BH   Bickel Hodges

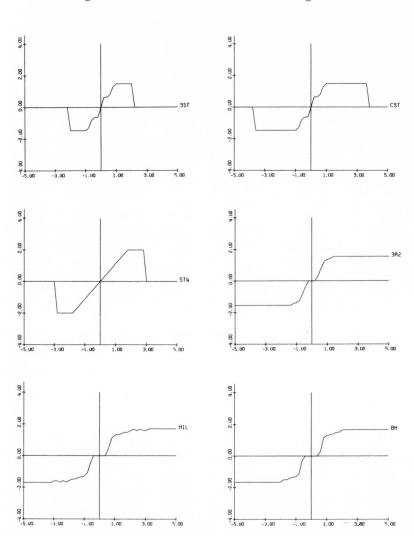

Exhibit 5-15e

Stylized Sensitivity Curves for

CML Cauchy maximum...
MEL Mean likelihood   SHO Shorth (for shortest...
BIC Bickel modified...

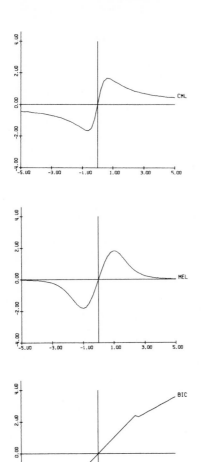

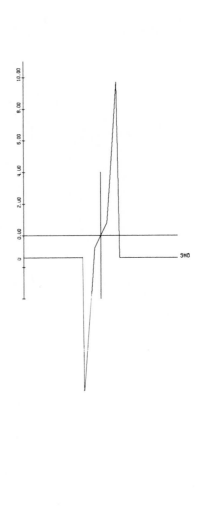

5F BREAKDOWN BOUNDS

To obtain some idea of the tolerance of the estimator to extremely aberrant data, breakdown bounds were calculated. These bounds give the largest percent of bad observations an estimator can tolerate.

More specifically, for sample size n, j sample points were 100, 200, ..., j00. The remaining n-j points were taken to be the n-j expected normal order statistics from a sample size n-j.

The estimator was said to break down if the resulting estimate was greater than 3. The numbers recorded in Exhibit 5-16 give the largest j for which the estimate is less than 3. A similar table was calculated as before but with the j 'outliers' all located at 100. The results were similar. In particular, one estimate has a bound greater than 50% while other estimates have bounds close to 50%.

It should be noted that under other circumstances an estimator may break down at a lower proportion of contamination. The tabled values should be viewed only as upper bounds.

For the corresponding asymptotic concept of "breakdown point", see Hampel (1968, 1971).

Exhibit 5-16

BREAKDOWN BOUNDS

The largest % contamination such that the estimate $T < 3$.
Data consists of $n-j$ expected normal order statistics
from sample of size $n-j$ together with j "bad" points
at $100, 200, \ldots, 100j$.

TAG \ n	5	10	20	40	TAG \ n	5	10	20	40
M	0	0	0	2.5	ADA	40	40	45	47.5
5%	0	0	5	7.5	AMT	40	40	45	47.5
10%	0	10	10	10	ØLS	20	20	20	25
15%	0	10	15	15	TØL	20	30	30	32.5
25%	20	20	25	25	MEL	20	10	10	10
50%	40	40	45	47.5	CML	40	40	45	47.5
GAS	20	30	30	32.5	LJS	20	20	30	32.5
TRI	20	20	20	22.5	SST	20	30	40	45
JAE	20	20	25	25	CST	20	20	30	35
BIC	20	10	20	20	33T	20	20	30	30
SJA	20	10	25	25	3T0	-	-	-	-
JBT	20	20	25	25	3T1	20	20	30	30
JLJ	-	-	25	-	5T1	20	30	30	27.5
H20	0	10	15	17.5	5T4	20	30	30	32.5
H17	20	20	20	20	CTS	20	30	30	32.5
H15	20	20	25	25	TAK	20	10	15	17.5
H12	20	30	30	27.5	DFA	40	50	50	62.5
H10	20	30	30	32.5	JØH	40	20	20	22.5
H07	20	30	35	37.5	HGP	0	20	25	25
M15	20	20	25	0	HGL	0	20	25	25
D20	20	20	25	25	THP	20	20	30	32.5
D15	20	20	25	25	THL	20	30	20	32.5
D10	20	20	25	25	JWT	20	20	30	35
D07	20	20	25	25	H/L	20	20	25	27.5
A20	20	30	35	37.5	BH	20	20	20	22.5
A15	40	40	40	40	2RM	0	10	10	10
P15	40	40	40	40	3RM	0	0	5	5
HMD	40	40	45	47.5	4RM	0	0	0	2.5
25A	40	40	45	47.5	3R1	20	10	10	10
22A	40	40	45	47.5	3R2	20	20	20	15
21A	40	40	45	47.5	CPL	40	40	45	47.5
17A	40	40	45	47.5	SHØ	40	50	50	50
12A	40	40	45	47.5					

- denotes not calculated.

5G   EASE OF COMPUTATION

Some of the estimates may be calculated easily by
hand, others only after considerable time, by a computer.
Exhibit 5-17 summarizes some measures of ease of computation. Under the heading "hand" we give the approximate sample size at which one author would prefer to
punch the data and use a computer program assumed available. Under "order of computation" we give the order of
the number of operations required to calculate the estimate. Since most estimates require ordering, in some
sense, the data, the minimal entry here is n log n.
There are a few estimates, notably the trimmed means,
50% (median), GAS, TRI, that do not require total ordering, but only a few selected order statistics. Partial
sorting algorithms are available which require only a
multiple of n operations. For very large "samples",
the order of 10^4 or more, these estimators must be
considered advantageous.

Under the heading computation time we give the time
in milliseconds to compute an estimate from an ordered
sample of size 40. At a charge of about $400 per hour
1 millisecond costs about $0.01 cents. Thus, excluding
the cost of sorting, even the most expensive estimate
costs less than 3 cents.

Exhibit 5-17

Estimate	Maximum size considered by hand	Order of operations by computer	Computation time in msec.
Mean	200	n	.1
Trimmed means	100		.1
GAS, TRI, 50% (median)	200	n	.1
Huber proposal 2:			
Scale iterated	10		.4
Scale fixed	20		.4
One step estimates	50		.2
Other M-estimates	< 10	$n \log n$	1.8
Adaptive estimates:			
JOH, JAE, HGP, HGL, THP, THL	0		
TAK	0	n^3	247.
DFA	0	n^2	17.
Folded estimates:			
BH, 2RM, 3RM, 4RM, 3R1, 3R2	50	$n \log n$	.6
H/L { numerically	5	(naively) n^2	1.5
graphically	20	good algorithm n	

We include here also an example of how to sort data quickly by hand using a stem and leaf display by Tukey (1970). If a hand operation is the writing of a number and an estimate can be subsequently calculated from the resulting display, then only n operations are required.

Exhibit 5-18

SORTING NUMBERS BY HAND
Stem and Leaf Display

Original 20 numbers: 32, 95, 11, 94, 36, 91, 62, 62,
74, 14, 63, 58, 87, 41, 10, 95, 99, 75, 57, 75

Stage 1		Stage 3		Stage 10		Stage 20 (final)	
0		0		0		0	
1		1	1	1	14	1	140
2		2		2		2	
3	2	3	2	3	26	3	26
4		4		4		4	1
5		5		5		5	87
6		6		6	22	6	223
7		7		7	4	7	455
8		8		8		8	7
9		9	5	9	541	9	54159

The sorted array may be easily read (if necessary) from the last stage:

10, 11, 14, 32, 36, 41, 57, 58, 62, 62, 63, 74,
75, 75, 87, 91, 94, 95, 95, 99.

5H   TRIMMED MEANS

The trimmed means form a family of estimators which
may be used as a basis of comparison for other estimators.
Many estimators (H07-H20, D07-D20, JAE) are asymptotically
equivalent to trimmed means. For finite sample-sizes,
the effective trimming proportion -- that trimmed mean
having the same variance and most highly correlated with
an estimate -- is one useful measure of variability.

Winsorized means were not included explicitly in the
survey. However, the variance of any Winsorized mean may
be obtained from the variances and covariances of two
trimmed means. If $W(k,n)$ denotes the Winsorized mean,
then

$$W(k,n) = \frac{1}{n} \left((k+1)x_{(k+1)} + \sum_{k+2}^{n-k-1} x_{(i)} + (k+1)x_{n-k} \right)$$

$$= \frac{1}{n} ((k+1)(n-2k)(A\%) - (n-2k-2)(B\%))$$

where $A = 100k/n\%$ and $B = 100(k-1)/n\%$ are two "integral" trimmed means. The variance of this linear combination may be easily calculated.

Exhibit 5-19 presents the variances of the trimmed
means for trimming proportions 0%(5%)50% and a variety of
sampling situations.

Exhibit 5-19

VARIANCES OF TRIMMED MEANS

	NORMAL				.75N(0,1) + .25N(0,9)/U(0,1)			
	n = 5	n = 10	n = 20	n = 40	n = 5	n = 10	n = 20	n = 40
0%	1.000	1.000	1.000	1.000	77.75	*	*	*
5%	1.004	1.009	1.022	1.025	55.35	637.41	3.838	2.103
10%	1.020	1.048	1.056	1.058	32.91	3.836	1.811	1.605
15%	1.060	1.070	1.098	1.094	13.14	2.875	1.636	1.558
20%	1.144	1.120	1.145	1.136	2.861	2.087	1.615	1.575
25%	1.156	1.148	1.199	1.184	2.587	1.848	1.640	1.617
30%	1.182	1.215	1.263	1.242	2.473	1.705	1.684	1.674
35%	1.248	1.248	1.334	1.312	2.361	1.738	1.753	1.750
40%	1.465	1.366	1.411	1.380	2.426	1.869	1.830	1.837
45%	1.465	1.366	1.498	1.478	2.426	1.869	1.940	1.945
50%	1.465	1.366	1.498	1.527	2.426	1.869	1.940	2.001

	CAUCHY				CONTAMINATED NORMAL (20-r)N(0,1) & rN(0,100)			
	n = 5	n = 10	n = 20	n = 40	r = 0	r = 1	r = 2	r = 5
0%	*	*	*	*	1.000	6.485	11.538	26.221
5%	*	*	23.984	15.787	1.022	1.232	2.902	14.931
10%	*	27.220	7.324	5.399	1.056	1.205	1.460	6.718
15%	542.188	17.056	4.566	3.645	1.098	1.215	1.431	3.288
20%	7.370	8.575	3.584	2.960	1.145	1.245	1.439	2.387
25%	7.072	5.777	3.108	2.613	1.199	1.288	1.472	2.182
30%	6.704	3.857	2.839	2.426	1.263	1.339	1.519	2.164
35%	6.306	3.730	2.740	2.337	1.334	1.399	1.587	2.210
40%	6.315	3.660	2.761	2.327	1.411	1.471	1.675	2.320
45%	6.315	3.660	2.876	2.389	1.498	1.555	1.804	2.485
50%	6.315	3.660	2.876	2.434	1.498	1.555	1.804	2.485

CONTAMINATED NORMAL
(20-r)N(0,1) & rN(0,9)

	r = 0	r = 1	r = 2	r = 3	r = 5	r = 10	r = 15
0%	1.000	1.420	1.883	2.259	3.007	5.019	6.994
5%	1.022	1.156	1.389	1.637	2.270	4.448	6.860
10%	1.056	1.166	1.308	1.471	1.926	3.977	6.658
15%	1.098	1.194	1.317	1.443	1.799	3.561	6.450
20%	1.145	1.230	1.354	1.459	1.774	3.290	6.248
25%	1.199	1.271	1.406	1.495	1.790	3.131	6.106
30%	1.263	1.316	1.470	1.540	1.845	3.078	5.993
35%	1.334	1.374	1.540	1.600	1.929	3.128	5.974
40%	1.411	1.435	1.620	1.664	2.037	3.223	6.083
45%	1.498	1.516	1.704	1.747	2.156	3.369	6.428
50%	1.498	1.516	1.704	1.747	2.156	3.369	6.428

* denotes variances > 1000.

5I ASYMMETRIC SITUATIONS

Except in a few instances there may be no reason to believe the underlying distribution is symmetric. It is essential then to know how sensitive our estimates are to asymmetry. The breakdown bounds of 5F give some idea of the dependence for asymmetric contamination in one extreme tail of the distribution. Much more detailed information is required.

To give some idea of the effects of asymmetry two situations were examined and the average and variance of the estimates calculated. These are recorded in Exhibit 5-20.

Much more must be learned but a beginning has been made.

Exhibit 5-20

ASYMMETRIC SITUATIONS
MEANS AND VARIANCES OF THE ESTIMATES
SAMPLE SIZE 20 VARIANCE MULTIPLIED BY 20
AVERAGE MULTIPLIED BY $\sqrt{20}$

	18N(0,1) & 2N(2,1)		18N(0,1) & 2N(4,1)	
	Average	Variance	Average	Variance
M	.933	1.000	1.789	1.000
5%	.818	1.039	1.309	1.051
10%	.752	1.092	.893	1.153
15%	.713	1.145	.790	1.192
25%	.659	1.264	.684	1.301
50%	.653	1.605	.656	1.655
GAS	.653	1.301	.675	1.345
TRI	.701	1.212	.757	1.258
JAE	.710	1.175	.735	1.263
BIC	.752	1.130	.934	1.229
SJA	.724	1.163	.842	1.255
JBT	.700	1.177	.754	1.331
JLJ	.683	1.267	.589	1.544
H20	.859	1.023	1.258	1.115
H17	.812	1.048	1.060	1.139
H15	.779	1.071	.954	1.152
H12	.728	1.118	.829	1.180
H10	.695	1.167	.758	1.217
H07	.656	1.266	.683	1.308
M15	.769	1.084	.923	1.192
D20	.838	1.042	1.143	1.159
D15	.769	1.082	.939	1.172
D10	.695	1.161	.764	1.212
D07	.663	1.233	.699	1.273
A20	.832	1.044	1.120	1.155
A15	.764	1.092	.918	1.179
P15	.763	1.092	.917	1.181
HMD	.588	1.369	.261	1.453
25A	.770	1.103	.785	1.346
22A	.690	1.209	.441	1.455
21A	.719	1.155	.610	1.380
17A	.673	1.212	.518	1.374
12A	.635	1.291	.452	1.417

Exhibit 5-20 (cont.)

ASYMMETRIC SITUATIONS
MEANS AND VARIANCES OF THE ESTIMATES
SAMPLE SIZE 20 VARIANCE MULTIPLIERS BY 20
AVERAGE MULTIPLIED BY $\sqrt{20}$

	18N(0,1) & 2N(2,1)		18N(0,1) & 2N(4,1)	
	Average	Variance	Average	Variance
ADA	.740	1.177	.674	1.446
AMT	.731	1.148	.565	1.455
OLS	.628	1.435	.542	1.469
TOL	.574	1.533	.234	1.632
MEL	.505	1.597	.220	1.349
CML	.541	1.833	.317	1.763
LJS	.833	1.050	1.039	1.209
SST	.514	1.725	.086	1.753
CST	.665	1.285	.403	1.536
33T	.874	1.540	.575	2.137
3T0	.735	1.227	.413	1.615
3T1	.690	1.225	.377	1.525
5T1	.687	1.239	.366	1.571
5T4	.685	1.237	.364	1.561
CTS	.522	1.595	.107	1.620
TAK	.837	1.100	.885	1.337
DFA	.681	1.533	.563	1.714
JOH	.724	1.211	.603	1.456
HGD	.903	1.083	1.164	1.565
HGL	.822	1.131	.848	1.485
THP	.755	1.271	.453	1.656
THL	.736	1.256	.407	1.649
JWT	.681	1.284	.429	1.717
H/L	.761	1.106	.921	1.187
BH	.809	1.078	1.033	1.188
2RM	.877	1.029	1.278	1.152
3RM	.906	1.010	1.751	1.032
4RM	.919	1.002	1.716	1.004
3R1	.782	1.067	.992	1.148
3R2	.727	1.130	.823	1.188
CPL	.589	1.438	.404	1.477
SHO	.317	4.836	.065	4.320

5J NORMALITY OF THE ESTIMATES

The normality of the distribution of the estimates may be examined from the percentage points of this distribution. Normal probability plots are useful in this regard. Since (i) the estimates are all scale invariant and (ii) most distributions are symmetric, for these cases half normal probability plots or equivalently X^2 probability plots will serve. These latter plots may be summarized by

$$ s = \left(\frac{F^{-1}(0.01)}{\Phi^{-1}(0.01)} \right) \Big/ \left(\frac{F^{-1}(0.25)}{\Phi^{-1}(0.25)} \right) \quad . $$

If F, the distribution of the estimate, is normal, the quantity s will be close to 1. If the distribution has longer tails, s will be greater than 1. If the estimate has shorter tails, s will be less than 1. Exhibit 5-21 presents values of s for a selection of situations.

EXHIBIT 5-21A: NON-NORMALITY INDEX: (1% PSEUDO-VAR)/(25% PSEUDO-VAR)

	CODE	NAME	NORMAL 5	NORMAL 10	NORMAL 40	1C% 3N 10	10% 3N 20	50% 3N 20
1	M :	MEAN (AVERAGE)	0.999	0.999	0.999	0.991	0.974	0.976
2	5%:	5% SYM TRIM MEAN	0.999	0.999	0.999	0.964	1.002	0.997
3	10%:	10% SYM TRIM MEAN	0.999	0.999	0.999	0.988	0.996	1.019
4	15%:	15% SYM TRIM MEAN	0.999	0.999	0.999	0.991	0.998	1.033
5	25%:	MIDMEAN	1.001	1.000	0.996	0.998	1.006	1.037
6	50%:	MEDIAN	1.009	1.007	1.010	1.008	1.028	1.043
7	GAS:	GASTWIRTH	1.001	1.002	0.997	1.004	1.010	1.040
8	TRI:	TRIMEAN	1.000	1.001	0.997	0.996	1.000	1.053
9	JAE:	JAECKEL ADAPT TRIM	1.000	1.004	1.003	0.991	1.000	1.121
10	BIC:	BICKEL ADAPT TRIM	1.005	1.004	1.002	0.988	0.994	1.060
11	SJA:	SYM JAECKEL AD TRIM	1.005	1.004	1.003	0.990	0.997	1.055
12	JBT:	2-CHOICE ADAPT TRIM	1.000	1.005	1.002	0.992	1.002	1.128
13	JLJ:	ADAPT TRIM LINCOM					1.015	1.195
14	H20:	HUBER PROP 2, K=2.0	0.999	0.999	0.999	0.969	0.992	1.009
15	H17:	HUBER PROP 2, K=1.7	1.001	0.999	0.999	0.978	0.993	1.026
16	H15:	HUBER PROP 2, K=1.5	1.004	1.000	0.999	0.983	0.994	1.042
17	H12:	HUBER PROP 2, K=1.2	1.005	1.001	0.999	0.990	0.998	1.054
18	H10:	HUBER PROP 2, K=1.0	1.004	1.002	0.999	0.994	1.000	1.055
19	H07:	HUBER PROP 2, K=0.7	1.003	1.004	0.998	1.002	1.010	1.040
20	M15:	1ST-HU 1.5*IQS MEAN	1.041	1.001	0.999	0.986	0.995	1.077
21	D20:	1ST-HU 2.0*IQS MED	1.004	1.000	0.999	0.972	0.992	1.057
22	D15:	1ST-HU 1.5*IQS MED	1.005	1.000	0.999	0.980	0.994	1.083
23	D10:	1ST-HU 1.0*IQS MED	1.004	1.001	0.999	0.990	0.997	1.080
24	D07:	1ST-HU 0.7*IQS MED	1.002	1.000	0.997	0.996	1.004	1.051
25	A20:	HUBER 2.0*ADS	1.014	1.001	0.999	0.977	0.992	1.034
26	A15:	HUBER 1.5*ADS	1.020	1.002	0.999	0.986	0.995	1.045
27	P15:	1S-HU 1.5*ADS MED	1.019	1.002	0.999	0.986	0.996	1.051
28	HMD:	M (2.0,2.0,5.5)*AD	1.082	1.032	1.009	1.029	1.031	1.065
29	25A:	M (2.5,4.5,9.5)*AD	1.060	1.009	1.000	0.993	0.998	1.043
30	22A:	M (2.2,3.7,5.9)*AD	1.081	1.026	1.004	1.012	1.010	1.052
31	21A:	M (2.1,4.0,8.2)*AD	1.069	1.014	1.001	1.003	1.001	1.050
32	17A:	M (1.7,3.4,8.5)*AD	1.065	1.015	1.001	1.008	1.007	1.052
33	12A:	M (1.2,3.5,8.0)*AD	1.059	1.017	1.000	1.013	1.015	1.046
34	ADA:	M (ADA,4.5,8.0)*AD	1.073	1.024	1.003	1.005	1.010	1.081
35	AMT:	M-TYPE (SINE)	1.083	1.018	1.001	1.004	1.001	1.050
36	OLS:	"OLSHEN'S"	0.998	0.997	0.996	1.001	1.015	1.043
37	TOL:	T-SKIP "OLSHEN'S"	1.015	1.000	0.991	1.005	1.013	1.052
38	MEL:	MEAN LIKELIHOOD	0.979	0.976	0.981	1.015	1.022	1.048
39	CML:	CAUCHY MAX LIKELIHD	0.999	1.007	1.001	0.996	1.049	1.059
40	LJS:	LEAST FAVORABLE	1.005	1.000	0.999	0.979	0.993	1.064
41	SST:	ITER S-SKIP TRIMEAN	1.016	1.041	1.018	1.016	1.046	1.086
42	CST:	ITER C-SKIP TRIMEAN	1.012	1.009	0.999	1.001	1.005	1.059
43	33T:	MX(3T,2)-SKIP MEXTR	1.026	1.013	0.994	1.068	1.107	1.123
44	3TO:	3T-SKIP MEAN	1.044	1.024	1.011	1.000	1.008	1.068
45	3T1:	MX(3T,2)-SKIP MEAN	1.024	1.018	1.010	1.006	1.006	1.059
46	5T1:	MX(5T,2)-SKIP MEAN	1.024	1.023	1.011	1.005	1.006	1.069
47	5T4:	MN(5T,.6)-SKIP MEAN	1.025	1.020	1.011	1.003	1.004	1.072
48	CTS:	C&T&S-SKIP TRIMEAN	1.184	1.026	1.002	1.010	1.020	1.076
49	TAK:	TAKEUCHI ADAPTIVE	1.012	1.001	1.000	0.986	1.002	1.142
50	DFA:	MAXIMUM EST LIKE	1.057	1.039	1.022	1.034	1.078	1.072
51	JOH:	JOHNS ADAPTIVE	1.009	1.008	1.003	1.001	1.011	1.205
52	HGP:	HOGG 67 ON KURTOSIS	1.008	1.011	1.003	0.971	1.001	1.074
53	HGL:	HOGG 69 ON KURTOSIS	1.000	1.003	1.005	0.987	1.001	1.134
54	THP:	T-SKIP HOGG 67	1.008	1.022	1.010	1.013	1.011	1.130
55	THL:	T-SKIP HOGG 69	1.024	1.024	1.010	1.009	1.014	1.112
56	JWT:	ADAPT FROM SKIPPING	1.023	1.021	1.015	1.010	1.023	1.080
57	H/L:	HODGES-LEHMANN	1.003	1.001	0.999	0.990	0.998	1.057
58	BH :	BICKEL-HODGES	1.003	1.000	1.000	0.987	0.994	1.103
59	2RM:	2 TIMES FOLDED MED	0.999	0.999	0.999	0.973	0.995	1.017
60	3RM:	3 TIMES FOLDED MED	0.999	0.999	0.999	0.959	1.043	0.998
61	4RM:	4 TIMES FOLDED MED	0.999	0.999	0.999	0.959	0.985	0.986
62	3R1:	3-FOLDED 1-TRIM MED	1.000	0.999	0.999	0.988	0.995	1.033
63	3R2:	3-FOLDED 2-TRIM MED	1.000	1.000	0.999	0.996	0.999	1.055
64	CPL:	CAUCHY PITMAN (LOC)	1.002	1.000	0.996	1.000	1.012	1.053
65	SHO:	SHORTH	0.844	0.836	0.863	0.802	0.893	1.058

EXHIBIT 5-21B: NON-NORMALITY INDEX: (1% PSEUDO-VAR)/(25% PSEUDO-VAR)

CODE	NAME	25% 1/U 5	25% 1/U 10	25% 1/U 20	25% 1/U 40	25% 3/U 20	10% 1ON 20
1	M : MEAN (AVERAGE)	26.59	13.97	5.82	10.67	4.49	0.85
2	5%: 5% SYM TRIM MEAN	20.98	14.43	2.02	1.23	7.16	1.42
3	10%: 10% SYM TRIM MEAN	15.06	1.74	1.18	1.04	2.39	0.98
4	15%: 15% SYM TRIM MEAN	7.91	1.37	1.07	1.02	1.81	0.99
5	25%: MIDMEAN	1.66	1.10	1.04	1.02	1.21	0.99
6	50%: MEDIAN	1.26	1.05	1.05	1.02	1.13	1.00
7	GAS: GASTWIRTH	1.66	1.07	1.04	1.02	1.19	0.99
8	TRI: TRIMEAN	1.56	1.13	1.07	1.02	1.55	0.99
9	JAE: JAECKEL ADAPT TRIM	1.57	1.15	1.06	1.02	1.25	0.99
10	BIC: BICKEL ADAPT TRIM	7.44	1.59	1.27	1.07	1.81	1.16
11	SJA: SYM JAECKEL AD TRIM	7.39	1.31	1.19	1.05	3.21	1.10
12	JBT: 2-CHOICE ADAPT TRIM	1.57	1.13	1.06	1.03	1.23	0.99
13	JLJ: ADAPT TRIM LINCOM				1.05	1.13	1.01
14	H20: HUBER PROP 2, K=2.0	26.59	2.40	1.27	1.07	2.55	0.96
15	H17: HUBER PROP 2, K=1.7	2.51	1.48	1.15	1.05	2.30	0.97
16	H15: HUBER PROP 2, K=1.5	2.28	1.25	1.11	1.04	2.00	0.98
17	H12: HUBER PROP 2, K=1.2	2.03	1.16	1.07	1.03	1.51	0.99
18	H10: HUBER PROP 2, K=1.0	1.91	1.12	1.06	1.02	1.34	0.99
19	H07: HUBER PROP 2, K=0.7	1.82	1.08	1.04	1.02	1.18	0.99
20	M15: 1ST-HU 1.5*IQS MEAN	4.20	61.83	222.86	155.61	136.05	1.57
21	D20: 1ST-HU 2.0*IQS MED	2.51	1.35	1.14	1.05	1.80	0.97
22	D15: 1ST-HU 1.5*IQS MED	2.26	1.23	1.10	1.03	1.66	0.98
23	D10: 1ST-HU 1.0*IQS MED	2.09	1.16	1.06	1.02	1.40	0.99
24	D07: 1ST-HU 0.7*IQS MED	1.79	1.13	1.05	1.02	1.26	0.99
25	A20: HUBER 2.0*ADS	2.03	1.22	1.13	1.04	1.51	0.97
26	A15: HUBER 1.5*ADS	1.75	1.15	1.09	1.03	1.40	0.98
27	P15: 1S-HU 1.5*ADS MED	1.76	1.15	1.09	1.03	1.40	0.98
28	HMD: M (2.0,2.0,5.5)*AD	1.36	1.10	1.04	1.02	1.08	1.01
29	25A: M (2.5,4.5,9.5)*AD	1.77	1.11	1.09	1.03	1.33	1.00
30	22A: M (2.2,3.7,5.9)*AD	1.55	1.12	1.07	1.02	1.23	1.01
31	21A: M (2.1,4.0,8.2)*AD	1.58	1.10	1.07	1.02	1.26	1.00
32	17A: M (1.7,3.4,8.5)*AD	1.44	1.09	1.05	1.02	1.19	1.00
33	12A: M (1.2,3.5,8.0)*AD	1.33	1.07	1.04	1.02	1.14	1.00
34	ADA: M (ADA,4.5,8.0)*AD	1.40	1.09	1.06	1.02	1.20	1.01
35	AMT: M-TYPE (SINE)	1.72	1.11	1.08	1.02	1.28	1.00
36	OLS: "OLSHEN'S"	3.07	1.07	1.03	1.02	1.12	1.00
37	TOL: T-SKIP "OLSHEN'S"	1.42	1.06	1.03	1.02	1.07	1.00
38	MEL: MEAN LIKELIHOOD	2.11	1.46	1.36	1.71	4.46	1.00
39	CML: CAUCHY MAX LIKELIHD	1.18	1.02	1.01	1.03	1.04	1.00
40	LJS: LEAST FAVORABLE	2.11	1.21	1.09	1.03	1.65	0.97
41	SST: ITER S-SKIP TRIMEAN	1.35	1.09	1.04	1.04	1.09	1.03
42	CST: ITER C-SKIP TRIMEAN	1.41	1.09	1.05	1.01	1.23	1.01
43	33T: MX(3T,2)-SKIP MEXTR	1.58	1.21	1.11	1.05		1.04
44	3TO: 3T-SKIP MEAN	1.73	1.17	1.06	1.02		1.01
45	3T1: MX(3T,2)-SKIP MEAN	1.46	1.14	1.06	1.02	1.26	1.01
46	5T1: MX(5T,2)-SKIP MEAN	1.46	1.10	1.06	1.02	1.18	1.01
47	5T4: MN(5T,.6)-SKIP MEAN	1.46	1.10	1.05	1.02	1.17	1.00
48	CTS: C&T&S-SKIP TRIMEAN	4.71	1.09	1.04	1.02	1.60	1.02
49	TAK: TAKEUCHI ADAPTIVE	2.13	1.31	1.07	1.02	1.36	1.00
50	DFA: MAXIMUM EST LIKE	1.64	1.10	1.13	1.09	1.15	1.03
51	JOH: JOHNS ADAPTIVE	1.26	1.10	1.04	1.02	1.12	1.00
52	HGP: HOGG 67 ON KURTOSIS	19.23	1.59	1.09	1.04	1.49	1.00
53	HGL: HOGG 69 ON KURTOSIS	39.83	1.40	1.05	1.03	1.28	0.99
54	THP: T-SKIP HOGG 67	2.67	1.27	1.09	1.03	1.52	1.02
55	THL: T-SKIP HOGG 69	2.51	1.20	1.08	1.02	1.43	1.02
56	JWT: ADAPT FROM SKIPPING	1.50	1.12	1.07	1.04	1.09	1.00
57	H/L: HODGES-LEHMANN	2.17	1.16	1.08	1.04	1.41	0.98
58	BH : BICKEL-HODGES	2.17	1.22	1.09	1.04	1.94	0.97
59	2RM: 2 TIMES FOLDED MED	14.66	2.95	1.62	1.10	5.54	0.97
60	3RM: 3 TIMES FOLDED MED	14.66	14.48	3.89	2.39	15.37	2.98
61	4RM: 4 TIMES FOLDED MED	14.66	14.48	6.14	8.51	4.75	1.05
62	3R1: 3-FOLDED 1-TRIM MED	1.56	1.74	1.20	1.06	2.39	0.98
63	3R2: 3-FOLDED 2-TRIM MED	1.56	1.11	1.07	1.04	1.60	0.98
64	CPL: CAUCHY PITMAN (LOC)	1.30	1.05	1.03	1.02	1.05	1.00
65	SHO: SHORTH	1.04	0.87	0.89	0.87	0.89	0.87

EXHIBIT 5-21C: NON-NORMALITY INDEX: (1% PSEUDO-VAR)/(25% PSEUDO-VAR)

	CODE	NAME	ALL 1/U 10	ALL 1/U 20	CAUCHY 5	CAUCHY 10B	CAUCHY 20	CAUCHY 40
1	M :	MEAN (AVERAGE)	19.86	9.44	8.39	1.47	15.14	1.47
2	5%:	5% SYM TRIM MEAN	17.64	3.65	8.16	1.47	3.33	2.49
3	10%:	10% SYM TRIM MEAN	3.23	1.78	7.74	3.80	1.75	1.29
4	15%:	15% SYM TRIM MEAN	2.48	1.40	7.06	3.04	1.38	1.23
5	25%:	MIDMEAN	1.66	1.15	4.07	2.23	1.29	1.18
6	50%:	MEDIAN	1.38	1.15	3.73	1.76	1.30	1.14
7	GAS:	GASTWIRTH	1.53	1.17	4.07	1.89	1.30	1.17
8	TRI:	TRIMEAN	1.99	1.32	3.92	2.80	1.37	1.19
9	JAE:	JAECKEL ADAPT TRIM	2.08	1.23	3.95	2.95	1.41	1.21
10	BIC:	BICKEL ADAPT TRIM	3.21	1.42	5.61	4.75	1.51	1.25
11	SJA:	SYM JAECKEL AD TRIM	2.57	1.32	5.65	3.51	1.44	1.21
12	JBT:	2-CHOICE ADAPT TRIM	1.99	1.23	3.95	2.79	1.35	1.22
13	JLJ:	ADAPT TRIM LINCOM		1.27			1.36	
14	H20:	HUBER PROP 2, K=2.0	4.46	2.01	8.39	5.33	1.65	1.28
15	H17:	HUBER PROP 2, K=1.7	2.46	1.75	5.81	3.13	1.53	1.23
16	H15:	HUBER PROP 2, K=1.5	2.19	1.54	5.13	3.19	1.47	1.22
17	H12:	HUBER PROP 2, K=1.2	2.13	1.29	4.88	3.04	1.41	1.24
18	H10:	HUBER PROP 2, K=1.0	1.86	1.21	4.61	2.32	1.35	1.22
19	H07:	HUBER PROP 2, K=0.7	1.61	1.17	4.37	2.01	1.30	1.20
20	M15:	1ST-HU 1.5*IQS MEAN	12.85	112.52	7.65	42.99	131.25	158.05
21	D20:	1ST-HU 2.0*IQS MED	2.75	1.50	5.69	3.88	1.59	1.25
22	D15:	1ST-HU 1.5*IQS MED	2.51	1.41	5.35	3.77	1.56	1.28
23	D10:	1ST-HU 1.0*IQS MED	2.37	1.27	5.16	3.51	1.44	1.26
24	D07:	1ST-HU 0.7*IQS MED	1.96	1.17	4.27	2.73	1.33	1.20
25	A20:	HUBER 2.0*ADS	1.90	1.29	4.44	2.50	1.38	1.18
26	A15:	HUBER 1.5*ADS	1.81	1.24	4.11	2.32	1.35	1.22
27	P15:	1S-HU 1.5*ADS MED	1.82	1.24	4.17	2.35	1.35	1.21
28	HMD:	M (2.0,2.0,5.5)*AD	1.61	1.21	3.56	1.82	1.31	1.18
29	25A:	M (2.5,4.5,9.5)*AD	1.75	1.34	4.21	2.05	1.31	1.30
30	22A:	M (2.2,3.7,5.9)*AD	1.65	1.30	3.87	1.89	1.34	1.26
31	21A:	M (2.1,4.0,8.2)*AD	1.70	1.29	3.99	1.93	1.31	1.28
32	17A:	M (1.7,3.4,8.5)*AD	1.64	1.25	3.76	1.83	1.29	1.24
33	12A:	M (1.2,3.5,8.0)*AD	1.55	1.21	3.56	1.72	1.26	1.20
34	ADA:	M (ADA,4.5,8.0)*AD	1.59	1.26	3.85	1.90	1.32	1.26
35	AMT:	M-TYPE (SINE)	1.81	1.34	4.28	2.05	1.36	1.27
36	OLS:	"OLSHEN'S"	1.52	1.16	9.61	2.07	1.28	1.15
37	TOL:	T-SKIP "OLSHEN'S"	1.47	1.16	3.85	1.85	1.29	1.14
38	MEL:	MEAN LIKELIHOOD	2.79	1.99	7.09	3.24	2.58	3.94
39	CML:	CAUCHY MAX LIKELIHD	1.36	1.13	3.38	1.59	1.25	1.15
40	LJS:	LEAST FAVORABLE	2.32	1.52	4.99	3.35	1.57	1.30
41	SST:	ITER S-SKIP TRIMEAN	1.67	1.21	3.50	1.91	1.25	1.12
42	CST:	ITER C-SKIP TRIMEAN	1.74	1.24	3.60	2.14	1.38	1.13
43	33T:	MX(3T,2)-SKIP MEXTR	1.96	1.63	4.33	3.28	1.79	1.36
44	3TO:	3T-SKIP MEAN	2.00	1.43	4.59	3.02	1.56	1.28
45	3T1:	MX(3T,2)-SKIP MEAN	1.88	1.42	3.82	2.63	1.54	1.28
46	5T1:	MX(5T,2)-SKIP MEAN	1.79	1.32	3.82	2.39	1.44	1.21
47	5T4:	MN(5T,.6)-SKIP MEAN	1.85	1.35	3.73	2.41	1.46	1.21
48	CTS:	C&T&S-SKIP TRIMEAN	1.68	1.22	7.39	1.88	1.28	1.16
49	TAK:	TAKEUCHI ADAPTIVE	2.58	1.37	5.09	3.85	1.41	1.24
50	DFA:	MAXIMUM EST LIKE	1.46	1.26	4.34	2.02	1.38	1.18
51	JOH:	JOHNS ADAPTIVE	1.75	1.20	3.73	2.26	1.33	1.18
52	HGP:	HOGG 67 ON KURTOSIS	3.88	1.52	8.82	6.33	1.86	1.16
53	HGL:	HOGG 69 ON KURTOSIS	2.80	1.27	7.88	5.28	1.50	1.17
54	THP:	T-SKIP HOGG 67	2.65	1.63	6.57	3.82	1.53	1.21
55	THL:	T-SKIP HOGG 69	2.54	1.58	6.16	3.63	1.54	1.36
56	JWT:	ADAPT FROM SKIPPING	1.74	1.15	3.85	2.35	1.39	1.15
57	H/L:	HODGES-LEHMANN	1.97	1.26	5.22	2.96	1.42	1.20
58	BH :	BICKEL-HODGES	2.97	1.71	5.21	4.19	1.82	1.50
59	2RM:	2 TIMES FOLDED MED	5.68	3.81	7.77	6.56	3.42	2.34
60	3RM:	3 TIMES FOLDED MED	17.85	8.33	7.77	1.47	6.56	7.32
61	4RM:	4 TIMES FOLDED MED	17.85	8.89	7.77	1.47	14.34	10.13
62	3R1:	3-FOLDED 1-TRIM MED	3.23	1.83	3.92	3.80	1.82	1.61
63	3R2:	3-FOLDED 2-TRIM MED	1.79	1.36	3.92	2.42	1.44	1.39
64	CPL:	CAUCHY PITMAN (LOC)	1.38	1.16	3.04	1.56	1.24	1.15
65	SHO:	SHORTH	1.38	1.16	3.02	1.75	1.37	1.05

6. A DETAILED ANALYSIS OF THE VARIANCES

This chapter endeavors to analyze a substantial part of the detailed results, which include, in addition to what has been presented earlier:

- covariances between every pair of estimates for every situation.

- indications of variance for unbiased linear combinations of every pair of estimates for every situation. (Coefficients step by 1/4, providing three internal combinations. One external combination is available at each end.)

There has been no attempt to analyze covariances and linear combinations in any detail. Selected instances have been looked at with interesting results. We have uncovered enough leads for progress by altering estimates to make an attempt at complete analysis of the results for the first 65 estimates unnecessary.

The structure of this chapter is as follows:

6A   Introductory considerations
     6A1 Looking at two situations at once
     6A2 Efficiencies and deficiencies
     6A3 Classification of situations
     6A4 Notionality of the Cauchy (practical
             deficiencies)
     6A5 Outdoing and outperforming
     6A6 Kinds of criteria
     6A7 Ways to use a criterion

6B   Selected families of estimates
     6B1 Hubers
     6B2 Hampels
     6B3 Trims
     6B4 Sitsteps and alternates
     6B5 Families summarized
     6B6 Extending the hampels
     6B7 Global linear combinations
     6B8 Sitstep ease
     6B9 Hybrid vigor

Note: In this chapter, "Gaussian" will replace "normal". Similarly, "10% of 3N" and "25% of 10N" will become "10% of 3G" and 25% of 10G", and so on.

6A INTRODUCTORY CONSIDERATIONS

Our first task is to deal with various introductory matters so that we can begin to look hard at the calculated variance.

The quantitative examination of robustness of variance is a very multiresponse affair. There are many possible situations, of which we have variances for only a selected few. What we would like to do would be to pay adequate attention to all situations -- where adequate means neither too much nor too little.

There is much to be said for pictures. So we begin by preparing to use pictures.

6A1 Looking at two situations at once

The first step -- perhaps the only one we will really succeed in taking -- toward a clear visualization of robustness as reflected in variances in different situations is taken when we can clearly grasp what is going on for several estimates in two situations at once. (It is convenient to distinguish the two as the conventional situation and the threatening situation respectively.) The case of one situation at a time is so simple that we usually feel a table of numbers is enough -- that we can avoid the effort of making a picture. (We may not always be right in feeling this way.)

To look at two situations it is natural to look at two coordinates in the plane. Since we are to analyze "variances" we should probably start with the case where the coordinates are variances. Exhibit 6-1 shows qualitatively what can happen. Here estimates A,B,C,D, E,F,G,H and J have their variances plotted for two situations. (Look only at the inner scale on each axis, for the moment.) Estimate A, for example, has variance 3.5 in the conventional situation and 3.0 in the threatening situation.

The curved segment represents the boundary between what estimates can do and what they cannot. In this hypothetical example, the lowest variance in the conventional situation is 2.0. K has this variance, but its variance in the threatening situation is high at 7.0. The picture indicates that 2.0 and 6.0 or 3.0 and 3.0 (the two ball-shaped ends of the curved segment) could be attained if we knew what estimate to use.

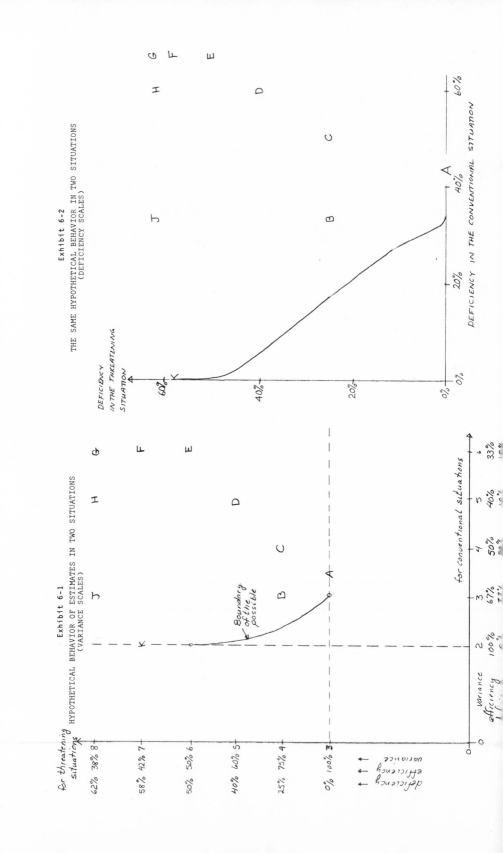

Exhibit 6-1

HYPOTHETICAL BEHAVIOR OF ESTIMATES IN TWO SITUATIONS
(VARIANCE SCALES)

Exhibit 6-2

THE SAME HYPOTHETICAL BEHAVIOR IN TWO SITUATIONS
(DEFICIENCY SCALES)

We can attain 2.0 in the conventional situation and 3.0 in the threatening one. It by no means follows that we can attain both simultaneously. For finite sample sizes, the boundary of the possible is a curve. (See Appendix 13 for further discussion.) True, as n —> ∞, this curve shrinks down to a vertical portion and horizontal portion. As n —> ∞, we can have the best of both worlds (but real data fails to have n —> ∞).

The problem of estimate synthesis for two situations involves (1) learning to get reasonably close to the boundary over a reasonable part of its length and (2) reducing the effort required to calculate estimates that get close.

6A2 Efficiencies and deficiencies

Variances are not the nicest numbers to use. In our example where variances under 10 are common, if estimate X had variances 500 and 800 but Y had variances 1000 and 1600, it would be clear that Y was worse than X, but we would not care much about the difference. Both would be horridly variable, not to be considered by any rational man.

The use of efficiencies, defined by

$$\text{efficiency} = \frac{\text{variance of standard}}{\text{variance}}$$

instead of variances is familiar and rather effective. It has two disadvantages in this chapter. First, numbers are too big. Watching a few % change in numbers between 90% and 98% is inefficient. Second, efficiencies go up when variances go down, and vice versa.

The easy way to avoid all the difficulties just mentioned is to use the deficiency, where

$$\text{deficiency} = 1 - \text{efficiency}$$

so that an efficiency of 95% translates to a deficiency of 5%.

121

Exhibit 6-2 shows the same hypothetical estimates in terms of deficiencies. Clearly we have:

- NOT disturbed order relations.

- put the corner near (0%,0%) under greater relative magnification than the rest of the picture.

It is easier to see numerically that, for small deficiencies, the deficiency scale is closely proportional to the variance scale. In fact, we have:

|deficiency|	|relative variance|	|(ratio of changes)|
0%	1.00	***
5%	1.053	(1.05)
10%	1.111	(1.11)
15%	1.176	(1.17)
20%	1.250	(1.25)

All our two-situation pictures in this chapter will be in deficiency terms.

6A3 Classification of situations

In practice, calculations are finite. As a consequence, the diversity of situations investigated is finite -- and usually small. For n = 20 we have a reasonable variety of situations; for n = 10, fewer; for n = 5 or 40, fewer still.

There are three ways in which it is natural to break down our situations:

- as gentle or as vigorous.

- as modified Gaussians or as alternative.

- as reasonable or as notional.

Let us illustrate.

The mixtures of some observations from a basic Gaussian with varying numbers of others from a Gaussian with the same center and three times the scale (notated below as "--% from 3G") and the t-distribution on 3 degrees of freedom are all gentle.

122

The remainder, including mixtures drawing on a Gaussian distribution with 10 times the scale (notated below as "--% from 10G"), samples from distributions in which the Gaussian is contaminated with varying amounts of Gauss/rectangular (notated below as "--% of 1/U" or "--% of 3/U", depending on whether the rectangular is (0,1) or (0,1/3)), and the two pure Cauchy-tailed alternatives (pure 1/U and pure Cauchy) are vigorous.

Both the contaminations of a Gaussian by suitable amounts of Gauss/rectangular and the mixtures of Gaussians of different spread are modified Gaussians. The pure Student's t on 3 d.f. (t_3), the pure Gaussian/rectangular (1/U), and the pure Cauchy are alternatives.

Two modified Gaussians and one alternative of those we calculated for are notional (i.e. unreasonable); all others are reasonable. The unreasonable modifications are 50% and 75% of 3G. They are unreasonable in two overlapping ways:

- it is hard to think of a situation where such situations would arise in practice.

- it is unreasonable to expect high performance (as compared to what is possible) from a reasonable, general-purpose estimate when applied to a locally peaked situation (in large part, not exclusively, because such situations do not seem to show up in practice).

The Cauchy distribution is unusual (or, if you prefer, very unGaussian) in TWO ways:

- its tails are long, falling off like x^{-2}.

- its center is peaked.

We argue in more detail below how the second makes it notional (i.e. unrealistic).

The Gaussian/rectangular was introduced to provide the first unusualness without the second.

6A4 Notionality of the Cauchy

There are many circumstances where data with long tails arises according to a simple, general pattern: different observations are made with different precision. In such cases there is almost always a best precision. Usually, something close to this best precision occurs much of the time. (The case of the one octane tester who is twice as precise as all others provides an exception to the second remark, but not to the first.)

It is well-known that a Cauchy variate is representable as

$$\frac{\text{Gaussian}}{|\text{another Gaussian}|}$$

Since the denominator is unbounded, this represents such a Cauchy variate as a mixture of Gaussians in which the precision is sometimes arbitrarily high, as well as arbitrarily low. The convenient way to avoid this unrealism is to replace the denominator by a non-negative variate that is bounded. A rectangularly distributed denominator is a convenient choice.

If it made a large difference, the writer would be prepared to forget the Cauchy entirely in making detailed comparisons of estimates. Its presence in this survey would then have to be excused on account of the desirability of comparison between our results and those of others.

Fortunately, we need not be so drastic. The qualitative behavior of the Cauchy is not very different from those of other vigorous situations, though it is a little "more so". What is very noticeable is that a few estimates (CML, TOL, and perhaps OLS; and also CPL which is not scale invariant), no one of which perform satisfactorily over any broad diversity of situations, perform up to 15% better for the Cauchy than do estimates of wider utility. In assessing the Cauchy quantitatively, it presently seems reasonable to neglect these estimates in judging the deficiency of other estimates. Such a view does not affect the relative comparison of any estimates, it merely keeps us from being "scared" by unrealistically large deficiencies. We call such a deficiency -- one corresponding to an efficiency greater than 100% for the neglected estimates -- a practical deficiency.

6A5 Outdoing and outperforming

Since we like to make both deficiencies smaller, our desire in such a picture is to move south or west or both. If one point lies both south or west of another there is no possible question but that both its variances are better. In this situation the first estimate will be said to outdo the second.

It is usually reasonable to say more. We are unlikely to be willing to increase the deficiency for the conventional or mild situation by more than 1% in order to reduce the deficiency for threatening situation by only 1%. Similarly, if we really believe in the threat we are unlikely to be willing to increase the threatening deficiency by more than 5% to gain only 1% reduction in the conventional deficiency.

If we would have to make an unlikely exchange, as so defined, to move from one estimate to another, the first will be said to outperform the other.

Exhibit 6-3 shows, for a rather unreasonable pair of situations, how the relationship of another estimate to a given one depends upon location in the plane.

* comparisons of estimates -- and of families *

Not infrequently, we are concerned with (one-parameter) families of estimates. We are likely to want to compare, at one time or another:

- one estimate with another (as just discussed).

- one estimate with a family.

- one family with another.

So long as we stick to a single pair of situations, the comparison of families turns out to be the easiest. For it is easy to indicate what part of the deficiency-deficiency plane is outdone -- or outperformed -- by some one of the estimates of a family. Exhibit 6-4 shows an example.

In later pictures, we will usually omit the horizontal and vertical boundaries -- shown dotted in exhibit 6-4 -- since the reader's eye can supply these when needed.

Because family-family comparison seems easiest, we will go through our detailed comparison in this order:

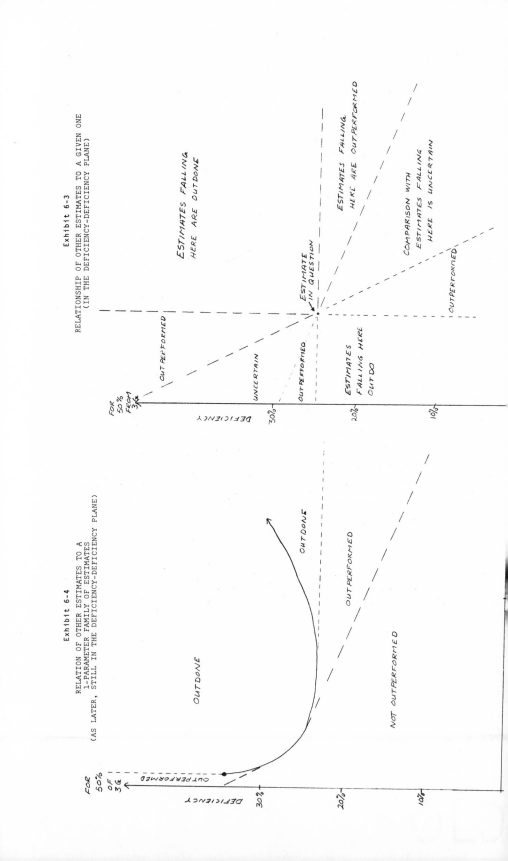

Exhibit 6-3

RELATIONSHIP OF OTHER ESTIMATES TO A GIVEN ONE
(IN THE DEFICIENCY-DEFICIENCY PLANE)

Exhibit 6-4

RELATION OF OTHER ESTIMATES TO A
1-PARAMETER FAMILY OF ESTIMATES
(AS LATER, STILL IN THE DEFICIENCY-DEFICIENCY PLANE)

- families compared with families.

- individual estimates compared with selected families.

- choice of estimate within families.

6A6 Kinds of criteria

We have to face at least three more-or-less formalized approaches to the choice of an estimate on the basis of robustness of variance:

- acceptance of a specific insurance premium (of a chosen deficiency in the conventional situation) and "minimization" of the deficiency for threatening situations. ("Minimization" is in quotes because there are many alternative situations and the best choice for one may not be the best for another.)

- adoption of exchange ratios among deficiencies for pairs of situations (we have essentially constrained these ratios in our definition of "outperforming"), and rejection of estimates which are usually less satisfactory than others.

- adoption of multiple exchange ratios, applying to all situations, so that any estimate's deficiencies can be combined across situations into a single weighted whole. (The basis for this choice may be Bayesian, but surely need not be.)

In view of the limited number of situations for which variances are actually available, the third choice seems at best to be an impractical counsel of perfection. The first choice corresponds to semi-formal working with the notion of "outdone" after some restriction to a vertical strip; the second to the same with "outperformed".
Our discussions will try to give appreciable weight to both the second choice and the first choice, and -- at least implicitly -- to some mixed choices.

127

6A7 Ways to use a criterion

Even with agreement on a kind of criterion, a number of different attitudes may seem rational. Three deserve names, those of:

- Messrs. One-toe-in-the-water, where contamination is admitted, but only gentle contamination (represented in this study by 3G) and probably only reasonable and gentle contamination.

- the wary classicist, who takes pure Gaussian as the conventional situation and faces both gentle or vigorous modifications or alternatives, so long as these are reasonable.

- the cautious realist, who believes that the situations he faces always involve some contamination and who, consequently, takes the less favorable of the available 5% situations as the conventional situation, again faces either gentle or vigorous modifications or alternatives, so long as these are reasonable.

6B SELECTED FAMILIES

We have selected four families -- one in two
versions -- for our initial stage, comparison of
families. In each case we have selected a specific
one-parameter family.

6B1 Hubers

A one-parameter family of estimates can arise in
various ways. The simplest, of course, is to involve
a single adjustable parameter, as in the estimates here
called "hubers" being Huber's proposal 2 with different
values of k. (Realized: H07, H10, H12, H15, H17 and
H20. Interpolation by linear combination.)

6B2 Hampels

The family, so far lightly explored, here called
"hampels", come from a larger family with three parameters.
The three treated here as defining a one-parameter
family are 3 of the 4 so far tried out. Like the hubers,
they can be converted into a continuous-parameter family
by considering unbiased linear combinations

$$\phi T + (1-\phi)V$$

of adjacent estimates. (Realized: 25A, 21A, 17A.
Plotting of curve through 12A is sometimes done.)

6B3 Trims

The process of local linear combination makes any
sequence of estimates into a one-parameter family. Its
apotheosis, in one sense, is the continuous family of
trimmed means, in which the result of trimming a
fractional number of observations at each end is always
a linear combination of the results of trimming the two
adjacent integer numbers of observations.

6B4 Sitsteps and alternates

A sit mean, where "sit" stands for skip-into-trim, is found by the following sort of procedure:

- sort the observations.

- skip (i.e. set aside) observations that are unusually far out, as indeed from a simple assessment of scale.

- trim a number of observations from each end of the remaining set -- the number trimmed depending upon the total number skipped.

Only a few sit means have so far been tried.
Sitstep estimates are global linear combinations of a sit mean with a one-step estimate of a quite different character. Our concern is often with a whole family of linear combinations.
Because of the advantages in adapting to sample size, we consider two families of sitstep estimates:

- sitsteps, that are unbiased linear combinations of 5T4 and D20.

- alternates, that are unbiased linear combinations of 5T4 and P15.

The advantages of such families will be discussed in 6B8.

6B5 Families summarized

These then are our four families:

- trims. (The first estimates suggested for robustness of variance [Harris and Tukey 1949].)

- hubers, actually local linear combinations of H20, H17, H15, H12, H10, H07. (Until the developments toward the close of the present study,

one only one estimate, TAK, which is very laborious
to compute, outperformed the hubers at all consistently.)

- hampels, actually local linear combinations
of 25A, 21A, 17A. (Only part of a family, so far,
with hopes for even better performance after
"tuning" the parameters and extending the range.)

- sitsteps, actually linear combinations of
5T4 and D20. (Again with much possibility for further improvement by "tuning" both parameters and
choice of basic estimate.)

- alternates, actually linear combinations of
5T4 and P15. (Of more interest for samples of 10
and less.)

6B6 Extending the hampels

The first step in exploring the hampels further
would seem to be their extension to larger values of k.
(A simple choice for the three parameters that is
relatively consistent with 25A, 21A and 17A is
(k, k+2, k+7).) It might then pay to explore the effect
of changing relationships among the parameters.

6B7 Global linear combinations

As we shall soon learn, it is much easier to pick
out a family of estimates than to choose an estimate
out of a family. Both realism (in facing up to the
existence of a hard problem) and flexibility are increased when all the estimates of the family are routinely
before us. This happens, whether we will or no, when
the family consists of all linear combinations of two
estimates. Once we have these two numbers, we are so
close to all their linear combinations to make avoiding
the question of which to use difficult.
Further, this situation often affords useful
flexibility. If we have several batches, each to be
summarized and a straight-line or two-way fit to be
made to the summaries, we can often choose our linear
combination to improve the quality of the fit. Experience
with such linear-combination families should turn up
other instances of using information extraneous to the
batch to guide a choice within the family.

6B8 Sitstep ease

At the present time the ease of computing a sitstep estimate is greater than that of other high-performance estimates (mainly of hampel type).

D20 begins with a median, and then makes one step of a Huber proposal 2 iteration with k = 2.0. For n = 20, 5T4 alone is better than the median alone for all the situations considered except for the three considered notional (i.e. unreasonable). Define E20 as the result of starting with 5T4 and making one step of a Huber proposal 2 iteration. There is ground for hope that E20 & 5T4 will perform slightly better than D20 & 5T4 . Its computation will certainly be even simpler to describe.

Define Q15 similarly with respect to P15. The linear combinations Q15 & 5T4 may perform better than P15 & 5T4.

6B9 Hybrid vigor

The high performance of intermediate unbiased linear combinations of 5T4 and D20 is an extreme example of "hybrid vigor", in which unbiased linear combinations of two estimates perform better than might be naively expected from the performance of the individual estimates. This even makes the local linear combinations used for trims (by definition) and for hubers (for convenience) perform somewhat better than the estimates from which they are combined.

This effect is usually too small to notice, but, as exhibit 6-5 shows for both families, it occasionally becomes quite noticeable.

Hybrid vigor, as shown most strongly between good estimates of quite different character, is an important tool in improving estimates empirically, which is the natural first step toward a deeper understanding and the invention of new formulations.

6C   COMPARISON OF FAMILIES FOR n = 20

In comparing the four families, as in later comparisons, we go through the situations in the order gentle (reasonable), vigorous (reasonable), unreasonable, thus meeting the needs of Messrs. one-toe-in-the-water (first group only) and wary classicists (two or three groups). Then we turn to the cautious realists, and their different views as to what is a conventional situation.

Exhibit 6-6 provides a key applicable to the later exhibits.

6C1   Gentle and reasonable situations

Exhibits 6-7 to 6-9 show the four families for 5% of 3G, 25% of 3G and t_3. (10% and 15% of 3G fall in between 5% and 25%.) The first things to notice in these pictures are the double-headed arrows labelled "both 1%" that show the size of a 1% simultaneous change in both deficiencies. The differences between the families is less than this in all the meaningful places.

The curve of sitsteps does "bottom out" higher than the others, but the dashed line, a tangent of slope 1   to this sitstep curve, remains close to both the curves and the corresponding slope-1 lines for the other families (not drawn). Thus so long as we do not value a reduction of threatening deficiency more than a reduction of conventional deficiency, we are talking of differences of less than 1%.

There is no need to make a choice in the gentle and reasonable case.

If we were forced into a choice, we might well take hubers as an overall choice. Hampels just do not go far enough, as yet, to be competitive at 5% from 3G. It is not until t_3 that they begin to do a little better than the hubers.

If we want ease of computation, we need to choose between trims and sitsteps. The trims are better at 5% from 3G, and worse for 25% from 3G and t_3. The choice between them clearly rests upon:

- how deviant gentle situations we expect.

- how important we judge the special advantages of a family of global linear combinations.

exhibit 6-6
Standard designations on pictures comparing families
(including exhibits 6-7 to 6-23)

Solid curves:

 trims   trimmed means

 hubers   huber proposal 2; H20, H17, H15, H12,
        H10, H07 and local linear combinations
        (if H07 appears, it is marked —◀).

 hampels   selected examples from a three-parameter
        family: 25A, 21A, 17A, (sometimes 12A)
        and local linear combinations (25A is
        marked ◀— to emphasize the value of
        extension in the indicated direction)
        17A is marked ╬ or +).

 sitsteps   global linear combinations of 5T4, and
        D20 (5T4 is marked with an open circle,
        D20, if present, is marked —•).

Dashed curves:

 alternates global linear combinations of 5T4 and
        P15. (5T4 is marked with an open
        circle, P15 is marked —•).

Dash-and-dot curves:

        Boundaries of outperformance of sitsteps
        when 1:1 and 1:5 limits are placed on
        exchange ratios. (see exhibit 6-4)

Dotted curves:

        Boundaries of outperformance of alter-
        nates when 1:1 and 1:5 limits are placed
        on exchange ratios. (see exhibit 6-4)

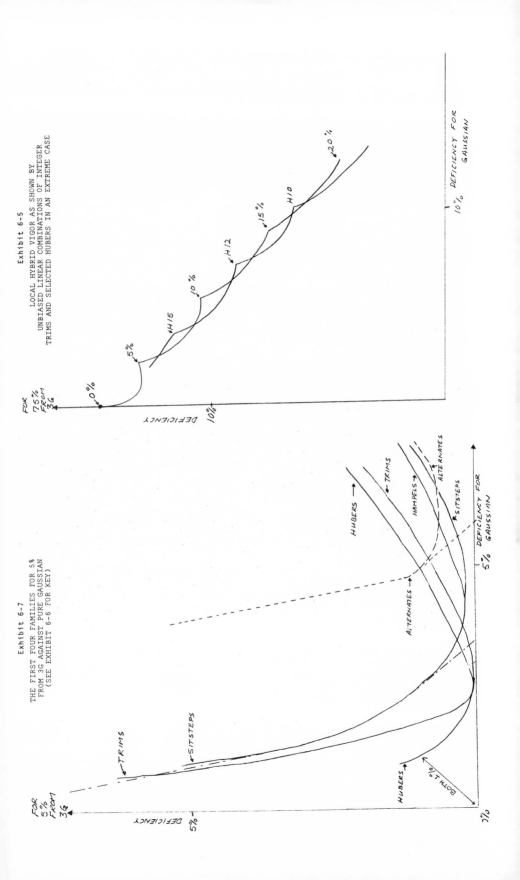

Exhibit 6-5

LOCAL HYBRID VIGOR AS SHOWN BY
UNBIASED LINEAR COMBINATIONS OF INTEGER
TRIMS AND SELECTED HUBERS IN AN EXTREME CASE

Exhibit 6-7

THE FIRST FOUR FAMILIES FOR 5%
FROM 3G AGAINST PURE GAUSSIAN
(SEE EXHIBIT 6-6 FOR KEY)

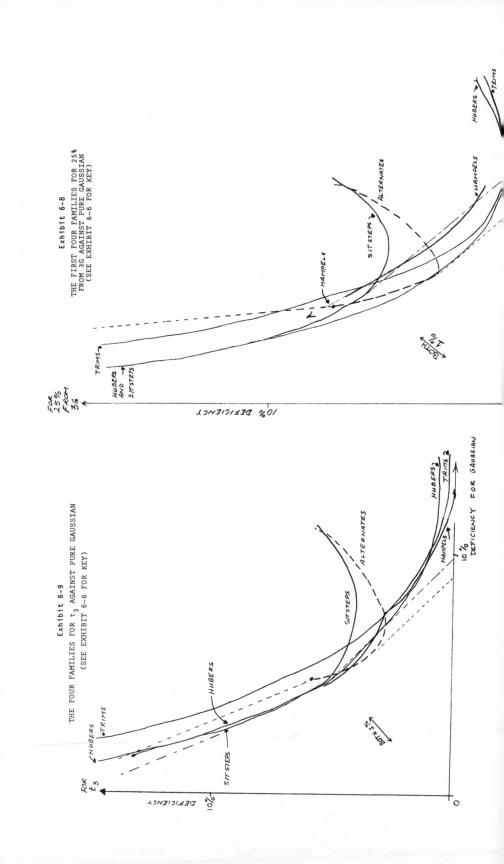

Exhibit 6-9

THE FOUR FAMILIES FOR t_3 AGAINST PURE GAUSSIAN
(SEE EXHIBIT 6-6 FOR KEY)

Exhibit 6-8

THE FIRST FOUR FAMILIES FOR 25%
FROM 3G AGAINST PURE GAUSSIAN
(SEE EXHIBIT 6-6 FOR KEY)

It is fair to say, for all choosers with broader
concern, that gentle situations do not speak strongly
against any one choice.

Alternates fall down badly for 5% from 3G, but
are fully competitive for 25% from 3G and t_3. They do
well on outperforming or nearly outperforming, but are
inferior on outdoing. For gentle situations for n = 20
they are close, but should be passed by.

6C2 Vigorous and reasonable situations

Exhibits 6-10 to 6-14 show the four families for 5%
from 10G, 10% of 1/U, 10% of 3/U, 25% from 10G, and
pure 1/U.

In every case the hampels at least partially outperform the others. For 5% from 10G, 10% of 1/U and
10% of 3/U (and for 10% from 10G and 25% from 1/U),
extending the hampels to larger values of k will clearly
pay substantial dividends. For 25% from 10G (and for
25% of 3/U) extension seems not to be needed, while for
pure 1/U the situation is somewhat uncertain. (In this
latter instance, 17A also outperforms the sitsteps.)

For 5% from 10G and 10% of 1/U, hubers outperform
sitsteps for Gaussian deficiencies below 3% or 2%
respectively. Otherwise the hubers appear to be outperformed by the sitsteps throughout. In these two
instances, the hubers outperform the sitsteps in only
part of the range and by less than 1%, as the double-
headed arrows show.

Our general position, then, if both gentle reasonable and vigorous reasonable situations are considered,
is that:

> - hampels, often after extension, seem to be
> the family of choice when only decreasing the
> variance is sought.

> - if other considerations are taken into
> account, and reasonably vigorous situations are
> to be feared, sitsteps are the family of choice.

Again, alternates do not match sitsteps for the
gentler situations (5% from 10G and 10% of 1/U) but
do quite well for the more vigorous (especially 25%
from 10G and pure 1/U).

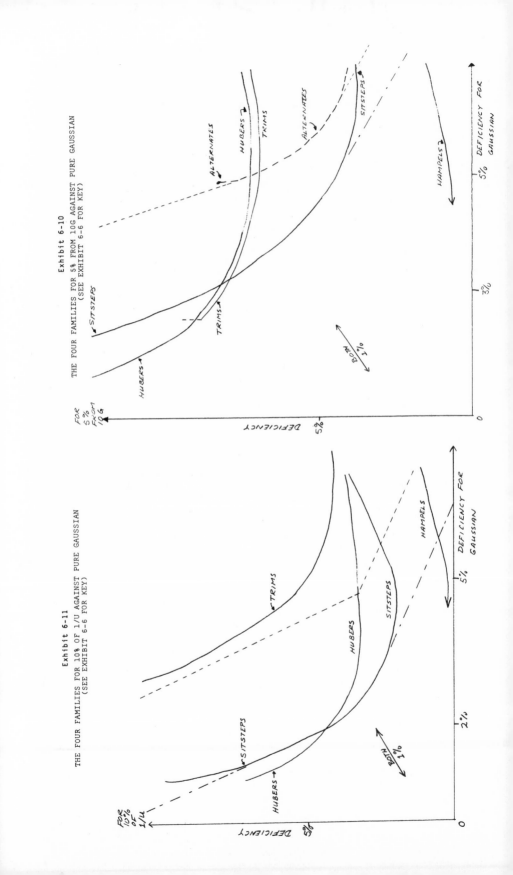

Exhibit 6-10

THE FOUR FAMILIES FOR 5% FROM 10G AGAINST PURE GAUSSIAN
(SEE EXHIBIT 6-6 FOR KEY)

Exhibit 6-11

THE FOUR FAMILIES FOR 10% OF 1/U AGAINST PURE GAUSSIAN
(SEE EXHIBIT 6-6 FOR KEY)

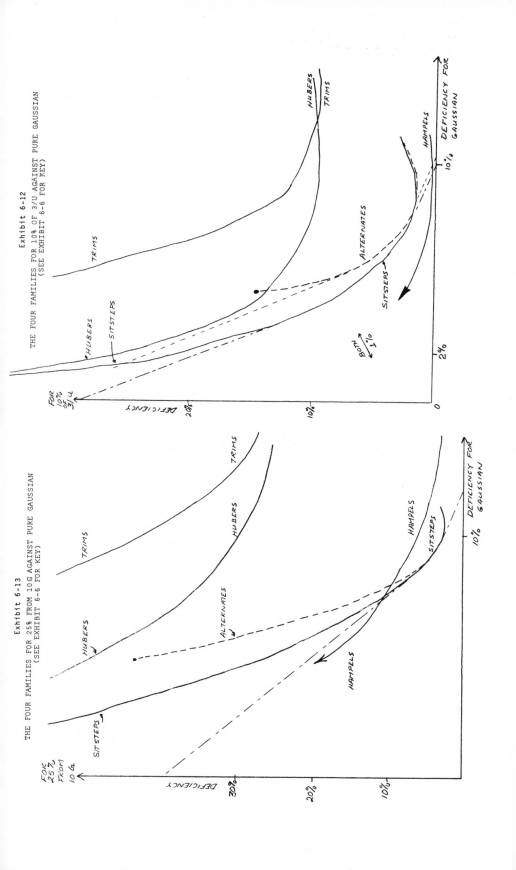

Exhibit 6-12

THE FOUR FAMILIES FOR 10% OF 3/U AGAINST PURE GAUSSIAN
(SEE EXHIBIT 6-6 FOR KEY)

Exhibit 6-13

THE FOUR FAMILIES FOR 25% FROM 10G AGAINST PURE GAUSSIAN
(SEE EXHIBIT 6-6 FOR KEY)

In going on to compare individual estimates with families it would suffice to carry over sitsteps and 2 hampels (25A and 21A) and, occasionally, 2 hubers (H20 and H17).

6C3 Unreasonable situations

Exhibit 6-15 shows our one vigorous but unreasonable situation, the pure Cauchy. Behavior is much like the pure 1/U, except that the hampels greatest advantage over the sitsteps is considerably reduced.

Exhibits 6-16 and 6-17 show the two gentle but unreasonable situations, 50% from 3G and 75% from 3G. If we believe in the boundary exchange ratios of 1:1 and 5:1, then at 75% from 3G sitsteps (in the form of D20) outperform everything else except trims with less than 2% trimmed from each end. At 50% from 3G, a most unreasonable situation, the other three families outperform sitsteps for many exchange ratios between 1:1 and 2:1.

All in all, the unreasonable situations do not cause us to change any of our conclusions based on the examination of reasonable situations only.

Again the alternates are close to the sitsteps, but not quite competitive with them.

6C4 The cautious realist

We turn next to the pictures appropriate for those who feel that most, if not all, of the data they see is contaminated. For them we make two changes in our procedure:

- the conventional situation becomes the poorer of 5% from 3G and 5% from 10G.

- the Gaussian alternative is now marked vigorous and unreasonable.

Exhibits 6-18 to 6-21 show four representative situations, 10% from 3G, 15% from 3G, 25% of 3/U and pure 1/U. The other reasonable situations behave similarly. In all, the hampels win hands down, without extension. The second choice in every situation is the alternates, followed closely by the sitsteps. No doubt ever arises.

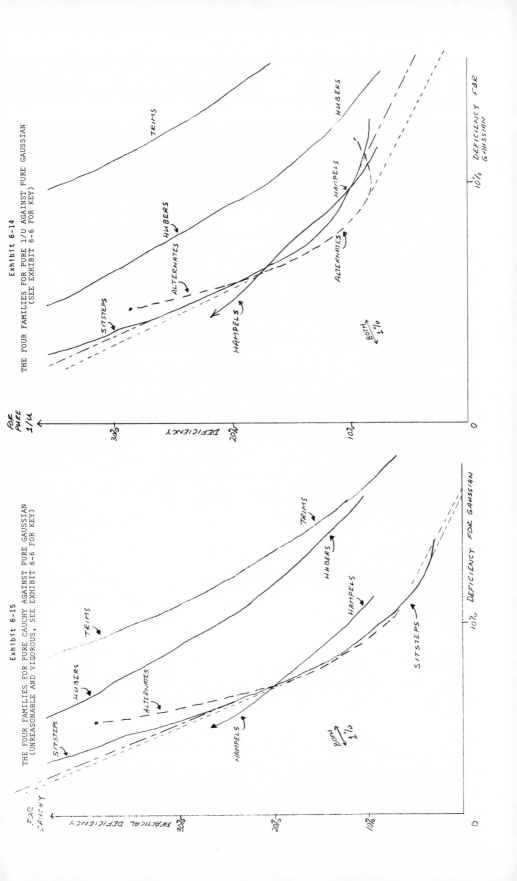

Exhibit 6-14

THE FOUR FAMILIES FOR PURE 1/U AGAINST PURE GAUSSIAN
(SEE EXHIBIT 6-6 FOR KEY)

Exhibit 6-15

THE FOUR FAMILIES FOR PURE CAUCHY AGAINST PURE GAUSSIAN
(UNREASONABLE AND VIGOROUS, SEE EXHIBIT 6-6 FOR KEY)

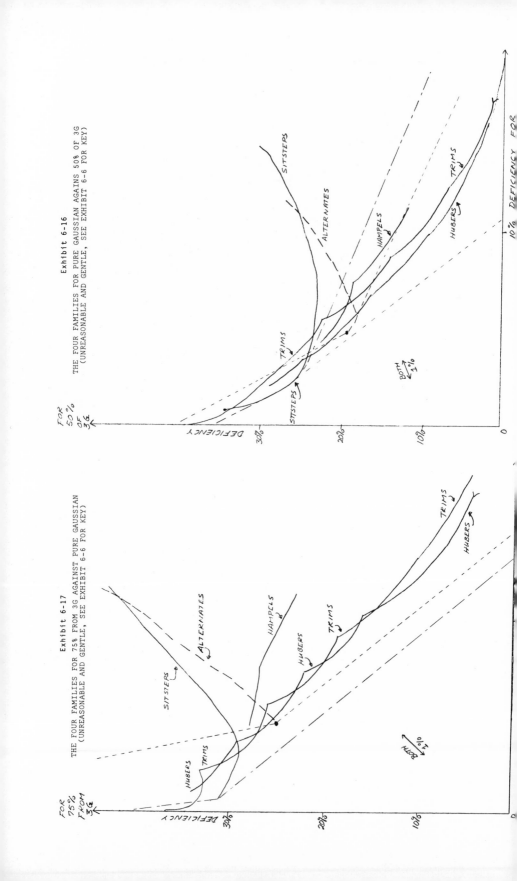

Exhibit 6-16

THE FOUR FAMILIES FOR PURE GAUSSIAN AGAINS 50% OF 3G
(UNREASONABLE AND GENTLE, SEE EXHIBIT 6-6 FOR KEY)

Exhibit 6-17

THE FOUR FAMILIES FOR 75% FROM 3G AGAINST PURE GAUSSIAN
(UNREASONABLE AND GENTLE, SEE EXHIBIT 6-6 FOR KEY)

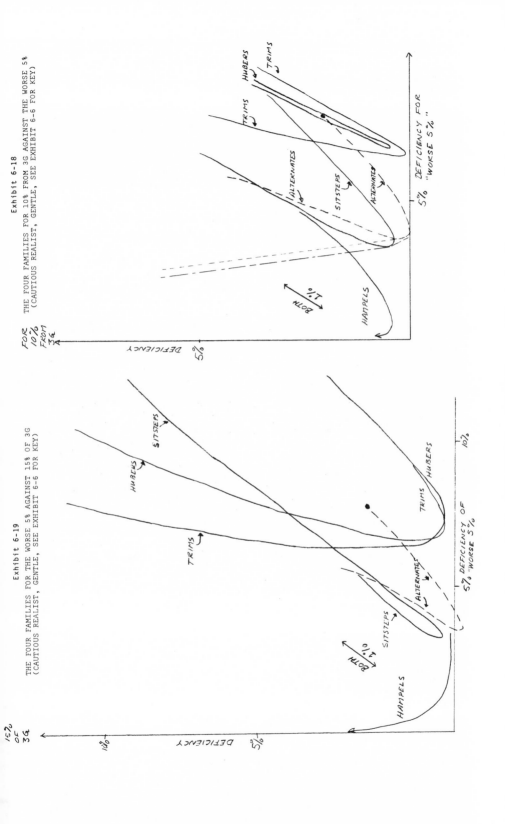

Exhibit 6-18

THE FOUR FAMILIES FOR 10% FROM 3G AGAINST THE WORSE 5%
(CAUTIOUS REALIST, GENTLE, SEE EXHIBIT 6-6 FOR KEY)

Exhibit 6-19

THE FOUR FAMILIES, FOR THE WORSE 5% AGAINST 15% OF 3G
(CAUTIOUS REALIST, GENTLE, SEE EXHIBIT 6-6 FOR KEY)

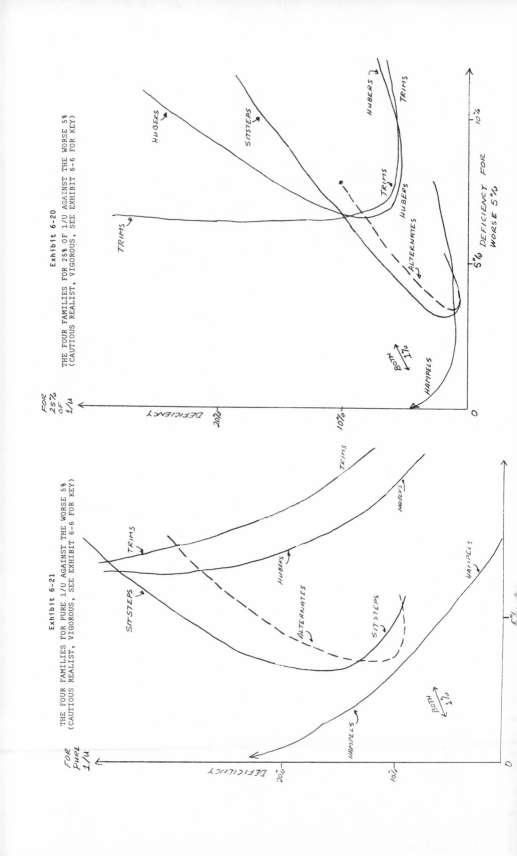

Exhibit 6-20

THE FOUR FAMILIES FOR 25% OF 1/U AGAINST THE WORSE 5%
(CAUTIOUS REALIST, VIGOROUS, SEE EXHIBIT 6-6 FOR KEY)

Exhibit 6-21

THE FOUR FAMILIES FOR PURE 1/U AGAINST THE WORSE 5%
(CAUTIOUS REALIST, VIGOROUS, SEE EXHIBIT 6-6 FOR KEY)

Exhibits 6-22 and 6-23 show similar pictures for two unreasonable situations, pure Cauchy and pure Gaussian. For the pure Cauchy, the sitsteps outperform the hampels by just a shade. For the Gaussian, extending the hampels would be profitable. These differences are not large enough to alter the views we based on the reasonable situations.

Whether Frank Hampel intended to respond to such wishes or not, the first hampels are clearly beau ideal estimates in the eyes of a cautious realist with a sample of size near 20.

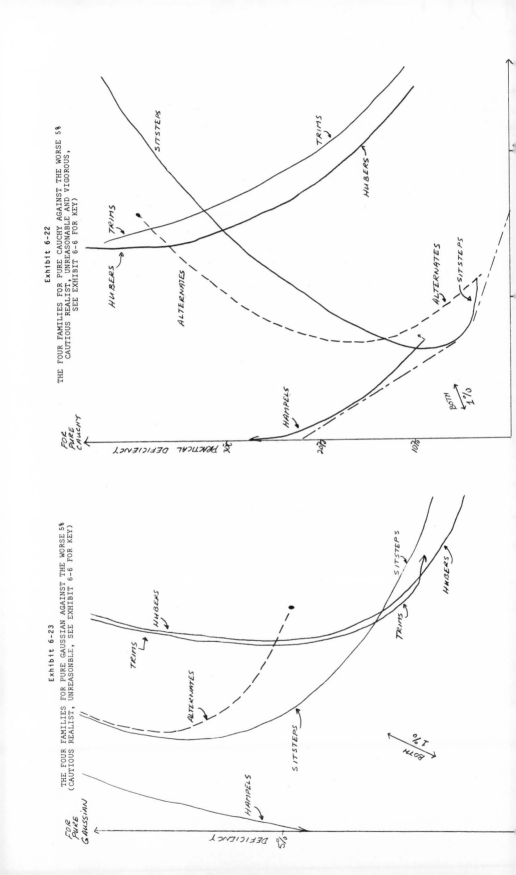

Exhibit 6-22

THE FOUR FAMILIES FOR PURE CAUCHY AGAINST THE WORSE 5%
CAUTIOUS REALIST, UNREASONABLE AND VIGOROUS,
SEE EXHIBIT 6-6 FOR KEY)

Exhibit 6-23

THE FOUR FAMILIES FOR PURE GAUSSIAN AGAINST THE WORSE 5%
(CAUTIOUS REALIST, UNREASONBLE, SEE EXHIBIT 6-6 FOR KEY)

6D COMPARISON OF INDIVIDUAL ESTIMATES
WITH FAMILIES FOR n = 20

We now turn to the performance of individual
estimates in achieving low variance. Exhibit 6-24
provides a key for use in connection with later
exhibits.

6D1 Gentle and (reasonable) situations

Exhibits 6-25 to 6-27 show -- for 5% from 3G, 25%
from 3G, and pure t_3 -- (1) a variety of individual
estimates, (2) selected trims, hubers and hampels, (3)
the curve of sitsteps, with 5T4 shown as an open circle,
(4) the curve of alternates shown dashed. The other
gentle and reasonable situations, 10%, 15% from 3G,
fall in between.

As an encouragement to optimism, mainly in the region
that will undoubtedly be penetrated by extended hampels,
certain hybrids are shown as *. Similarly, certain
other sitsteps are shown as small squares. Since both
are only of passing interest, they are not identified.

We return to a discussion of the individual estimates in 6E below.

6D2 Vigorous (reasonable) situations

Exhibits 6-28 to 6-32 show the same information
for 5% from 10G, 10% of 3/U, 25% of 1/U, 25% from 10G,
and pure 1/U. The other vigorous situations -- 10%
of 1/U, 10% from 10G, and 25% of 3/U -- fall in between
(often near one of those shown).

6D3 Unreasonable situations

Exhibit 6-33 shows the same information for the
Cauchy, using practical deficiencies. Exhibit 6-34
goes further, giving a squeezed picture that includes
estimates that we believe to be taking advantage of
unrealistic center peakedness. Since the sitsteps
outperform OLS, TOL and CML, we have only to be
responsive to the fine performance of TAK.

We have not shown pictures for 50% and 75% from 3G,
on the ground that they ought not to influence our
analyses.

exhibit 6-24
Standard designations on pictures comparing
individual estimates with families
(including exhibits 6-25 to 6-34)

Solid curve:

  sitsteps: global linear combinations of 5T4 and
             D20. (5T4 is marked with an open
             circle. Weights in quarters give
             A, B and C.)

Dashed curve:

  alternatives: global linear combinations of 5T4
             and P15. (Weights in quarters give
             D, E and F.)

Enclosed characters:

  ◇   estimates of huber type (including
           modifications).

  ⬭   estimates of hampel type (including
           a variety of related estimates).

  —O—   sitsteps (if empty = 5T4).

Special characters:

  ✱   miscellaneous hybrids.

  ▢   other instances of sitsteps.

Among sitsteps and among alternates, three particular
linear combinations from each family are located and
tagged, namely

  A, which is 3/4 of 5T4 plus 1/4 of D20
  B, which is 1/2 of 5T4 plus 1/2 of D20
  C, which is 1/4 of 5T4 plus 3/4 of D20
  D, which is 3/4 of 5T4 plus 1/4 of P15
  E, which is 1/2 of 5T4 plus 1/2 of P15
  F, which is 1/4 of 5T4 plus 3/4 of P15.

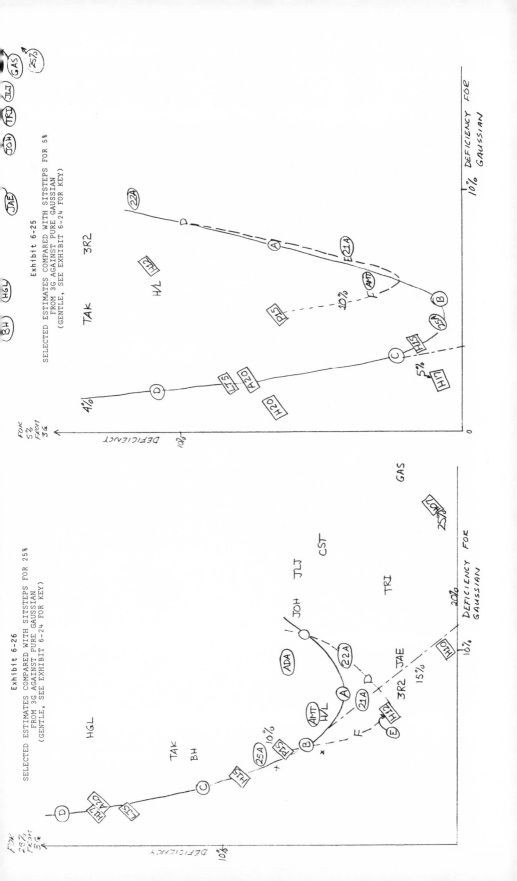

Exhibit 6-25

SELECTED ESTIMATES COMPARED WITH SITSTEPS FOR 5%
FROM 3G AGAINST PURE GAUSSIAN
(GENTLE, SEE EXHIBIT 6-24 FOR KEY)

Exhibit 6-26

SELECTED ESTIMATES COMPARED WITH SITSTEPS FOR 25%
FROM 3G AGAINST PURE GAUSSIAN
(GENTLE, SEE EXHIBIT 6-24 FOR KEY)

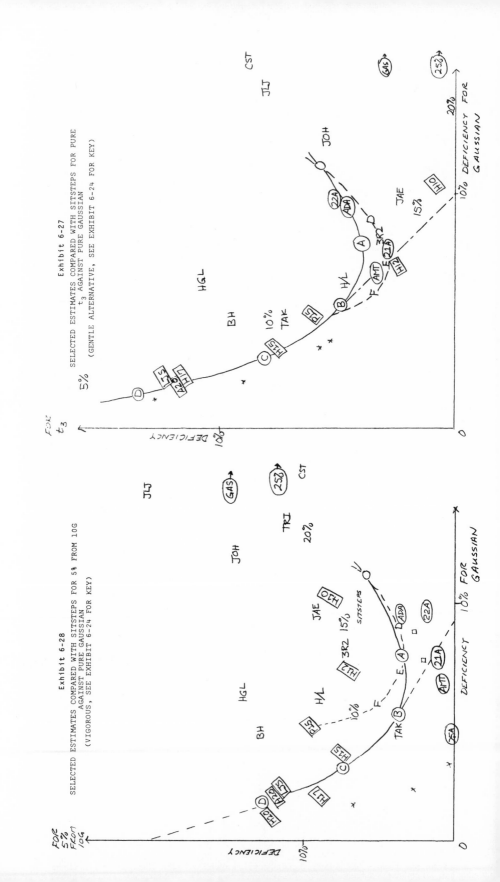

Exhibit 6-27

SELECTED ESTIMATES COMPARED WITH SITSTEPS FOR PURE
t_3 AGAINST PURE GAUSSIAN

(GENTLE ALTERNATIVE, SEE EXHIBIT 6-24 FOR KEY)

Exhibit 6-28

SELECTED ESTIMATES COMPARED WITH SITSTEPS FOR 5% FROM 10G
AGAINST PURE GAUSSIAN

(VIGOROUS, SEE EXHIBIT 6-24 FOR KEY)

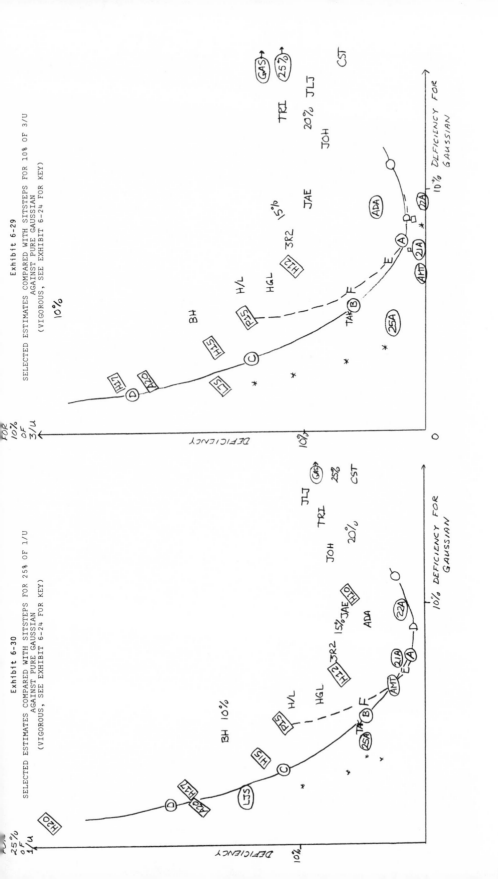

Exhibit 6-29

SELECTED ESTIMATES COMPARED WITH SITSTEPS FOR 10% OF 3/U
AGAINST PURE GAUSSIAN
(VIGOROUS, SEE EXHIBIT 6-24 FOR KEY)

Exhibit 6-30

SELECTED ESTIMATES COMPARED WITH SITSTEPS FOR 25% OF 1/U
AGAINST PURE GAUSSIAN
(VIGOROUS, SEE EXHIBIT 6-24 FOR KEY)

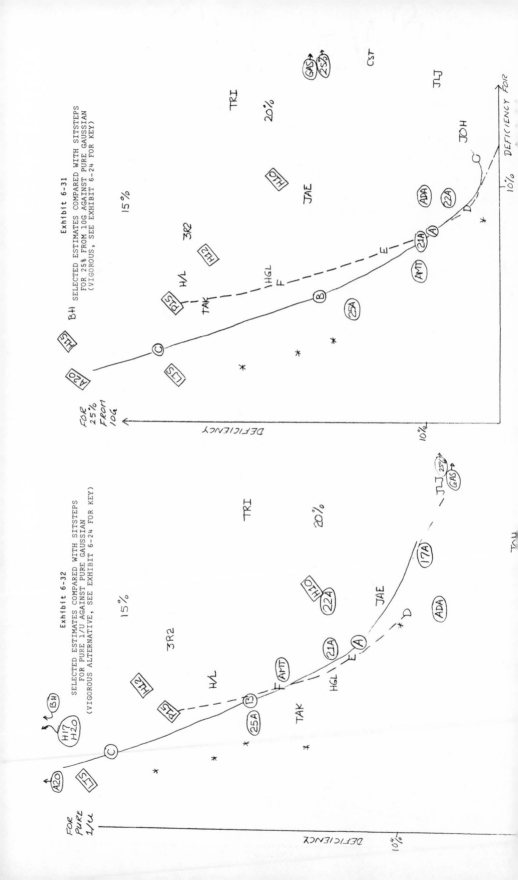

Exhibit 6-31

BH SELECTED ESTIMATES COMPARED WITH SITSTEPS
FOR 25% FROM 10G AGAINST PURE GAUSSIAN
(VIGOROUS, SEE EXHIBIT 6-24 FOR KEY)

Exhibit 6-32

SELECTED ESTIMATES COMPARED WITH SITSTEPS
FOR PURE 1/U AGAINST PURE GAUSSIAN
(VIGOROUS ALTERNATIVE, SEE EXHIBIT 6-24 FOR KEY)

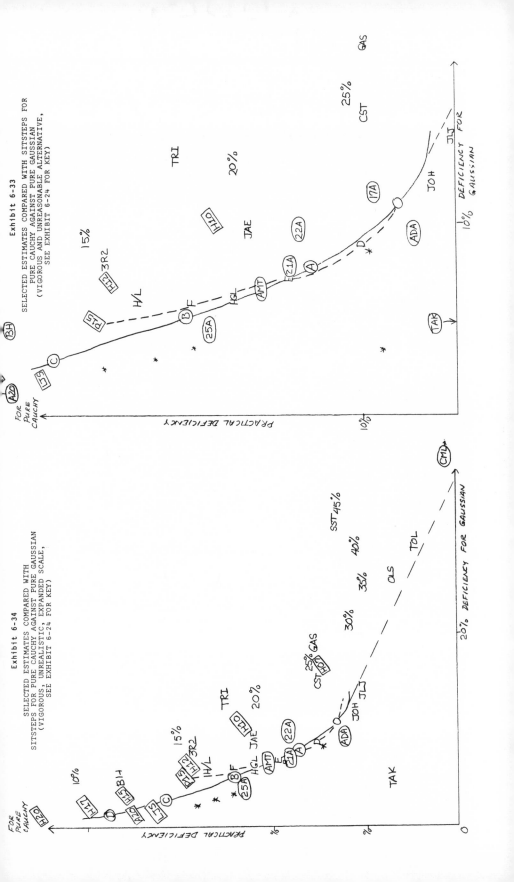

Exhibit 6-33

SELECTED ESTIMATES COMPARED WITH SITSTEPS FOR
PURE CAUCHY AGAINST PURE GAUSSIAN
(VIGOROUS AND UNREASONABLE ALTERNATIVE,
SEE EXHIBIT 6-24 FOR KEY)

Exhibit 6-34

SELECTED ESTIMATES COMPARED WITH
SITSTEPS FOR PURE CAUCHY AGAINST PURE GAUSSIAN
(VIGOROUS, UNREALISTIC, EXPANDED SCALE,
SEE EXHIBIT 6-24 FOR KEY)

6D4 The cautious realist

The outstanding performance of (first) hampels and (second) alternates and sitsteps for the cautious realist (see 6C4) is so clear and definite as to make comparisons of individual estimates with these families unnecessary. (If comparisons are needed, they can be made indirectly via the pictures of 6D1 to 6D3.)

6E SPECIFIC ESTIMATES FOR n = 20

6E1 Takeuchi's estimate

The main disadvantage of Takeuchi's estimate is its great computing effort. In comparison with sitstep B, which it resembles in Gaussian deficiency, it does less well for the gentle situations, about the same for the vigorously contaminated, and rather better for pure 1/U. For the realistic situations it does no better than break even.

For the pure Cauchy, TAK performs strikingly well. If this is only a matter of being "adaptive" then it does a good job for n = 20 once we go as far as the Cauchy. We might then expect to do better for n = 40 and worse for n = 10. If, on the other hand, it is a matter of seizing the unrealistic peakedness, one would, unfortunately, expect the same sort of dependence on n. This leaves us a challenge -- what is going on?

6E2 Estimates of hampel type

ADA was an attempt at an adaptive hampel. Planned before any hampels had run, it clearly used a poor choice of parameters. Perhaps HMPL D(ADA, ADA+2, ADA+7) would perform better.

22A deviates from the 25A-21A-17A curve in the same direction as ADA. Its parameters are k, k+1.5, k+3.7 for k = 2.2. Perhaps HMPL D(k, k+3, k+10) might do even better than HMPL D(k, k+2, k+7).

AMT performs somewhat less well for gentle situations than 25A-21A curve, but for increasingly vigorous contamination does increasingly better than that curve. (It falls back a little for pure 1/U.) Clearly rounding some of the corners in a hampel adjustment may help.

6E3 Folded estimates

H/L, the Hodges-Lehmann estimate, is now of mature years. Its performance is classically good, usually falling between the trims and the hubers. (H13, a Huber estimate with k = 1.3, would probably outdo it slightly.)

BH, the simplified version, has a lower Gaussian deficiency, but a higher contaminated deficiency. For pure 1/U (or pure Cauchy) its performance is bad.

3R2, the retrimmed and refolded estimate, performs in the trims-hubers region with a higher Gaussian deficiency than H/L. If we didn't have easier and better estimates, we might consider it.

6E4 Estimates of Jaeckel type

JAE, the prototype Jaeckel estimate, runs close to the trims-hubers band.

JLJ, a modified Jaeckel, outperforms the sitsteps for the pure Cauchy, but its Gaussian deficiency is very large, and its performance for contaminated situations is discouraging.

6E5 Johns's estimate

JOH performs like 5T4 for gentle situations, but for mildly vigorous contamination does less well, coming closer to 5T4 for 25% from 10G and outperforming it for pure 1/U and pure Cauchy. There is a small challenge here to save the good without the bad.

6E6 Hogg estimates

HGL, the most promising of the Hogg estimates, is outperformed by H/L for the gentler situations but outperforms H/L for the more vigorous ones. For pure 1/U, it outperforms the sitsteps. On balance, those who fear vigorous deviations from Gaussianity would prefer HGL to H/L -- but would prefer a sitstep to both for n=20.

6E7 Simple estimates

In addition to 5T4, which some would not quite consider simple, and 25% (the midmean), which involves a fair amount of addition and division, we ought to consider TRI, CST, GAS, SST and 45% (the median).

For gentler situations, 25% behaves best, with CST coming next. For vigorous contaminations, CST outperforms 25%. For pure 1/U (but not for pure Cauchy) 25% outperforms CST. Throughout sitstep B outperforms these leading easy estimates.

Note that 25% (the midmean) uniformly outdoes GAS for this sample size.

6E8   Summary

Besides sitsteps, then, we might want to consider the following estimates for practical use (considering the Cauchy as unrealistic):

- 25A, 21A and extended hampels (together with modifications or outgrowths of AMT).

- H20 and H17, for very gentle situations only.

- CST and 25% for even simpler calculation.

Exhibit 6-35 shows the deficiencies for the eleven resulting estimates for the realistic alternatives. These can then be summarized as in exhibit 6-36. As panel B indicates, we might well concentrate on two sitsteps -- B and C -- and two hampels -- 25A and 21A -- for performance and on one of the 5T4, CST and 25% for varying degrees of ease.

For further development, besides extending and tuning sitsteps and hampels, we need to understand how to keep the good in TAK and JOH without the bad.

Techniques for using the behavior of several samples to adjust the estimate for each -- perhaps using the mean number T-skipped in 5T4's to select among 5T4, A, B, C and D20 -- offer promise of further reduction of the already small deficiencies.

exhibit 6-35

Performance of 11 selected estimates for the
reasonable situations individually

Gaussian deficiency	Tag	5% of 3G	10% of 3G	15% of 3G	15% pure t₃	25% of 3G	Deficiencies 5% of 10G	10% of S	10% of 3S	10% of 10G	25% of S	25% pure S	25% of 10G	25% of 3S
1	H20	1%	7%	14%	17%	23%	12%	9%	39%	30%	29%	69%	83%	79%
2	H17	0	3	7	12	15	9	5	25	21	18	59	70	67
2	D20	1	6	11	13	17	13	6	24	27	19	50	59	59
3	C	0	2	4	8	11	7	3	15	16	10	39	44	46
4	25A	0	1	3	5	8	0	0	2	1	2	28	18	25
5	B	0	0	1	5	6	3	2	6	7	3	28	24	31
8	A	1	1	0	4	5	3	4	1	3	0	19	6	15
8	21A	1	1	1	3	4	1	1	0	0	1	22	8	12
11	5T4	4	4	3	6	6	6	9	2	5	1	16	0	7
15	CST	9	6	5	4	6	10	12	6	11	5	16	5	14
17	25%	9	7	4	1	1	13	11	11	15	6	13	22	19

exhibit 6-36
Performance of 11 selected estimates
for the reasonable situations, summarized

A) FULL TABLE

| | | Deficiencies | | | |
Tag	Gaussian	4 gentle*	5 less vigorous*	3 more vigorous*	(Notes)
H20	1%	23%	30%	83%	
H17	2	15	25	70	
D20	2	17	27	59	(sitstep)
C	3	11	16	46	(sitstep)
25A	4%	8%	2%	28%	
B	5	6	7	31	(sitstep)
A	8	5	4	19	(sitstep)
21A	8	4	1	22	(sit mean)
5T4	11%	6%	9%	16%	
CST	15	9	12	16	(simple)
25%	17	9	15	22	(describable)

*The % given is the largest for the group. (The 5 less
vigorous are 5% from 10G, 10% of 1/U, 10% of 3/U, 1% from 10G
and 25% of 1/U. The 3 more vigorous are pure 1/U, 25% from
10G, and 25% of 3/U.

B) FURTHER SELECTED

C	3%	11%	16%	46%	
25A	4	8	2	28	
B	5	6	7	31	
21A	8	4	1	22	
5T4	11%	6%	9%	16%	(minimax*)
CST	15	9	12	16	(minimax*)
25%	17	9	15	22	

*among those listed here

6F COMPARISON BETWEEN VARIANCES AND PSEUDOVARIANCES

As soon as we extend our analysis to smaller samples -- to n = 10 and n = 5 -- and try to do a careful job of comparison with n = 20, we find signs of complications. At first weak, these signs strengthen as we look at more vigorous situations. By the time we reach pure Cauchy at n = 10, the indications are much too strong to be left alone. Not only are jumpy estimates, such as those using skipping, misbehaving, but those which we might naively consider almost unrockable boats, like the hubers, show clearly anomalous behavior.

As a matter of first aid, we ran a new set (n = 10B) of 1000 configurations for pure Cauchy at n = 10. The variances were quite different from those at n = 10A, but fell much more nearly in line with those for other values of n. This put us to further analysis -- in particular to the comparison of variances and pseudo-variances. The results of this comparison were rather striking, and appear here, since the results will be of considerable use as we go on to n less than 20.

6F1 In which interval?

Recall that

(1) pseudovariances are the results of con-verting % points by asking what would be the variance of a Gaussian distribution matching a given % point. (If the pseudovariance appears constant as the % changes, the distribution of the estimate appears to be Gaussian.)

(2) pseudovariances were calculated for 50%, 25%, 10%, 2.5%, 1%, 0.5% and 0.1%.

The simplest thing to do is to look at the variance and ask "In which interval defined by calculated pseudo-variances does it fall?". For well-behaved situations, the answer is almost always: Between the 10% pseudo-variance and the 2.5% pseudovariance!

We will discuss why this ought to be so in 6F2. For the present, note that this occurrence guided us, after further exploration, to classify estimates by their variance relative to the 2.5% pseudovariance and the 10% pseudovariance as follows:

X)   Between the 2.5% and 10% pseudovariances, or within a factor of 1.001 of being there.

E)   Between 1.001 and 1.01 times the 2.5% pseudovariance.

D)   Between 1.01 and 1.1 times the 2.5% pseudovariance.

C)   More than 1.1 times the 2.5% pseudovariance but less than 2 times the 10% pseudovariance.

B*)   More than 2 times the 10% pseudovariance but less than 1.1 times the 2.5% pseudovariance.

B)   More than 2 times the 10% pseudovariance and more than 1.1 times the 2.5% pseudovariance but less than 2 times the 2.5% pseudovariance.

A)   More than 2 times the 2.5% pseudovariance.

Most would be willing to include E's with X's. Some, probably many, would be willing to include D's as well -- as "not very far out of line". A's, and even B's, represent cases where variances are far out of line with pseudovariances.
Some estimates are often out of line. Exhibit 6-37 shows the classification -- for each available situation -- of 12 of these "often out of line" estimates. Exhibit 6-38 does the same for another set of 12. Exhibits 6-39 and 6-40 complete the record.
We should notice that:

1)   For some situations, variances are quite frequently quite different from 2.5% and 10% pseudovariances.

2)   For only 6 cases -- 25% of 1/U and pure Cauchy at n = 5, 25% of 1/U and pure Cauchy (A and B) at n = 10, and 25% of 3/U at n = 20 -- are these differences of outstanding size for many estimates.

3)   For another 6 to 10 situations, the differences are large for all or part of the first 12 or so estimates (exhibit 6-37) and not for other estimates.

exhibit 6-37
Relation of variances to 2.5% and 10% pseudovariances
(first 12 estimates, see text for coding)

	M	M15	4RM	3RM	2RM	SJA	BIC	MEL	CTS	5%	10%	15%
n=10, Cauchy A	A	A	A	A	A	B	A	B	A	A	A	B
", 25% of 1/U	A	A	A	A	A	A	A	C	X	A	B	C
n=5, 25% of 1/U	A	B	A	A	A	A	A	C	A	A	B	B
", Cauchy	A	B	A	A	A	A	A	A	A	A	A	A
n=10, Cauchy B	A	B	A	A	B	B	B	A	X	A	B	C
n=20, 25% of 3/U	A	B	A	A	B	A	C	B	C	B	C	D
n=10, pure 1/U	A	A	A	A	B	B	B	C	X	A	C	C
n=20, 10% of 3/U	A	A	A	B	B	A	C	C	X	C	D	E
", 10% of 1/U	A	A	A	C	D	A	A	P	X	D	E	X
", 25% of 1/U	A	A	A	B	C	C	C	D	X	C	X	X
", Cauchy	A	B	A	B	B	X	C	B	X	B	B	X
", pure 1/U	A	X	A	B	X	X	D	E	X	X	X	X
n=40, 25% of 1/U	A	A	B	D	X	X	X	C	X	X	X	X
", Cauchy	A	B	B	B	D	X	X	C	X	C	X	X
n=20, 25% from 10G	X	A	X	X	X	X	X	X	X	X	X	X
", 10% from 10G	X	C	X	X	X	X	X	X	X	X	X	X
", 5% from 10G	X	C	X	X	X	X	X	X	X	X	X	X
", t₃	C	C	X	X	X	X	X	X	X	X	X	X
n=20, 75% from 3G	X	X	X	E	X	X	X	X	X	X	X	X
", 50% from 3G	X	X	X	E	X	X	X	X	X	E	X	X
", 25% from 3G	X	X	X	X	X	X	E	X	X	X	X	X

The following are X for all estimates:

n=20: 15% or 10% or 5% from 3G; Gauss;
n=10: 20% or 10% from 3G; Gauss;
n=5 or 40: Gauss.

162

exhibit 6-38

Relation of variances to 2.5% and 10% pseudovariances
(second 12 estimates, see text for coding)

	H20	H17	D20	HGP	HGL	THL	THP	3R1	BH	TAK	OLS	JOH
n=10, Cauchy A	A	A	A	A	A	A	A	A	A	A	C	B
", 25% of 1/U	A	A	A	C	C	A	C	B	C	C	C	D
n=5, 25% of 1/U	A	C	C	A	A	A	A	C	C	A	A	D
", Cauchy	A	B	B*	A	A	B*	B*	D	B*	B*	A	C
n=10, Cauchy B	B	C	B	A	D	X	C	B	B	B	A	B
n=20, 25% of 3/U	C	C	C	D	X	E	D	B	C	X	X	D
n=10, pure 1/U	B	X	X	B	X	X	X	C	X	X	X	D
n=20, 10% of 3/U	D	D	X	D	X	X	X	X	X	X	X	X
", 10% of 1/U	X	X	X	X	X	X	X	X	X	X	X	X
", 25% of 1/U	X	X	X	X	X	X	X	X	X	X	X	X
", Cauchy	D	X	X	X	X	X	X	X	X	X	X	X
", pure 1/U	X	X	X	D	X	X	X	X	X	X	X	X
n=40, 25% of 1/U	X	X	X	X	X	X	X	X	X	X	X	X
", Cauchy	D	X	X	X	X	X	X	X	X	X	X	X
n=20, 25% from 10G	X	X	X	D	C	X	E	D	D	X	X	X
", 10% from 10G	X	X	X	X	X	X	X	X	X	X	X	X
", 5% from 10G	D	X	X	X	X	X	X	X	X	X	X	X
", t_3	X	X	X	D	X	X	X	X	X	X	X	X
n=20, 75% from 3G	X	X	X	X	X	X	X	X	X	X	X	X
", 50% from 3G	X	X	X	E	X	X	X	X	X	X	X	E
", 25% from 3G	X	X	X	X	X	X	X	X	X	X	X	X

163

exhibit 6-39

Relation of variances to 2.5% and 10% pseudovariances
(remaining estimates, see text for coding and exhibit 6-42 for further details)

	X	E	D	C	B	A
n=10, Cauchy A	(22)	·	·	22A,3R2,H/L (7)	TOL (4)	D10,LJS,P15 (17)
" , 25% of 1/U	(27)	A20	25% A15,P15,A20 (18)	(the other 39)	·	·
n=5, 25% of 1/U	(none)	21A,H07	(7)	SST,HMD (4)	(6,5*)	·
" , Cauchy	(9)	5T1,H12	(11)	H15)	·	·
n=10, Cauchy B	(30)	JAE,H10	·	·	·	·
n=20, 25% of 3/U	(30)	·	·	·	·	·
n=10, pure 1/U	(all 43)					
n=20, 10% of 3/U	(all 44)					
" , 10% of 1/U	(all 44)					
" , 25% of 1/U	(all 44)					
" , Cauchy	(all 44)					
" , pure 1/U	(43)		HGP			
n=40, 25% of 1/U	53		7.5%	2.5%		
" , Cauchy	53					
n=20, 25% from 10G	40	(4)				
" , 10% from 10G	(all 44)					
" , 5% from 10G	(all 44)					
" , t3	(all 44)					
n=20, 75% from 3G	(42)	CST,JWT				
" , 50% from 3G	(all 44)					
" , 25% from 3G	(all 44)					

164

exhibit 6-40
Further details to exhibit 6-39
(estimates not X in situation indicated)

Situation	Class	Estimates
n=10, Cauchy A	A	ADA, 5T1, 5T4, CST, JWT, 3T1, 3T0, D07, TRI, JBT,
	"	JAE, SST, D10, H15, LJS, D15, HGL
n=10, 25% of 1/U	C	3T1, 3R2, D07, 5BT, TRI, 3T0, H/L,
	B	JAE, HU2, 33T, H15
n=5, Cauchy	D	ADA, CST, 17A, 5T4, TOL, 5T1, 3T1, JWT, TRI, 3R2,
	"	JBT, JAE, 22A, 33T, D07, 3T0, D10, D15
	B*	H12, LJS, D10, H/L, H15
	B	CML, JLJ, 50%, 12A, SHO, DFA
n=10, Cauchy B	D	3T1, JBT, D10, TRI, JAE, H/L, 33T
	C	D10, LJS, H15, D15
n=20, 25% of 3/U	D	5T1, CST, 3T1, D07, D10, TRI, H/L, H12, 3R2, LTS, D15

165

4) For all the regularities we see, the difference between Cauchy A and Cauchy B, at n=10, shows that just what happens in sensitive situations is much a matter of how much lightening happened to strike in a particular set of configurations.

5) The two extreme situations at n=5 have few estimates classed X -- 9 for pure Cauchy and 0 for 25% of 1/U.

6) The estimates most subject to difference either resemble the mean or are overtly adaptive.

To this last point we ought to give some detailed evidence. Let us then list the 24 more often discrepant estimates according to type. We have:

Close to the mean: M, M15, 4RM, 3RM, 2RM, 5%, 10%.
Rather close to same: 15%, H20 (matches M for n=5!)
H17, D20.
Overtly adaptive: SJA, BIC, MEL, CTS, HGP, HGL,
TAL, THP, TAK, JOH.
Other: 3R1, BH, OLS.

The analysis here (in 6F1) has shown us the outlines of the empirical facts, it is time to turn to attempts toward understanding.

6F2 Some Pearsonian considerations

One of the outstanding instances of apparent serendipity in modern statistics was the development of Karl Pearson of the family of Pearson curves (on the basis of the local properties of the hypergeometric distribution) and the later discovery that many distributions of functions of Gaussian samples followed Pearson curves either exactly (mainly through the work of R. A. Fisher) or approximately. This was followed by Egon Pearson's discovery that other 4-parameter families of distributions (noncentral t, Pearson and Morrington 1958; Johnson curves, Pearson 1963) behaved -- as wholes -- like the Pearson curves.

When the behavior of Pearson curves is examined in standard measure (sketchily, Table 42 of Pearson and Hartley 1954 to 1966; in more detail, Johnson, Nixon,

Amos, and Pearson 1963) it is found that the % points
most stable are those near (one-sided) 2.5% or 5%.
Later work (Pearson and Tukey 1965) led to more precise
approximations. If we had one-sided 5% points available,
then (Pearson and Tukey, page 540) it would be effective
to use the square of

$$\max\left\{\frac{5\% \text{ point}}{1.645} , \frac{2.5\% \text{ point}}{1.990}\right\}$$

for symmetric cases. (An unsymmetric extension is given
there.) Since we do not, the asymptotic alignment
suggested by N. L. Johnson (Pearson and Tukey, page 544),
which leads to an approximate match to 4.2% points, is
probably our best guide.

In view of this history, it seems:

1) natural to look toward approximately the
4.2% pseudovariance as having some primacy among
pseudovariances.

2) appropriate to call such a pseudovariance
a Pearsonian pseudovariance, (and, perhaps, even to
call it a variance).

3) important to understand what equality (or
inequality) of variance and Pearsonian pseudovariance
does and does not mean.

Estimate distributions need not be -- and are not
always -- close to Gaussian. For samples of 1 from
situations not purely Gaussian this is a triviality. In
many cases these distributions deviate from Gaussianity
toward long tails. When this happens, there can be --
and are -- two different sorts of long-tailedness:

1) more or less consistent long-tailedness;

2) unusually discrepant estimates for a few
quite unusual sorts of configurations.

When the first of these occurs, the distribution
is likely to be Pearsonian -- to have variance and

167

Pearsonian pseudovariance nearly equal -- when the second happens, the distribution is not likely to be Pearsonian. To be apparently nonPearsonian is not a matter of just being long-tailed (or even being longer tailed). Rather it is a matter of having been affected seriously, at least so far as the variance goes, by one configuration (or a very vew configurations) of the 640 to 1000 that were used. To be apparently Pearson indicates that, at least among the 640 to 1000 configurations used, no one turned up to which that estimate was extremely sensitive. (In using these remarks, we will have to keep in mind the disproportion of sampling. Even with perfect random numbers, rare events expected far less than 1 per thousand will occur, and less rare events, expected far more than 1 per thousand will fail to occur.)

6F3 Pearsonian pseudovariances and pseudodeficiencies

As we turn to the Pearsonian pseudovariances, it is important to be clear about the sampling situation on which all our pseudovariances are based. We are not concerned with a binomial situation. When we assess the fraction of an estimate distribution beyond some fixed value, the contributions from different configurations are not all 0 or 1. Each configuration produces an intermediate value, the result of integrating over the configuration.

If we are sampling from a pure Gaussian situation, we will get the same fraction (except for roundoff and approximate function evaluations) for each configuration, and our sampling error will be zero (with the minor exceptions noted.) As we move to situations far from the pure Gaussian this advantage will be reduced.

Since our distributions are all symmetric, upper and lower tails have been combined by averaging. Thus, no configuration can ever contribute a fraction greater than 0.5 as its contribution. Accordingly, our variance of estimating a fraction can never exceed 1/4 that for the binomial case, and is quite unlikely to reach as much as 1/10 the binomial for any reasonably resistant estimate.

For situations close to the Gaussian, our apparent Pearsonian pseudovariances will be as well determined as our apparent variances. Very far from the Gaussian, our apparent Pearsonian pseudovariance will be better determined than our apparent variances, perhaps very much so.

If all we seek is stable criteria, then, we ought to turn to Pearsonian pseudovariances whenever these disagree with variances. But is stability our only criterion?

Let us think hard about a situation where they differ. How much attention do we want to pay to the impact of very unusual configurations? If we say: "They are rare, and when they occur we are likely to be ruined anyway" we can be quite comfortable in using Pearsonian pseudovariance as our criterion. If we say "They are rare, but we must respond when they occur. At least until we know how to recognize them and treat them separately (when we will need to make allowance for separate treatment) we must allow for their presence!" we will not be satisfied with a pseudovariance criterion.

In this latter case we are unlikely to be satisfied with a simple variance criterion either. The increment of the variance over the Pearsonian pseudovariance is a matter of where and how often the lightening strikes. Chance plays a large part in both where and how often. Presumably we need to combine

1) the Pearsonian pseudovariance.

2) an indication, "averaged" over various situations where a substantial difference is likely, of how much larger (than the Pearsonian pseudovariance) the variance seems to run.

Just how to make the combination is far from clear. If we are lucky, it may suffice to avoid estimates for which indication (2) is large.

We are used to working in deficiencies. We should like to continue. So we define

$$\text{Pearsonian pseudodeficiency} = \frac{2}{3}(2.5\% \text{ pseudodeficiency})$$
$$+ \frac{1}{3}(10\% \text{ pseudodeficiency})$$

where for each %,

$$\text{pseudodeficiency} = 1 - \frac{\text{best available pseudovariance}}{\text{actual pseudovariance}}$$

Here the choice of 2/3 and 1/3 comes from the spacing c
the Gaussian 2.5%, 4.2% and 10% (one-sided) % points, and
the "best available" pseudovariance is used since an
optrim estimate may show nonPearsonian behavior, while
the best available -- in the sense of minimum pseudovariance -- estimate does not seem to do this enough
to bother us seriously. (If we actually had 5% or 4.2%
pseudovariances we would feel free to redefine the
Pearsonian pseudovariance accordingly.)

6F4 Some numerical comparisons

It is now time to look at some numbers. We begin
by looking at a moderately broad spectrum of estimates
(including original and modified hubers) for a sequence
of situations of increasing sensitivity. Exhibit 6-41
shows the increase of deficiency over Pearsonian pseudo
deficiency for 24 estimates and 9 situations (including
both pure Cauchys at n=10).

The first two columns correspond to the 10th and
11th lines of exhibits 6-37 to 6-39, where only the most
susceptible estimates fall in class A. The other 7
columns correspond to the top 7 lines of those exhibits,
where the greatest effects occur. Toward the left, the
entries in this table are small. As we move to the center
and right, large values appear, beginning with H20, HGL
and TAK and spreading to most estimates.

If we leave n=5, especially the pure Cauchy situation, aside for a moment, we see that:

> - the hampels proper (25A, 21A, 17A, 12A) and
> AMT are never affected appreciably.

> - the A/P modified hubers (A20, A15, P15) and
> the median (50%) are only affected quite moderately.

> - 5T4 is only affected moderately (except for
> Cauchy A at n=10 where it is "bombed").

On the basis of these insights, 9 estimates were
selected for relatively small differences and the differences rearranged as in the first panel of exhibit 6-42.
When adjusted for column medians, as in panel B of that
exhibit, the exceptional character of 3 situations for
A20, A15 and P15 became clear, as did the exceptional

exhibit 6-41

Deficiencies minus Pearsonian pseudodeficiencies (in permille)

	n=20 25% of 1/U	n=20 Cauchy	n=10 pure 1/U	n=20 25% of 3/U	n=10 Cauchy B	n=5 pure Cauchy	n=5 25% of 1/U	n=10 25% of 1/U	n=10 Cauchy A
H20	6	26	113	45	84	34	123	379	116
H17	0	18	28	55	53	0	108	407	178
H15	1	17	15	79	72	-37	189	411	228
H12	-2	19	2	39	58	-20	212	119	5
H10	-1	20	2	28	46	-26	199	12	-8
H07	-1	24	0	0	23	-38	176	2	-17
D20	-2	27	15	65	75	-19	191	440	204
D15	-3	26	13	69	91	2	252	495	247
D10	-2	29	8	61	102	10	280	480	289
D07	-1	21	-1	31	90	28	174	331	207
A20	-1	15	5	8	19	-84	5	23	-10
A15	-2	16	2	5	23	-97	13	12	-9
P15	-3	16	2	8	25	-97	13	13	-8
5T4	-2	4	19	34	61	-28	131	2	519
TRI	4	11	2	44	71	-47	99	303	290
CST	-2	-7	1	71	40	-13	71		397
SST	0	1	17	0	37	-22	55	-1	498
50%	-2	-2	-2	-2	22	-99	-14	-1	0
25A	1	5	-2	1	12	-66	34	1	-10
21A	1	4	-4	3	11	-49	18	1	-8
17A	0	1	0	6	14	-23	8	0	-5
12A	-3	-1	1	0	12	0	0	-2	-7
TAK	5	23	35	17	135	-4	308	207	287
HGL	-1	37	56	16	75	30	43	131	236
ADA	3	0	-5	10	24	-55	12	-1	1
JLJ	-1	11		29					
JOH	-1	16	4	4	90	-99*	-14*	34	350
JWT	-4	2	17	8	53	-18	155	4	521
AMT	3	4	6	8	20	-79	49	1	-7

exhibit 6-42
Part of exhibit 6-41 rearranged and adjusted for column
and row medians

	n=20			n=10				n=5	
	25% of 1/U	pure Cauchy	25% of 3/U	pure 1/U	Cauchy B	25% of 1/U	Cauchy A	pure Cauchy	25% of 1/U
A. AFTER REARRANGEMENT									
A20	-1	15	8	5	19	23	-10	-84	5
A15	-2	16	5	2	23	12	-9	-97	13
P15	-3	16	8	2	25	13	-8	-97	13
25A	1	5	-2	-2	12	1	-10	-66	34
21A	2	4	-4	-4	11	1	-8	-49	18
17A	0	1	0	0	14	0	-5	-23	8
12A	-3	-1	1	1	12	-2	-7	0	0
AMT	3	4	6	6	20	1	-7	-79	49
50%	-2	-2	2	-2	22	-1	0	-99	-14
(med)	(-1)	(4)	(2)	(1)	(19)	(1)	(-8)	(-79)	(13)

B. ADJUSTED FOR COLUMN MEDIANS (†) (*

	25% of 1/U	pure Cauchy	25% of 3/U	pure 1/U	Cauchy B	25% of 1/U	Cauchy A	pure Cauchy	25% of 1/U
		(**)				(**)		(**)	
A20	0	11	6	4	0	22	-2	-5	-8 (4) (
A15	-1	12	3	1	4	11	-1	-18	0 (3) (
P15	-2	12	6	1	6	12	0	-18	0 (6) (
25A	2	1	-4	-3	-7	0	-2	-13	21 (-2)
21A	2	0	-6	-5	-8	0	0	30	5 (0)
17A	0	-3	-2	-1	-5	-1	3	56	-5 (-1)
12A	-2	-5	-1	0	-7	-3	1	79	-13 (-2)
AMT	4	0	4	5	1	0	1	0	36 (1)
50%	-1	-6	0	-3	3	-2	8	-20	-27 (-1)

(†) (median of left 7)
(*) omitting columns labelled ** instead

	25% of 1/U	pure Cauchy	25% of 3/U	pure 1/U	Cauchy B	25% of 1/U	Cauchy A	pure Cauchy	25% of 1/U
C. ADJUSTED FOR (PARTIAL) ROW MEDIANS									
A20	0	(11)	6	4	0	(22)	-2	(-5)	(-8)
A15	-1	(12)	3	1	4	(11)	-1	(-18)	(0)
P15	-2	(12)	6	1	6	(12)	0	(-18)	(0)
25A	4	3	-2	-1	-5	2	0	(-11)	(23)
21A	2	0	-6	-5	-8	0	0	(30)	(5)
17A	0	-3	1	2	-5	-1	3	(81)	(-11)
AMT	3	-1	3	4	0	-1	0	(-1)	(35)
50%	0	-5	1	-2	4	-1	9	(-19)	(-26)
	(0)	(-2)	(3)	(1)	(0)	(0)	(0)		

	25% of 1/U	pure Cauchy	25% of 3/U	pure 1/U	Cauchy B	25% of 1/U	Cauchy A
D. READJUSTED FOR (PARTIAL) COLUMN MEDIANS							
A20	0		3	3	0		-2
A15	-1		0	0	4		-1
P15	-2		3	0	6		0
25A	4	5	-5	-2	-5	2	0
21A	2	2	-9	-6	-8	0	0
17A	1	0	-4	-1	-4	0	4
12A	0	-1	-2	1	-5	-1	3
AMT	3	1	0	3	0	-1	0
50%	0	-3	-2	-3	4	-1	9

character of the values for n=5. Accordingly, appropriate omissions were made in calculating row medians, whose use led to panel C. When the same cells were omitted again, both in finding column medians and in writing down adjusted results, we reached panel D.

Of 57 entries in panel D, the extremes are 9 and -9. Since these are in permille and since the values from which they were differenced came from direct division on a 10-inch slide rule, some allowance needs to be made for computational errors. Thus the actual agreement with one another is very good indeed.

What of the adjustments? The row adjustments amount to saying that A20, A15 and P20 seem to run 5‰ (0.5%) higher differences than the others. That there should be a small effect here is not surprising -- what is surprising, in view of the behavior of the other original and modified hubers, is that it is not larger.

The column adjustments are larger. But this should have been expected. All values differenced, deficiencies and Pearsonian pseudo-efficiencies alike, are based on just one estimate at each situation -- the "best available" estimate in the sense of having assessed variance. If this has, in fact, a true difference between variance and Pearsonian pseudovariance, the negative of this difference (when properly expressed) will propagate into each difference (for that situation) between deficiency and Pearsonian pseudodeficiency. Thus our column adjustments, which amount (both steps of adjustment combined) to -1, 2, 5, 2, 19, 1, and -8 for the situations at n=20 and n=10, respectively, presumably reflect such differences for "best available" estimates. (It is natural to ask what the "best available" estimates were. In order, they were: 17A, 12A, CML; CML, CML, 17A, CML; 12A, CML; a sequence that offers no clear suggestions.) None of these is really large, though the "19" for Cauchy B at n=10 seems a little worthy of note.

6F5 The situations at n=5

In our adjustments in exhibit 6-42, we avoided the
two situations at n=5. Our 9 selected estimates no longer
agree with one another as they did at n=20 and n=10. We
need to proceed more carefully here.

When we look at the values in panel C for these two
situations, we are led to believe that we will reflect
true differences best by bringing the more negative
values close to zero. A reasonable facsimile of an adjustment can be made by adjusting to the second lowest value
for each situation (-18 for pure Cauchy and -11 for 25%
of 1/U). Exhibit 6-43 shows the result, for all estimates
from exhibit 6-41, of making the column adjustments of
exhibit 6-42 and the two just mentioned.

We can now classify the estimates, at least roughly,
as follows:

never affected much: A20, A15, P15, 50% (JLJ)

affected even less except for n=5: 25A, 21A, 17A,
                                    12A, AMT, ADA

not affected much except for n=5: H07, H10 (?)

more or less subject to severe attack: the rest

The detailed reasons for the behavior thus classified
escape us, at least for the moment.

For the Cauchy situation at n=5 we have further
evidence. Exhibit 5-13 compares our Cauchy % points for
the median with exact values (which are relatively easily
calculated). At 10% and 2.5% the Monte Carlo results are
quite close, leading to a Pearsonian pseudovariance that
is perhaps $16/m$ low. At 0.1%, by contrast, the Monte
Carlo pseudovariance is about $460/m$ high. This suggests
that all estimates may be responding substantially to a
few "wild" samples in this situation, and that all
variances may be unduly high in our simulation. On the
other hand, we can probably trust our pseudovariances.

Since the differences for 25% of 1/U at n=5 are even
more violently perturbed, we will probably be wise to only
look at (Pearsonian) pseudovariances for n=5.

* /m = permille

exhibit 6-43
Differences of exhibit 6-41, adjusted as per text

	n=20			n=10				n=5	
	25% of 1/U	pure Cauchy	25% of 3/U	pure I/U	Cauchy B	25% of 1/U	Cauchy A	pure Cauchy	25% of 1/U
H20	7	24	40	111	65	378	124	131	121
H17	1	16	50	26	34	406	186	97	106
H15	2	15	74	13	53	410	236	60	187
H12	-1	17	34	0	39	118	13	77	210
H10	0	18	23	0	27	11	0	71	197
H07	0	23	-5	-2	4	1	-9	61	174
D20	-1	25	60	13	56	439	212	78	189
D15	-2	24	64	11	72	494	255	99	250
D10	-1	27	56	6	83	479	297	107	278
D07	0	19	26	-3	71	330	215	125	172
A20	0	13	3	3	0	22	-2	13	3
A15	-1	14	2	0	4	11	-1	0	11
P15	-2	14	3	0	6	12	0	0	11
5T4	-1	2	29	17	42	1	527	69	129
TRI	5	9	39	0	52	302	298	50	97
CST	-1	-9	66	-1	21	8	405	84	69
SST	1	-1	-5	15	18	-2	506	75	53
50%	-1	-4	-7	-4	3	-2	8	-2	-16
25A	2	3	-6	-4	-7	0	-2	29	32
21A	3	2	-2	-6	-8	0	0	48	16
17A	1	-1	1	-2	-5	-1	3	74	6
12A	-2	-3	-5	-1	-7	-3	1	97	-2
TAK	6	21	12	33	116	206	295	93	306
HGL	0	35	11	54	56	130	244	127	41
ADA	4	-2	5	-7	5	-2	9	42	10
JLJ	0	9	24						
JOH	0	14	0	2	71	33	358	-2*	-16*
JWT	-3	0	4	15	34	3	529	79	153
AMT	4	2	4	4	1	0	1	18	47
djusted for)	(-1)	(2)	(5)	(2)	(19)	(1)	(-8)	(-97)	(2)

*The same as 50% for this sample size.

175

exhibit 6-44

Calculating Pearsonian pseudodeficiencies from 10% and 2.5% pseudo-deficiencies
(all in permille)

	n=20, 25% of 1/U pseudodeficiencies				n=20, 25% of 3/U pseudodeficiencies				n=10, pure 1/U pseudodeficiencies			
	10%	2.5%	Pears.	(def)	10%	2.5%	Pears.	(def)	10%	2.5%	Pears.	(def)
H20	250	309	289	(295)	675	775	741	(786)	647	786	746	(853)
H17	161	195	184	(184)	536	664	621	(674)	567	670	636	(864)
H15	111	130	125	(124)	430	546	507	(586)	467	577	540	(555)
H12	60	70	67	(65)	286	361	336	(375)	341	476	431	(433)
H10	49	55	53	(52)	215	264	248	(276)	256	360	325	(327)
H07	74	70	71	(71)	163	175	171	(171)	130	199	176	(176)
D20	184	213	203	(201)	475	550	525	(590)	528	659	625	(640)
D15	105	120	115	(112)	354	433	407	(476)	453	594	547	(560)
D10	56	64	62	(60)	328	284	265	(326)	353	511	458	(466)
D07	60	61	61	(60)	178	209	199	(230)	261	387	345	(344)
A20	161	189	180	(129)	456	516	496	(504)	437	519	492	(497)
A15	89	103	98	(96)	329	384	366	(371)	341	425	397	(399)
P15	90	105	106	(97)	337	385	367	(315)	347	431	403	(405)
5T4	20	15	17	(15)	24	35	31	(65)	145	253	217	(236)
TRI	74	83	80	(76)	243	326	298	(342)	341	476	341	(433)
CST	52	54	53	(51)	50	80	70	(141)	150	247	215	(216)
SST	181	180	180	(180)	85	70	75	(75)	104	179	154	(171)
50%	214	211	212	(210)	241	237	238	(236)	102	106	105	(103)
25A	25	41	36	(37)	172	229	210	(211)	265	349	321	(319)
21A	5	11	8	(10)	89	129	116	(119)	195	285	255	(251)
17A	0	0	0	(0)	25	45	38	(46)	131	206	181	(181)
12A	25	24	24	(21)		0	0	(0)	61	117	99	(100)
TAK	35	35	35	(40)	158	220	199	(216)	378	536	483	(518)
HGL	90	94	93	(92)	200	236	224	(240)	425	562	516	(572)
ADA	34	31	32	(35)	50	74	55	(76)	125	192	170	(165)
JLJ	90	93	92	(91)	51	45	47	(76)	102	106	105	(103)
JOH	72	71	71	(70)	56	50	52	(56)	130	205	177	(181)
JWT	171	175	174	(170)	180	165	170	(176)	182	265	237	(254)
AMT	6	17	13	(16)	97	140	126	(134)	222	319	286	(292)

exhibit 6-45

Change from Cauchy A to Cauchy B for deficiencies,
Pearsonian pseudodeficiencies, and variances

	Deficiencies			Pears. pseudodefics.			variances	
	A	B	diff	A	B	diff	A	B
H20	939	916	-23	823	832	+9	49.78	32.23
H17	910	795	-115	732	742	+10	33.48	13.24
H15	872	735	-137	644	663	+19	27.70	10.27
H12	552	624	+72	547	566	+19	6.79	7.22
H10	429	500	+71	437	454	+17	5.30	5.42
H07	268	340	+72	285	297	+12	4.15	4.12
D20	921	807	-114	717	732	+15	38.03	14.03
D15	904	766	-138	657	675	+18	31.59	11.58
D10	868	700	-168	577	598	+21	22.98	9.05
D07	770	572	-198	463	482	+19	13.20	6.33
A20	588	636	+48	598	617	+19	7.36	7.49
A15	494	545	+51	503	522	+19	6.00	5.97
P15	497	554	+57	505	529	+24	6.04	6.07
5T4	828	420	-408	309	359	+50	17.70	4.70
TRI	774	586	-188	484	515	+31	13.31	6.56
CST	750	385	-365	353	345	-8	12.16	4.43
SST	762	293	-469	264	256	-8	11.80	3.88
50%	171	191	+20	171	169	-2	3.66	3.37
25A	386	458	+72	396	446	+50	4.95	5.00
21A	314	389	+75	322	378	+56	4.42	4.46
17A	244	311	+67	249	297	+48	4.01	3.95
12A	165	210	+45	172	198	+26		3.44
TAK	915	761	-154	628	626	-2	35.75	11.31
HGL	927	771	-156	691	696	+5	41.31	11.85
ADA	246	296	+50	245	272	+27	4.03	3.86
JLJ	171	191	+20	171	169	-2	3.66	3.37
JOH	510	382	-128	260	292	+32	6.19	4.40
JWT	853	405	-448	332	352	+20	20.59	4.57
AMT	345	430	+82	355	410	+55	4.65	4.77

6F6 Stretching across the gap

Before we leave our numbers, we ought to give some examples of how much difference between 10% and 2.5% pseudovariances is likely to be concealed in the Pearsonian pseudovariances. Exhibit 6-44 gives details for three moderately threatening situations: 25% of 1/U at n=20, 25% of 3/U at n=20, pure 1/U for n=10. In view of the size of many differences between 10% and 2.5% pseudodeficiencies, the agreement between deficiencies and Pearsonian pseudodeficiencies seems even better to us than it did before.

6F7 Relative precision of assessment

There is another question to which we should make some response: "What can be said about the relative precision with which variances -- or better deficiencies -- and Pearsonian pseudovariances -- or better Pearsonian pseudodeficiencies -- can be assessed?". The additional evidence we can gather easily comes from a comparison of Cauchy A with Cauchy B. Exhibit 6-45 shows, for our usual set of estimates how the differences in deficiencies compare with differences in Pearsonian pseudodeficiencies.

We see a great reduction in size of difference -- and, especially for the hubers, an almost corresponding increase in consistency of pattern. If we had no other reason for choosing between variances, and Pearsonian pseudovariances, we would choose the latter, in the interests of increased precision. (The last two columns compare raw variances, clearly not a thing to do if we can avoid it.)

6F8 Comments

It is time to say again where we stand, and why we were interested in the numbers we have been examining. If variance and Pearsonian pseudovariance are about the same for some situation:

- a user willing to hew to the "4.2%" points is glad to be guided by either.

- a user deeply concerned about the ill effects of rare configurations will be moderately happy; his happiness will increase as the number of situations of reasonably similar threat increases for which the approximate equality continues to hold.

- a user interested in very extreme % points of his estimates will not consider himself necessarily well guided by either variances or Pearsonian pseudovariances.

On the other hand, if variances are quite different from Pearsonian pseudovariances for some other situation,

- a user willing to hew to the "4.2%" points is glad to neglect variances and be guided by Pearsonian pseudovariances.

- a user deeply concerned about the ill effects of rare configurations will accept the variances (or deficiencies) as qualitative guidance, saying to himself "even these may well need to be larger". (He will not be very interested in the detailed values of the Pearsonian pseudovariances.)

- a user interested in very extreme % points will again lack any certain guidance.

For some, one sort of change. For others, another. For still others, no improvement either way.

These, then, are the alternatives we must face, subject to the further remark that:

- if we want acceptable numerical precision at n=5 or for the Cauchy case at n=10 we must stick to Pearsonian pseudovariances.

6G THE EFFECTS OF MODIFYING HUBERS

In one way or another, we make considerable use of
hubers, both original and modified. Their use as one
terminus of sitsteps is of course quite important to us.
Accordingly it is well worth while to look hard at the
behavior of the hubers under different kinds of modification, for different situations, and for different
sample sizes.
Four types have been explored at least minimally
in the present survey (not counting M15, a clear failure):

   H: An iterated M-estimate starting with a
scaling by the interquartile range (x_9-x_2 for

samples of 5, $x_{8.5}$-$x_{2.5}$ for samples of 10) and a

location by the median and iterating both location
and scale. (Explored: H20, H17, H15, H12, H10, H07)

   D: Like P, but scaling is by the interquartile
range.   (Explored: D20, D15, D10 and D07)

   A: A half-iterated half-M estimate in which
scaling is always by the median absolute deviation
from the median, but location is iterated.
(Explored: A20, A15)

   P: Like A, but only one step of location is
taken. (Explored: P15)

Four remarks deserve our attention:

   1) The locating function is z-shaped, flat
outside - k $\leq$ x $\leq$ k and linearly sloping within.
(The number used in tags are values of 10k.)

   2) The observed variances for P15 and A15 are
everywhere very close -- for reasonable situations,
even the deficiencies rarely differ by 1/50 of
themselves. Accordingly, we are not irresponsible
to take variances for A20 as reasonable facsimiles
of variances for P20, which has not been explored as
such. (We are much more interested in P's than in
A's.)

3) The behavior of pseudovariances for A and P shows equal closeness, and offers equal justification of the facsimile.

4) The behavior of P's relative to D;s seems to be more consistent than the behavior of either relative to H's, so that clarity of picture is served by the presentation described in exhibit 6-46.

6G1   Modification for n=20

We turn first to gentle situations, where we have a large variety for n=20 and two for n=10.   Exhibit 6-47 shows the situation for 0% (pure Gaussian), 5%, 10%, 15%, 25%, 50% and 75% from 3G for n=20.   Since t_3 behaves rather similarly, it is included.
Two things are rather rapidly clear when this picture is scanned:

1) Behavior is reasonably systematic, changing reasonably smoothly with either k or % of contamination.

2) The effects of modifying hubers are more along the curves -- corresponding to a change of k in the original form -- than they are across it.

Two more things emerge from harder looking:

3) The change from D to P, a change in the basis used for scaling (shown by the solid line segments in exhibit 6-47), is even more closely parallel to the curves. This suggests that intrinsic differences in the scaling bases have had very little effect.

4) With a few exceptions, corresponding to more or less extreme situations (k = 2.0 and k = 0.7 for 25%, k = 2.0 [not shown] and k = 1.5 for 50%, k = 2.0 and k = 1.5 for 75%) the one-steps perform less well than the originals. (The changes at k = 1.0, except at 50%, are too small to be judged.) The differences are, however, quite small.

exhibit 6-46
Key for pictures about huber modification
(u,v represent 2 digits, such as 20, 15, 10 or 07)

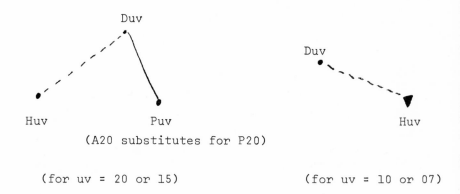

(A20 substitutes for P20)

(for uv = 20 or 15)   (for uv = 10 or 07)

Information for uv = 17 or 12 was used to draw smooth
curves through H's, but, since it cannot be used to
show effects of modification, no points are shown in
pictures.

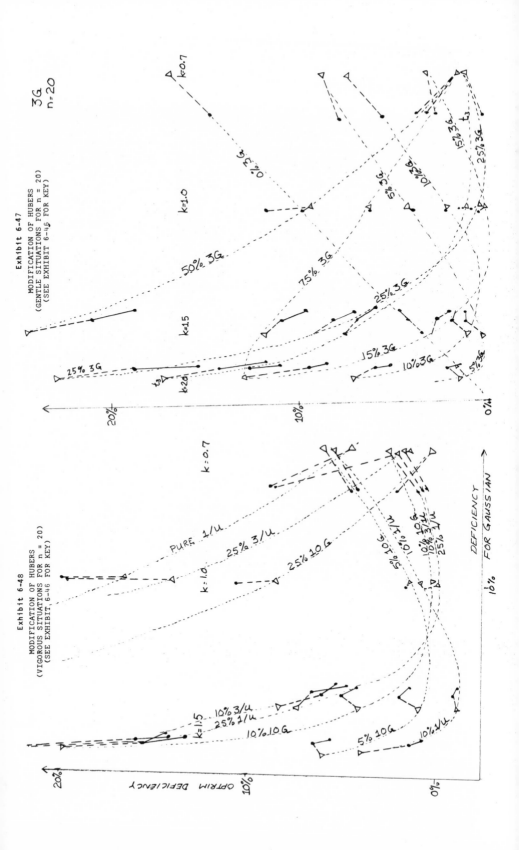

Exhibit 6-47

MODIFICATION OF HUBERS
(GENTLE SITUATIONS FOR n = 20)
(SEE EXHIBIT 6-46 FOR KEY)

Exhibit 6-48

MODIFICATION OF HUBERS
(VIGOROUS SITUATIONS FOR n = 20)
(SEE EXHIBIT 6-46 FOR KEY)

DEFICIENCY FOR GAUSSIAN

OPTRIM DEFICIENCY

Turning next to the vigorous situations for n=20, exhibit 6-48 shows similar information for 8 vigorous situations. Conclusions (1), (2) and (3) again hold, conclusion (4) is modified to:

4') For k = 2.0 and k = 1.5, use of the one-steps leaves us on the wrong side of the curve, indicating a small loss in efficiency. For k = 0.7, the situation is mixed, with apparent slight gains for a group of moderately vigorous situations (10% of 3/U, 10% from 10G, 25% of 1/U, 25% from 10G) balanced by apparent slight losses for others.

At 25% of 3/U, we have reached a situation where Pearsonian pseudodeficiencies have begun to differ from deficiencies. We ought to ask "By how much?". Exhibit 6-49 shows the answer. The differences are meaningful rather than large. Moreover, the picture is somewhat better behaved (in comparison with earlier pictures) in terms of Pearsonian pseudodeficiencies.

We feel no urge to go back and redo earlier exhibits, changing to Pearsonian pseudodeficiencies, but we do take note for the future, expecting to make such a change for more severe situations.

6G2 Modification for two gentle situations for n=10

Exhibit 6-50 shows the corresponding picture for 10% or 20% from 3G at n=10 and for comparison, 10% from 3G at n=20. We see very much the same thing going on, with 18/2 (10% from 3G at n=20) closer to 9/1 (10% from 3G at n=10) than to 8/2 (20% from 3G at n=10). (This suggests that % from 3G is more important than number from 3G.)

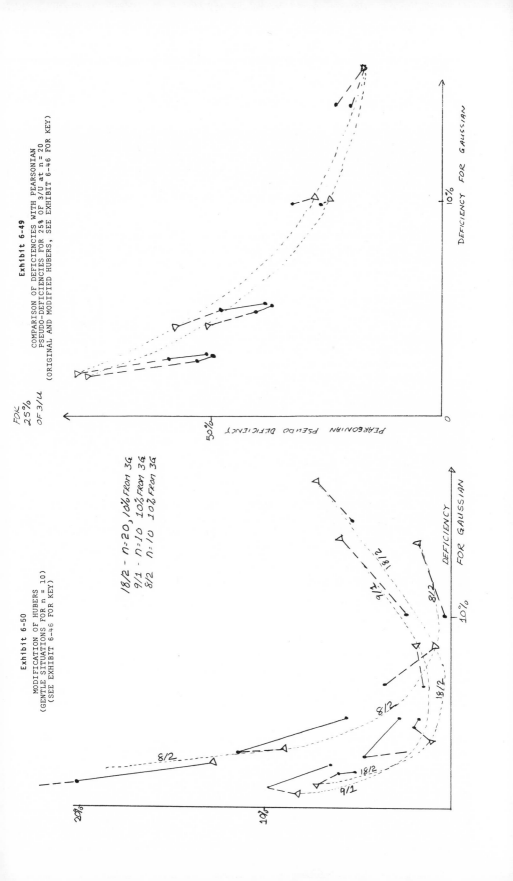

Exhibit 6-49

COMPARISON OF DEFICIENCIES WITH PEARSONIAN
PSEUDO-DEFICIENCIES FOR 25% OF 3/U at n = 20
(ORIGINAL AND MODIFIED HUBERS, SEE EXHIBIT 6-46 FOR KEY)

FOR 25% OF 3/U

PEARSONIAN PSEUDO DEFICIENCY

DEFICIENCY FOR GAUSSIAN

Exhibit 6-50

MODIFICATION OF HUBERS
(GENTLE SITUATIONS FOR n = 10)
(SEE EXHIBIT 6-46 FOR KEY)

18/2 - n=20, 10% FROM 3G
9/1 - n=10 10% FROM 3G
8/2 n=10 10% FROM 3G

DEFICIENCY FOR GAUSSIAN

6G3 Modification for 25% of 1/U at all n

The only vigorous contamination available for n=5, 10 or 40 is 25% of 1/U. Exhibit 6-51 shows the corresponding picture for n=40, n=20 and a part of n=5. (The omissions of n=10 and part of n-5 will be explained shortly.) Behavior for n=40 and n=20 is quite similar and closely consistent. Until asymptotic results are available, not much more needs to be said.

For n=5, the line segments connecting Duv with Puv for uv = 20 and 15 seem perhaps a little out of place, as does the entire behavior for uv = 10 and 07. It would be hard to argue that we have trouble from these values alone.

The fact that Huv is off the page for uv = 20 and uv = 15 at n=5 gives us pause, however. There is nothing in the behavior at n=20 and n=40 to suggest any such occurrence.

Slightly warned, then, we turn to Exhibit 6-52, which shows -- at 1/2 the vertical scale, note this carefully -- the whole picture for these uv at n=5 and for all uv at n=10. Clearly something is going on. There are strong indications of two regimes, one with vertical deficiencies of 5% to 15% and one with vertical deficiencies of 35% to 95%. There seem to be no transitional cases, each point belongs clearly to one regime or the other.

This is not so for one combination of uv and n. For uv=20 and uv=15 at n=10, we have Huv and Duv high and Puv low. For uv=10 and uv=07 at n=10, we have Huv low and Duv high. For uv=20 at n=5, we have Huv high, and both Duv and Puv low. The most of which we can be sure is that something is going on.

It is now time to turn to Pearsonian pseudodeficiencies. Exhibit 6-53 shows the result of plotting all four values of n at the same scale as exhibit 6-52. Except at n=5, things look reasonably clean and tidy.

So far as 25% of 1/U goes we have no grounds for fear down as far as n=10. At n=5, though, we shall probably have to think separately.

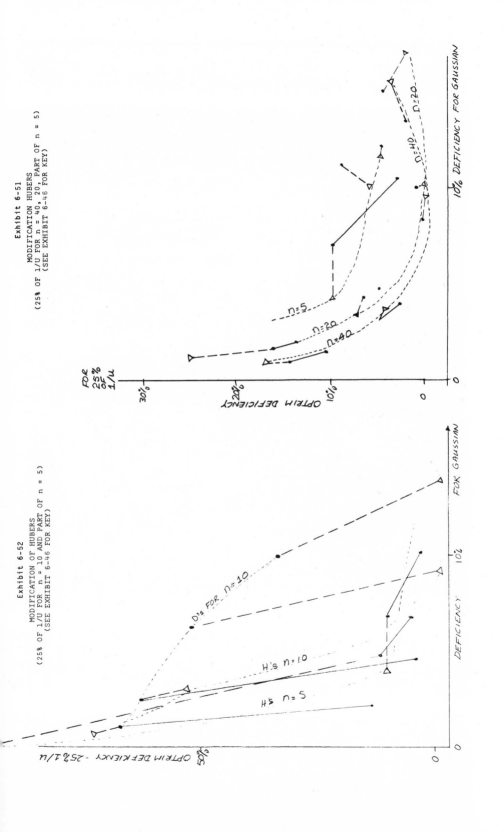

Exhibit 6-51
MODIFICATION HUBERS
(25% OF 1/U FOR n = 40, 20, PART OF n = 5)
(SEE EXHIBIT 6-46 FOR KEY)

Exhibit 6-52
MODIFICATION OF HUBERS
(25% OF 1/U FOR n = 10 AND PART OF n = 5)
(SEE EXHIBIT 6-46 FOR KEY)

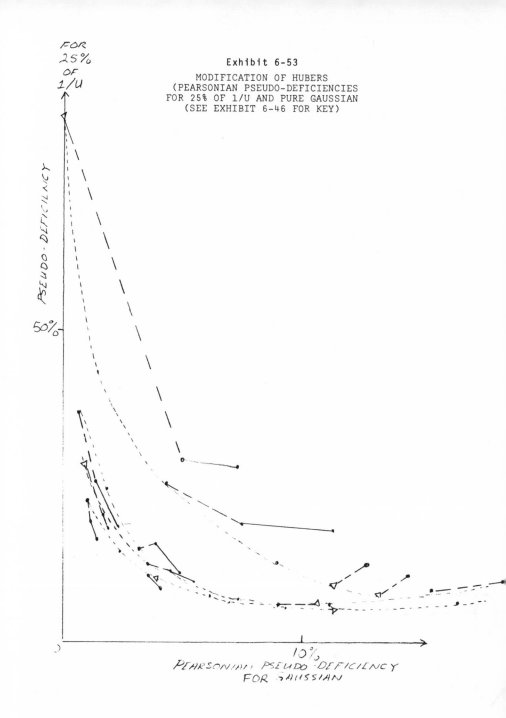

Exhibit 6-53

MODIFICATION OF HUBERS
(PEARSONIAN PSEUDO-DEFICIENCIES
FOR 25% OF 1/U AND PURE GAUSSIAN
(SEE EXHIBIT 6-46 FOR KEY)

FOR
25%
OF
1/U

PSEUDO-DEFICIENCY

50%

10%

PEARSONIAN PSEUDO-DEFICIENCY
FOR GAUSSIAN

6G4   Modification for pure 1/U (at n=10 and n=20)

Exhibit 6-54 shows, to the same scale, the behavior
of the Pearsonian pseudodeficiencies for pure 1/U, which
are only available at n=20 and n=10. The agreement is
reasonably close, so that we feel comfortable, for example,
in interpolating behavior as low as n=10. The keen eye,
sharpened by careful study of the previous exhibits, can
detect, however, traces of rapidly increasing trouble.
If we had data for n=5, we would expect to find a quite
considerab le shift.

6G5   Modification for pure Cauchy (at all n)

Exhibit 6-55 shows, to the same scale, the behavior
of the Pearsonian pseudodeficiencies at all 4 values of
n, incompletely for n=5. We see that the agreement between
n=40 and n=20 is good. At n=10 we have some deviations
(the Pearsonian pseudodeficiencies used are for the mean
of Cauchy A and Cauchy B) while the incomplete information
for n=5 seems to fit in quite well.

Exhibit 6-56 shows the details at n=5. We see that
the modifications behave quite neatly indeed.

Given that we have resolutely refrained from plotting
deficiencies -- after all, it was the Cauchy variances
at n=10 and n=5 that drove us to Pearsonian pseudo-
deficiencies in the first place -- these results are
respectably well behaved. We can be sure that something
small is going on for n=10, but we are probably justified
in passing it by, so far as general conclusions go.

6G6   Summary on modification

If we agree to look at Pearsonian pseudovariances --
through pseudodeficiency spectacles, of course -- when
these differ from variances, we find a relatively coherent
pattern of behavior from modifying hubers in essentially
all the situations we have examined. Among the high
points are these:

1) Behavior is reasonably systematic.

2) The effects of modification are more
along curves than across curves.

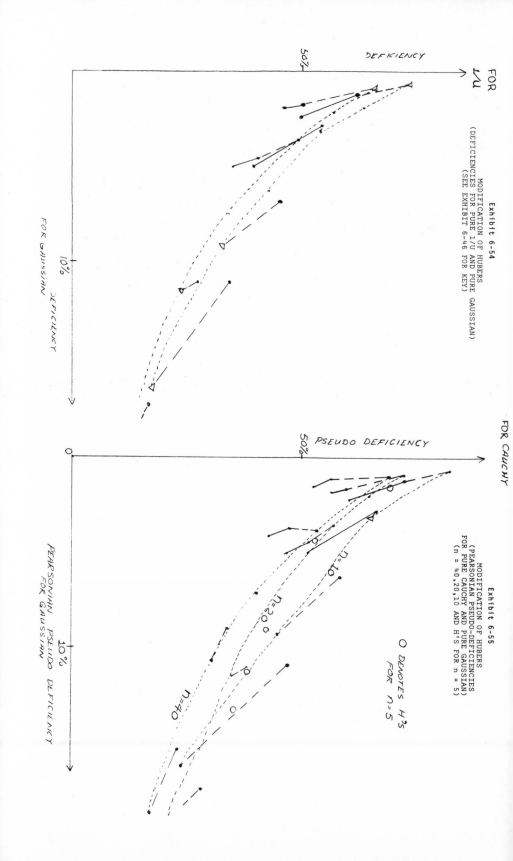

Exhibit 6-54

MODIFICATION OF HUBERS
(DEFICIENCIES FOR PURE 1/U AND PURE GAUSSIAN)
(SEE EXHIBIT 6-46 FOR KEY)

Exhibit 6-55

MODIFICATION OF HUBERS
(PEARSONIAN PSEUDO-DEFICIENCIES
FOR PURE CAUCHY AND PURE GAUSSIAN)
(n = 40,20,10 AND H'S FOR n = 5)

O DENOTES H'S
FOR n = 5

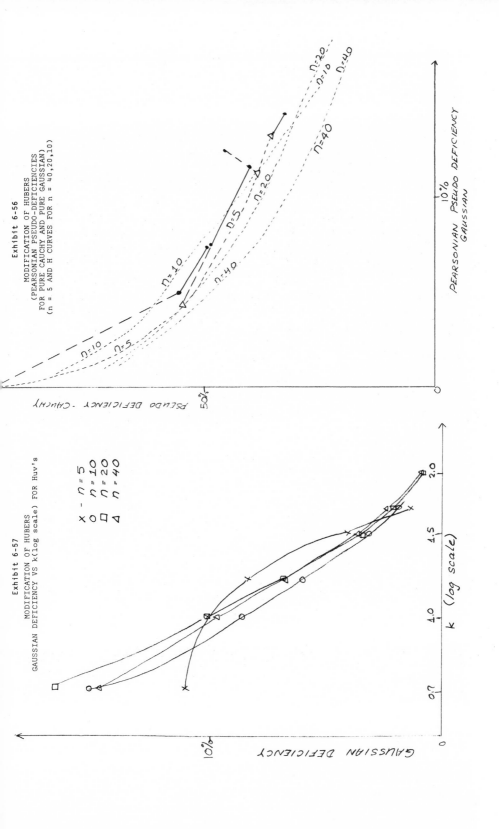

Exhibit 6-56

MODIFICATION OF HUBERS
(PEARSONIAN PSEUDO-DEFICIENCIES
FOR PURE CAUCHY AND PURE GAUSSIAN)
(n = 5 AND H CURVES FOR n = 40,20,10)

Exhibit 6-57

MODIFICATION OF HUBERS
GAUSSIAN DEFICIENCY VS k(log scale) FOR Huv's

3) The change from D to P (available only for k=2.0 and k=1.5) is still more along the curve than across the curve.

4) At k=0.7 and 1.0, where only D modifications are available, the result of modification is usually a net loss, the D points falling above the H curve.

5) At k=2.0 and 1.5, the effect of modification is often a loss in gentle situations but a gain in some vigorous ones (25% of 1/U at n=40 to 10 in exhibit 6-53, pure 1/U at n=20 to 10 in exhibit 6-54, Cauchy at n-40 to 10 in exhibit 6-55). (The other vigorous situations run close to the curve.)

6) The loss is often less, and the gain is often more, for A or P modification than for D modification.

The facts are far clearer than the reasons.

6H   SHOULD WE RESCALE WITH n?

When we look hard at the last several exhibits, we see two phenomena going on as n changes:

- the curves (for H's, for D's or for A's and P's) shift somewhat.

- the locations of specific estimates shift along the curve.

6H1   Scale families

All three of the families of estimates just considered are <u>scale families</u>, in the sense that

1)   each estimate is a scale-using estimate
and
2)   each estimate comes from each other by rescaling.

An estimate is <u>scale-using</u> if either their whole computation or their iterative step can be divided into two portions:

- in one we calculate a number, say s, that can be regarded as an indication of scale.

- in the other we use s to define a functional $f_s(\cdot)$ of samples which we then apply to get a location estimate.

Whenever we have a scale-using estimate, we can rescale it. We can introduce a third, middle, step in which:

- s is converted to ks, which is then used as s was before.

All hubers come from any one huber by rescaling with suitable constants. If the initial huber had k=1.0, then k above and k in huber definitions have the same meaning.

6H2 Rescaling with n

We are free to rescale. The result will always be a location estimate. Our freedom is at all n, better at each h. If we rescale by k(n), where k(n) -> 1 as n -> ∞, then the asymptotic behavior of our estimate will not change.

Consider an estimate T**, whose deficiency (or perhaps whose Pearsonian pseudodeficiency) does not change as n changes, both for a conventional situation and for a threatening one. For each n, T** will be represented by the same point in the dual deficiency plan. (We might say that T** is very asymptotic.)

Suppose further that T** is scale-using. Rescaling T** leads, for each n, to a family of estimates and hence to a curve in the dual deficiency plane. Suppose, further, that these curves are all the same, point for point. (Now all rescalings of T** are supposed very asymptotic.)

What if we rescale T** by k(n)? For each n we get an estimate represented by a point on the common curve, but now this point moves (along the curve) as n changes. The same is true when any fixed rescaling of T** is treated similarly.

Rescaling by k(n) all constant rescalings of T** gives us, for each n, the same curve that we had for all constant rescalings themselves. For all n, we get the same curve. But if we choose the constant in the constant rescaling as our parameter, we have different parametrizations for different n. This difference is a removable difference -- one way to remove it is to rescale again, this time by $(k(n))^{-1}$.

If all -- or even much -- of what we face as n changes is reparameterization of a fixed curve, we ought to try rescaling and see what can be done.

6H3 Rescaling for individual situations

So far we have been realistic. We have assumed, tacitly, that if we rescale we will rescale the same way for all situations for a given n. Nothing else can be realistic, since we never know in practice what situation we face. But if we want only to study the behavior of an estimate, we can perfectly well ask about the results of rescaling differently in different situations.

Suppose we find rescaling by k(n,Si) simplifies matters in situation Si, where k does vary from one S to another? What are the practical consequences? (We can assume k(n,Si) -> 1 as n -> ∞ without loss of generality:

    - if all k(n,Si) behave somewhat similarly (for different Si), we can improve simplicity of behavior -- specifically we can make asymptotic results better approximations in smaller samples -- if we rescale by a k(n) that resembles all the k(n,Si).

    - However the k(n,Si) behave, together they summarize (when they exist with exact properties) all that we need to know to convert asymptotic results into finite sample ones.

Clearly we gain knowledge and sometimes learn how to gain simplicity, by asking for k(n,Si)'s. In fact, of course, we still gain something, though not as much when the effect of k(n,Si) is a matter of approximation of competitive quality rather than of an exact match.

6H4 Rescaling the original hubers

Exhibit 6-57 shows the Pearson pseudodeficiency for a Gaussian situation, later called D, plotted against $\log k_o$ (we now use k_o for the huber parameter) for the original hubers at n=40, n=20, n=10, and n=5. (We introduce the logarithm in $\log k_o$ for a reason soon to appear.)

The curve for n=5 is clearly deviant, but the other three are not too far from being translates of one another.

If we rescale by k(n),

$$k_o \to k(n) \cdot k_o$$

and

$$\log k_o \to \log k(n) + \log k_o$$

the deficiency D remaining the same. Thus horizontal displacement in 6-57 is just what we hope to find.

To look more carefully, we may take any q(D) such that

$$\log k_o + q(D)$$

where D is the deficiency for k_o, varies less with D and plot

$$\log k_o + q(D)$$

against

$$D$$

for various n, looking now for a <u>vertical</u> constant displacement (of a curve that runs <u>much more</u> nearly at right angles to the shift.

Exhibit 6-58 continues the example with q(D) a simple quadratic, namely

$$q(D) = 2D - .95D^2$$

for $0 \leq D \leq 1$. We can now see what portion of the curves are roughly constantly spaced (in an increasing or decreasing order on n). Using these portions we are led to

$$q(D) = 2D - .95D^2 + \frac{.3}{n}$$

since .3 is log 2, the last term is (log2)/n. Such a shift would confirm rescaling the huber family by $2^{1/n}$.

Exhibit 6-59 shows the results of plotting

$$\log k + 2D - .95D^2 + \frac{.3}{n}$$

for all n. We see that:

196

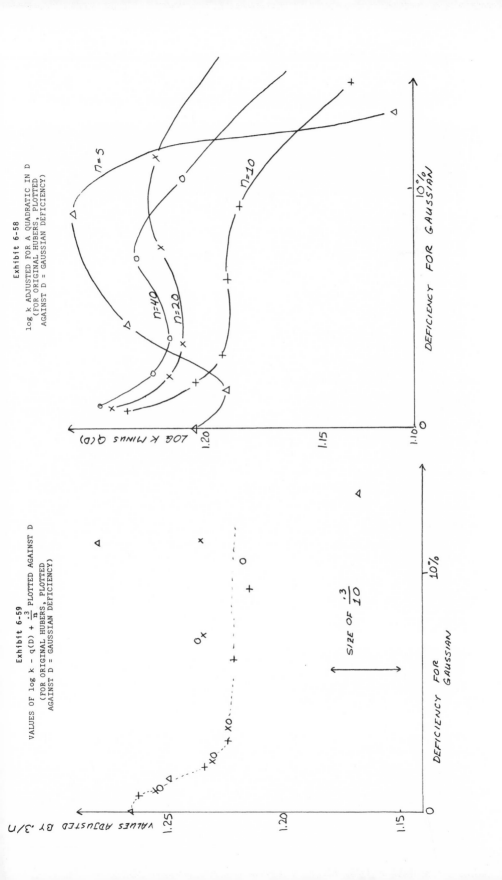

Exhibit 6-58

log k ADJUSTED FOR A QUADRATIC IN D
(FOR ORIGINAL HUBERS, PLOTTED
AGAINST D = GAUSSIAN DEFICIENCY)

Exhibit 6-59

VALUES OF log k - q(D) + $\frac{.3}{n}$ PLOTTED AGAINST D
(FOR ORIGINAL HUBERS, PLOTTED
AGAINST D = GAUSSIAN DEFICIENCY)

1) For low D's, the four curves correspond very closely.

2) For middle D's, things are more irregular.

3) For large D's, specifically for H07, the adjusted values (one for each n) are all about the same, though D changes appreciably.

On balance, rescaling by $2^{1/n}$ seems to simplify the behavior of the hubers for Gaussian situations.
Proceeding similarly for 25% of 1/U and the Cauchy, we find the following shifts:

$$25\% \text{ of } 1/U: \qquad \frac{1.2}{n} \text{ corresponding to } 2^{4/n}$$

$$\text{Cauchy:} \qquad \frac{1.5}{n} \text{ corresponding to } 2^{5/n}$$

Clearly the rescalings indicated for the three situations -- Gaussian, 25% of 1/U, and Cauchy -- are not the same. Since they are of the same sign, however, there is something to be gained from a compromise rescaling. Either .6/n or .9/n seem reasonable. They correspond to rescaling by $2^{2/n}$ or $2^{3/n}$ respectively.
The nature of these rescalings is that k should be replaced by $k \cdot 2^{2/n}$ or $k \cdot 2^{3/n}$. This means that the scale estimate used in the original hubers needs to be divided by $2^{2/n}$ or $2^{3/n}$ before being multiplied by the rescaled k.
The difference in rescaling indicated for Gaussian and long-tailed situations indicates a need for division by a bigger factor for the long-tailed situations. This would happen if we altered the scaling procedure to concentrate more on the central part of the sample.

6H5 Modified hubers

We can do the same for D-modified and A-modified hubers. The results are brought together in exhibit 6-60. We see that the results for the modified hubers resemble those for the original hubers, appearing somewhat more irregular.

All values in this exhibit are quantatively uncertain in detail, although the direction and general size of the values indicated are usually clear. For this reason, all exponents in panel B are held to integer multiples of 1/n.

6H6 Rescaling in general

Taken as a whole, both original and modified hubers indicate:

- a useful rescaling by about $2^{3/n}$ if we want asymptotics to be as good a compromise as possible.

- need for a scaling procedure that concentrates more on the center of batches if we are to make this compromise better. (Doing so may make the quality of estimate worse, something that may not be balanced out by close fitting of asymptotic results.)

exhibit 6-60
Rescaling -- for individual situations and overall --
suggested for original and modified hubers

A) CHANGES IN log k,

	H-family	D-family	A(P) family
pure Gauss	$\frac{.3}{n}$	$\frac{1}{n}$	$\frac{.8}{n}$
25% of 1/U	$\frac{1.2}{n}$	$\frac{.8}{n}$	$\frac{.4}{n}$
pure Cauchy	$\frac{1.5}{n}$	$\frac{2+}{n}$	$\frac{1.2}{n}$
(compromise)	$(\frac{.6}{n}$ or $\frac{.9}{n})$	$(\frac{.9}{n})$	$(\frac{.9}{n})$

B) RESCALING FACTORS

	H-family	D-family	A(P) family
pure Gauss	$(2)^{1/n}$	$2^{3/n}$	$2^{3/n}$
25% of 1/U	$2^{4/n}$	$2^{3/n}$	$2^{1/n}$
pure Cauchy	$2^{5/n}$	$2^{(6+/n)}$	$2^{4/n}$
(compromise)	$(2^{2/n} \text{ or } 2^{3/n})$	$(2^{3/n})$	$(2^{3/n})$

6I COMPARISON OF SITSTEPS AND RELATED
FAMILIES FOR SMALLER n

As we move to smaller n's -- below 20 -- the question
of choosing the ends of a global-linear-combination family
becomes more pressing. At n=20 we used D20, but for
lower n's we now see that the variance of D20 is untrust-
worthy for vigorous situations. If we were principally
concerned with pure D20, we could go, as we have just
seen, to pseudovariances. This is, of course, not our
position. We want to deal with linear combinations of
D20 with such antitheses as 5T4. To do this we need
covariances.

In planning the calculations, we were forethoughtful
enough to plan for (a) covariances and (b) pseudovariances.
We did not calculate pseudocovariances -- indeed it is not
quite clear whether such can be meaningfully defined.
Thus our interest has to be concentrated in P15, as a
one step and in A20, as a surrogate for the one step P20,
when we consider the left end of a family of global linear
combinations.

If we are to consider new and alternate left ends,
we may as well look at new and alternate right ends at
the same time.

6I1 Gentle situations (for n=10)

Exhibits 6-61 and 6-62 show, for 10% of 3G and 20%,
the behavior of the linear combinations joining:

- P15 or A20 on the one hand.

- 5T4, TRI, CST, 12A and SST on the other.

Rather clearly the 5T4 families

- outperform the families replacing 5T4 by the
other possibilities.

- but only by a fraction of 1%.

If we did not have 5T4, we could gain by using any of the
others. Since we have it, we may as well use it (though
the superior ease of TRI may tempt us.)

201

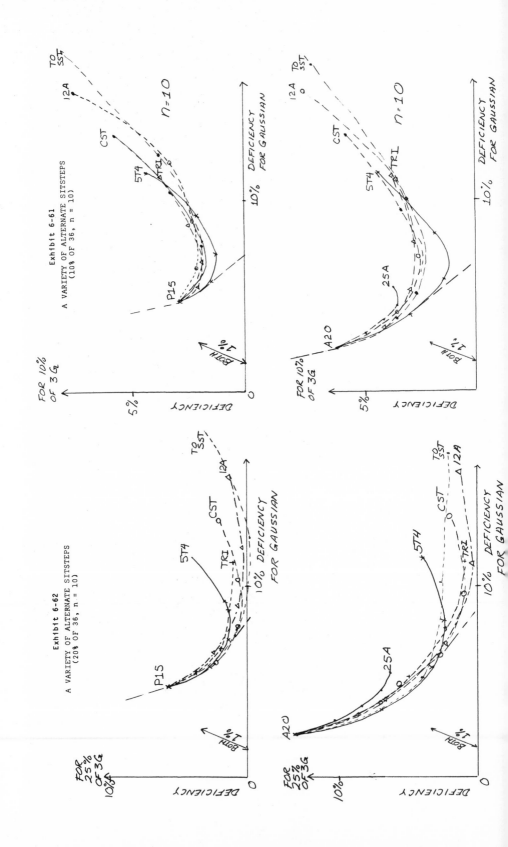

Exhibit 6-61

A VARIETY OF ALTERNATE SITSTEPS
(10% OF 36, n = 10)

Exhibit 6-62

A VARIETY OF ALTERNATE SITSTEPS
(20% OF 36, n = 10)

6I2 Vigorous situations (for n=10)

Exhibits 6-63 and 6-64 show the corresponding
pictures at n=10 for 25% of 1/U and pure 1/U, respectively. TRI and SST now fail to compete as right-
hand terminals (SST doing much better than TRI), but 12A
and CST continue to do well but not as well as 5T4.
We are skipping any corresponding picture for 25%
of 1/U at n=5, since we do not trust the deficiencies,
and do not have the Pearsonian pseudodeficiencies.
(We give no pictures for Cauchy at n=10 or n=5
for the same reason.)

6I3 Summary on these choices

Having stuck with 5T4 on the basis of comparison,
we are prepared to continue, for n≠20 a policy quite similar to the one we adopted for n=20, showing two families
of sitsteps:

- for n=40, as for n=20, we show 5T4 & D20 --
the old original, and 5T4 & P15.

- for n=10 and n=5 (here a large grain of
salt) we show 5T4 & A20 (hopefully like 5T4 & P20)
and 5T4 & P15.

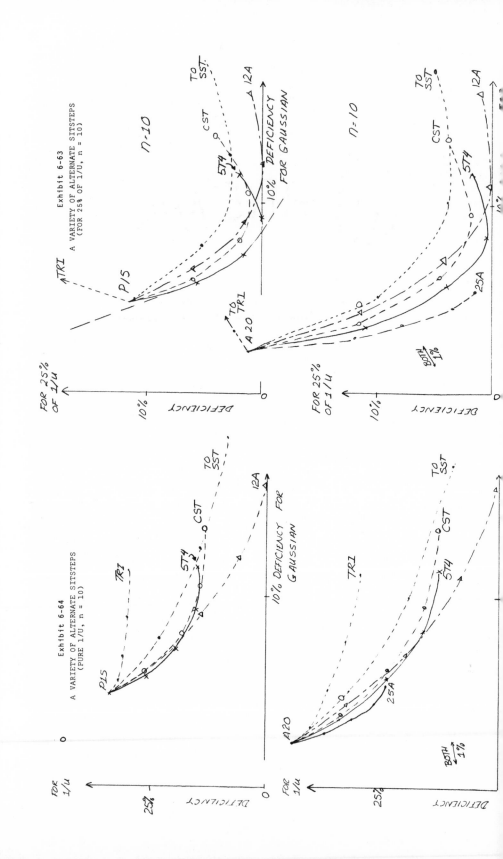

Exhibit 6-63

A VARIETY OF ALTERNATE SITSTEPS
(FOR 25% OF 1/U, n = 10)

Exhibit 6-64

A VARIETY OF ALTERNATE SITSTEPS
(PURE 1/U, n = 10)

6J COMPARISON OF FAMILIES AND SOME ESTIMATES
AT n'S OF 40, 10 AND 5

It is now time to look at the comparisons at n's
other than 20. At n=5, and for the Cauchy at n=10, we
dare look only at Pearsonian pseudovariances. Accordingly
our sitstep families are mostly matters of supposition.
At n=40, of course, all is simple. Other situations
fall in between. We shall compare families, often by
showing only the location of some of their members, with
each other and with certain interesting estimates.

6J1 The gentle situations (n=10)

Exhibits 6-65 and 6-66 are for 10% from 3G and for
20% from 3G at n=10. We see, that for 10% from 3G,
the gentlest contamination considered for n=10,
hubers do best, though both alternate sitsteps are only
a fraction of a percent behind. By 20% of 3G, the sitsteps have caught up with the hubers. In both cases
the hampels are about 1% (for both situations) behind
and the individual estimates are uninteresting.

6J2 Vigorous situations (all n)

The only vigorous situation available at n=5 and
n=40 is 25% of 1/U. Exhibit 6-67 shows Pearsonian
pseudodeficiencies for families and interesting estimates
at n=5. The dotted curves for 5T4 & P15 and 5T4 & A20
are quite hypothetical, but are unlikely to be too far
out. If so, they come quite close to the hubers. The
hampels fall appreciably behind.
Exhibit 6-68, for 25% of 1/U at n=10, shows
Pearsonian pseudodeficiencies for individual estimates
and multiestimate families, but curves based on deficiencies for the two alternate sitstep families. The sitstep alternates are now ahead of the hubers, and the
hampels are still further ahead of them.
Exhibit 6-69, for 25% of 1/U at n=20, uses ordinary
deficiencies throughout, and shows both sitstep alternates
as well as the original sitsteps. Both sitsteps and
hampels are well ahead of hubers. No individual estimate
is yet of interest.

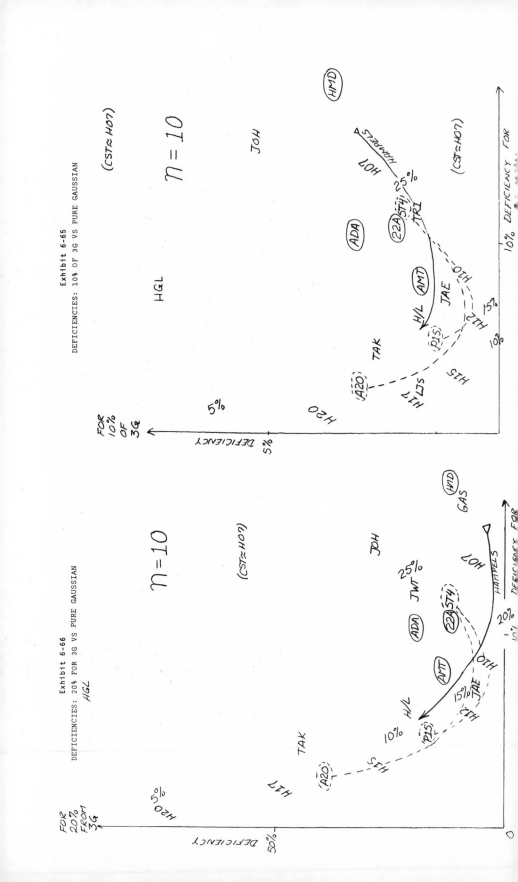

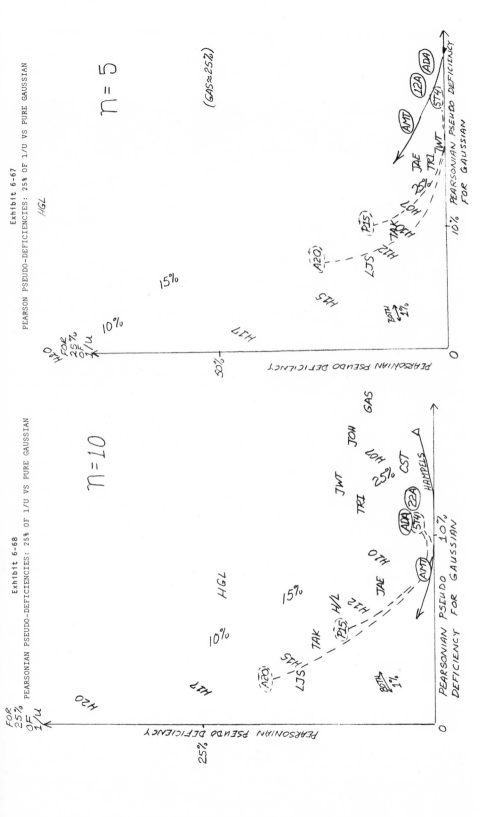

Exhibit 6-67

PEARSON PSEUDO-DEFICIENCIES: 25% OF 1/u VS PURE GAUSSIAN

Exhibit 6-68

PEARSONIAN PSEUDO-DEFICIENCIES: 25% OF 1/u VS PURE GAUSSIAN

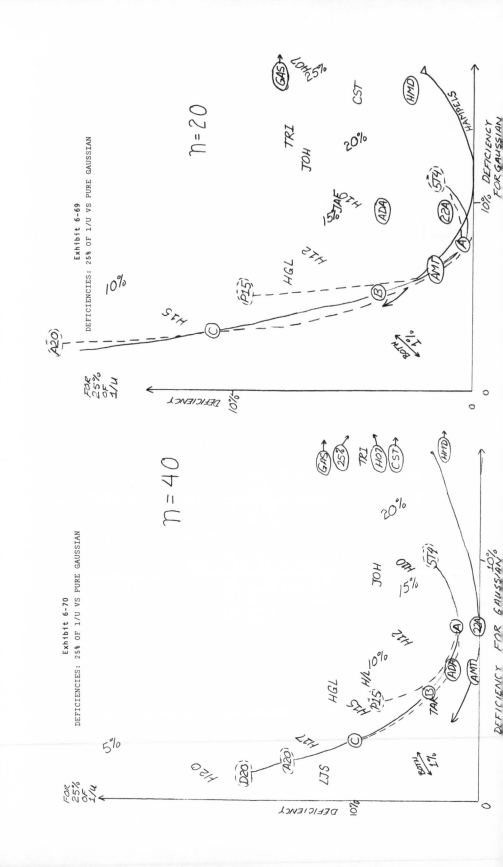

Exhibit 6-69

DEFICIENCIES: 25% OF 1/U VS PURE GAUSSIAN

$n = 20$

Exhibit 6-70

DEFICIENCIES: 25% OF 1/U VS PURE GAUSSIAN

$n = 40$

Exhibit 6-70, for 25% of 1/U at n=40, uses deficiencies throughout. Hampels are now well ahead of hubers. LJS does notably well, but we need not discuss the other individual estimates.

Turning next to pure 1/U, exhibit 6-71, at n=10, shows ordinary deficiencies as did exhibits 6-14 and 6-32, earlier, for pure 1/U at n=20.

At n=10, we now see sitstep alternates well ahead of hampels, and hampels well ahead of hubers.

As compared to n=20, the sitstep alternates do substantially better at n=10 and the individual estimates that did noticeably well in exhibit 6-32 (TAK, HGL, JOH) are not competitive for these smaller samples.

6J3 Unreasonable situations (all n)

Turning now to the Cauchy, exhibit 6-72 shows Pearsonian pseudodeficiencies at n=5, as does exhibit 6-73 at n=10. Exhibits 6-15 and 6-33, earlier, show ordinary deficiencies at n=20. Exhibit 6-74 shows ordinary deficiencies at n=40.

Comparing these, we see that the sitstep alternates do very well, being threatened only by:

- the hampels at n=20 (and, when extended, at n=40)

- TAK at n=20 and n=40 (why more strongly at n=20?)

- HGL, ADA, and JOH at n=40.

6J4 Summary of comparisons

We can do fairly well in summarizing the comparisons for all n by stating that:

- sitsteps or sitstep alternates do quite well throughout, provided we allow for n.

- hampels do better for vigorous situations with n $\geq$ 20.

- hubers do better for very gentle situations, but only by fractions of a percent.

209

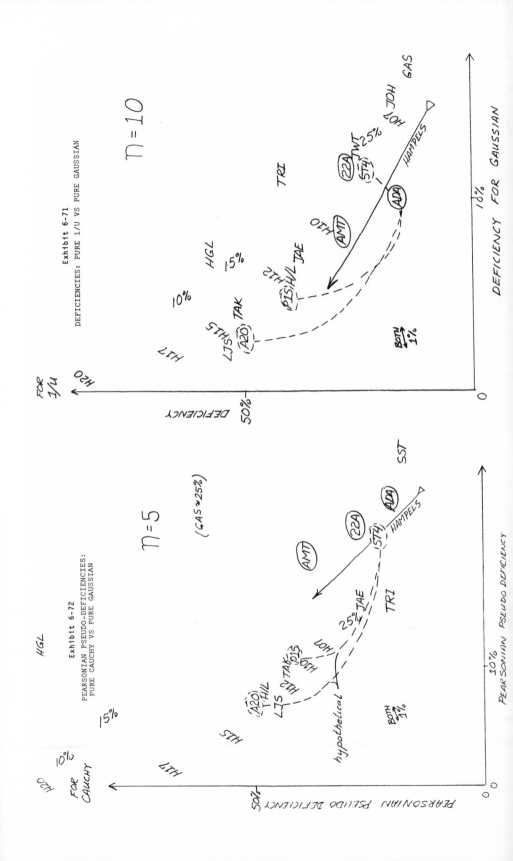

Exhibit 6-71

DEFICIENCIES: PURE 1/U VS PURE GAUSSIAN

Exhibit 6-72

PEARSONIAN PSEUDO-DEFICIENCIES:
PURE CAUCHY VS PURE GAUSSIAN

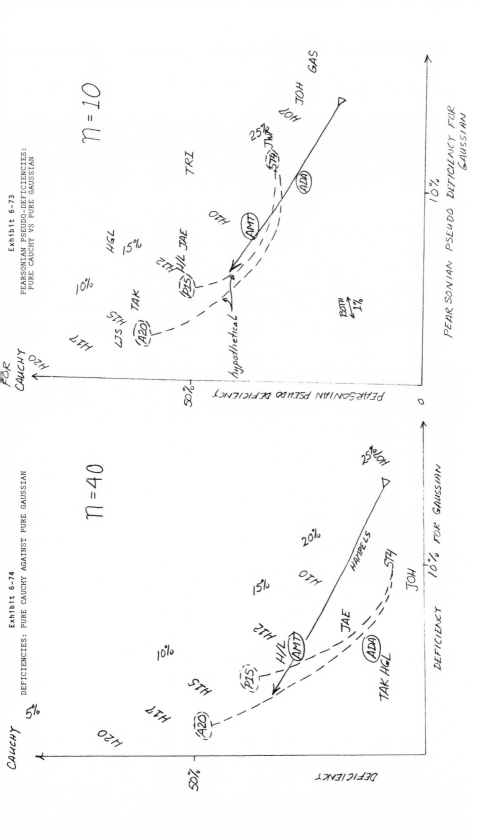

Exhibit 6-73

PEARSONIAN PSEUDO-DEFICIENCIES:
PURE CAUCHY VS PURE GAUSSIAN

$n = 10$

Exhibit 6-74

DEFICIENCIES: PURE CAUCHY AGAINST PURE GAUSSIAN

$n = 40$

- at n=40, some of the individual adaptive estimates, like TAK, HGL and JOH begin to come into their own for the Cauchy situation. (Perhaps they come into their own for more reasonable situations for $n \geq 60$, or for $n \geq 80$, or for $n \geq 100$.)

This leaves one writer with the view that one needs to consider for the future:

- sitsteps, sitstep alternates, and hampels, all of which are subject to considerable extension and tuning (which ought to lead to considerable further improvement).

- blatantly adaptive estimates for $n \geq 50$.

No doubt further studies will throw much light on the validity and completeness of this insight.

6K THE CONSERVATIVE APPROACH

We ought also to look at the approach furthest
from variances that is possible with our data. This is
to look at the assessed 0.1% points, and compare each
with the best available.
Exhibit 6-75 lists the results in terms of 10% steps
in 0.1% pseudovariance for ratios up to 1.7 times the
best available. (Since these ratios are of pseudovariances,
1.7 is only about a ratio of 1.3 for % points.)

6K1 When n=5

Here we have to note with some surprise the appearance
of A15 and P15. Somehow, P15 is a very conservative
estimate, even if a one-step, for this sample size.
Except for the median, the remaining 8 of the best
10 are of hampel type. We can do quite well with 12A,
whose performance for other n's is outstanding. A little
better may be done at this sample size by moving a little,
perhaps to 21A.

6K2 When n=10

Here only two estimates show good overall performance.
(The 1% pseudovariance used for pure Cauchy was the
arithmetic mean of Cauchy A and Cauchy B.)
12A does very well, and is by now the outstanding
candidate for the "one estimate for all n".
50% does very well, and is a strong candidate for
all sample sizes up to, perhaps, 12 to 14.

6K3 When n=20

12A is doing even better than before, along with 3
other estimates of hampel type. Since 12A has avoided
difficulty more thoroughly at other n, the lightening-
susceptibility argument suggests its use for this n,
even when we are free to change from one n to another.
Two more estimates of hampel type follow. The median,
along with GAS and 5T4, come another block later. The
median is still quite available for the conservative in a
hurry. (Given a little time and effort, the conservative
will favor 12A.)

exhibit 6-75
0.1% pseudovariances moderately close to best available
(parenthetic headings are ratios to best available,
order of listing within section not indicative of size)
estimates of the hampel type are on the left of each double
column

n=5		n=10		n=20		n=40	
(1.3 to 1.4)		(1.3 to 1.4)		(1.2 to 1.3)		(1.1 to 1.2)	
17A	A15	12A	50%	12A	.	.	JOH
21A	P15			ADA			
25A		(1.5 to 1.6)		OLS		(1.2 to 1.3)	
		.	GAS	TOL		12A	GAS
(1.4 to 1.5)							HGL
12A	50%	(1.6 to 1.7)		(1.3 to 1.4)			5T4
22A		17A	DFA	17A	.		5T1
ADA		ADA	CML	HMD			25%
AMT		HML					CST
				(1.4 to 1.5)			H07
(1.5 to 1.6)				.	50%		
HMD	.				DFA	(1.3 to 1.4)	
					JWT	17A	JWT
(1.6 to 1.7)					H07	ADA	HGP
.	CM.				GAS	HMD	TAK
					5T4	OLS	JBT
					5T4	TOL	JAE
					25%		
					JLJ	(1.4 to 1.5)	
							CTS
				(1.5 to 1.6)			D07
				.	SST		SJA
				(1.6 to 1.7)		(1.5 to 1.6)	
				(none)		21A	50%
						AMT	DFA
							SST
							3T1
							3T0
							H10
							BIC
						(1.6 to 1.7)	
						22A	CML
							TRI
							D10

6K4 When n=40

It appears that JOH has come into its own, at least for conservatives, when n=40. Its overall performance is indeed very good.

12A, along with GAS, 5T4, 25%, CST and HO7, to name a few, comes in the next block. 12A is no longer in the top block, but its performance for n=40 is better, in these terms, than the performance of any of our estimates for n=5 or n=10 (and similar to its own performance for n=20.

The median is down two more blocks, and is not highly attractive for n=40. The appearance of GAS and CST in the same block as 12A leaves 50% in a poor competitive situation.

6K5 Choices

The following four choices seem reasonable for a rampant conservative depending on his degree of hurry and willingness to vary his choice with n:

1) use 12A for all n.

2) use 21A for $n \leq 7$, use 12A for $8 \leq n \leq 30$, use JOH for $n \geq 31$.

3) use 50% for all n.

4) use 50% for $n \leq 20$ and GAS for $n \geq 21$.

6K6 Comment

TAK seems not yet close to coming into its own, for rampant conservatives, at n=40. At n=100 (where we shudder at the computation!) it may prove out.

5T4 does not do badly at n=20 and n=40. We have to be interested in learning about 0.1% pseudovariances for linear combinations from 5T4 & P15, 5T4 & P20 and from other sitstep families.

6L A RADICAL APPROACH

One antithesis of looking at 0.1% points -- in terms of 0.1% pseudovariances -- is to look at 25% points. Exhibit 6-76 lists estimates in blocks by their worst ratio of actual 25% pseudovariances to best available.

6L1 When n=5

The high performance of 17A and 21A is an incomplete repeat of the high performance of 17A, 21A and 25A at 0.1%. Either is a leading candidate, although the other 10 in the first block did at least as well. 12A is about 5% behind, not enough to be ruled out. The quick candidates are now TRI, GAS and 25% and no overlap with the 0.1% list occurs except for estimates of hampel type.

6L2 When n=10

12A now leads all the rest (by a small margin). 17A is one block back. The leading quick candidate is GAS, with 50% appreciably behind.

6L3 When n=20

The lead is now taken by ADA and JOH. (JLJ has independent identity, so far, only for n=20, which must be considered a possibly temporary defect.) 12A and 17A are close behind. The leading quicks are GAS and 25%.

6L4 When n=40

12A is now in the first block, as are 5T4 and TAK. The leading quicks are GAS and 25%.

6L5 Choices and comments

If one estimate is to be chosen for all n, the leader is undoubtedly 12A. Leading here as well as for 0.1% is an outstanding performance. (Indeed, except for 1.296 for pure Gaussian at n=5 and 1.201 for pure Gaussian at n=20, 12A is never worse than 1.18 among 25% pseudovariances.)

exhibit 6-76
25% pseudovariances moderately close to best available
(parenthetic headings are ratios to best available,
order within blocks not indicative)
estimates of the hampel type are on the left of each double
column

n=5		n=10		n=20		n=40	
(1.20 to 1.25)		(1.1 to 1.2)		(1.1 to 1.2)		(1.1 to 1.2)	
17A	5T4	12A	.	ADA	JLJ	12A	JOH
21A	3R2				JOH	ADA	5T4
OLS	3R1	(1.2 to 1.3)					HGL
	TRI	TOL	DFA	(1.2 to 1.3)			5T1
	JBJ	OLS	GAS	12A*	DFA		H07
	JAE	HMD	JOH	17A	5T4		TAK
	335	ADA					
	GAS	17A		(1.3 to 1.4)		(1.2 to 1.3)	
	25%			OLS	H07	17A	JWT
		(1.3 to 1.4)		21A	GAS		HGP
(1.2 to 1.3)		.	50%	22A	25%		GAS
12A	CST		H07	AMT	5T1		25%
	3T1		CTS		D07		JBT
	5T1		SST		JBJ		D07
	3T0		JWT		3T1		JAE
	1 to 7		5T1		TAK		SJA
	TAK		5T4		3T0		3T1
					JAE		3T0
(1.3 to 1.4)		(1.4 to 1.5)			HGL		
HMD	SST	22A	CST			(1.3 to 1.4)	
AMT	DFA			(1.4 to 1.5)		OLS	DFA
25A	A15			TOL	SST	HMD	CTS
	P15			HMD	JWT		BIC
	H10						LJS
	H12						THL
(1.4 to 1.5)						(1.4 to 1.5)	
ADA	50%					TOL	SST
	D10					21A	H10
	LJS					AMT	TRI
	MEL					22A	D10
	BH						
	H/L						
	D15						
	CTS						

*Note that 12A falls just out of the block above at 1.201

If we are free to vary our choice with n, we might again like to use 17A or 21A for small n.

(The relative regularity of behavior could suggest, to a suspicious mind, that:

- at n=10, something below 12A might do even better.

- the steady drift with n can probably be eliminated, and a single excellent choice found, by redefining the scaling algorithm slightly.)

If we want to be quick, we should use GAS. It does not seem to need to be changed as n changes.

We need to notice that 5T4 does well for all n. (This raises intriguing questions about the possible relation of this high performance to the qualities of linear combinations of 5T4 with modified hubers.)

While we find it hard to imagine a user who would be firm and consistent in adhering to a criterion of "worst 25% behavior", we have to admit that the results for 25% are, at least so far as 12A and the other hampels go, striking confirmation of what we found at 0.1%. For this they can hardly be less than very illuminating and substantially informative.

6M FURTHER COMMENTS

Before closing, we ought to do a little more to look at the results of our analysis as a whole.

6M1 The stages of analysis for variances

The first stages of this analysis, begun when there were only 40 estimates (not 65) and rather fewer situations, concentrated on the relation of variances to trimmed-mean variances, thus assigning to each estimate in each situation 1, 2, or 0 amounts of trimming which matched its variance. Deficiencies were introduced at this stage.
It was found that deficiencies were more easily looked at across situations when adjusted somewhat.
As a result the output of the main calculation included values of <u>modified optrim deficiencies</u>, defined to be:

$$(1-\sqrt{\text{contamination}})^2(1 - \frac{\text{optrim variance}}{\text{actual variance}})$$

where "optrim" refers to the minimum variance obtainable by trimming.
The development of dual-deficiency plotting then shifted our interest away from numerical comparisons across situations (graphical comparisons were more informative) so that any need for modification disappeared. Also, its emphasis on optrim variance rather than optavail variance, where "optavail" refers to minimum variance among those estimates tried, decreased our interests in optrim deficiency.
If the variance output were to be resigned at this writing, this writer would recommend using simply

$$1 - \frac{\text{optavail variance}}{\text{actual variance}}$$

as the working quantity.

6M2   What pseudovariances in the future?

On the whole, the behavior of the various % points, especially when converted into pseudovariances, was regular enough for us to not present tables in detail. It could well make sense to cut down on the number of % points calculated, perhaps to 0.1%, 4.2%, and 25%. The computing effort thus released could be used to recalculate "% outside" at the first approximations to these % points, thus greatly reducing dependence on interpolation and extrapolation.

6M3   Which estimates MUST we consider?

The discussion of 6J4, which focused, after we apply de minimis, on:

- sitsteps and sitstep alternates

- hampels

- blatantly adaptive estimates at n $\geq$ 50

was not disturbed by the very different analyses of 6K and 6L. The emphasis on hampels was, of course, strengthened.

It is both interesting and rewarding that estimates not considered by anyone before this study, as well as blatantly adaptive estimates published or invented since 1969, should prove the outstanding candidates.

In this context, we ought to point out that OLS, had it been rescaled widely and wisely, might have performed as well as (or maybe better than) the hampels. Our failure, at an early stage, to think in terms of families of estimates is responsible for this uncertainty of undervaluation.

6M4   Some estimates that were not expected to work

SHO was entered among our estimates, not because it was a likely candidate in the range of situations here considered, but rather in the hope that it might function well "beyond the Cauchy" and that, if it did, we would want to know about the penalties involved in its use in less violent situations. The $n^{-1/3}$ convergence is far from encouraging, as are its variances and pseudovariances.

4RM, 3RM and 2RM were selected to be "reasonable facsimiles" of the arithmetic mean. Since they were quite successful, they only appear in this study as examples of what to avoid when seeking robustness.

Because they do do a good job of matching the mean, they do lead us toward small-addition estimates that are robust. Replacing the mean in 5T4, where it follows a skip-into-trim process, by a refolded median should have little effect on quality of performance, while reducing arithmetic to many additions of pairs of numbers.

6M5 The case of small n

At some stage n becomes "small" in the sense that we learn more by forgetting asymptotic results. There are several reasons why the present study does not really pin down where this happens.

The most encouraging is that it appears to be true that "the better the estimate, the less difficulty it causes". It is the estimates which are good in other senses (hampels, A20, P15, 5T4, 50%) that keep variance close to pseudovariance the longest. The same estimates tend to perform well in 6L and so on.

Another is the severe limitation in variety of situations for n < 10.

A third is the absence of adequate understanding of what will happen if we rescale estimates to match performance (rather than some algebra that is always somewhat arbitrary) across n.

With all this taken care of, it is reasonably clear that the boundary can be pushed down below n=10. Whether it can be pushed all the way to n=5 is in doubt.

7. GENERAL DISCUSSION

This chapter consists of separate sections containing individual conclusions and remarks by some of the participants. The contributors are

7A   WHAT HAVE WE LEARNED?

Those who have participated actively in the study here reported are convinced that both science and art have been advanced significantly. It is time to try to make this clear to all -- and, particularly, to make clear the diversity of ways of advance.

7A1   What Have We Learned About Good Estimates?

We have learned that hampels are extremely promising, and deserve both extension and tuning.

We have learned that sitsteps do very well in view of the limited computational effort required, and deserve careful tuning at both ends. The use of the sit mean as the start of the one-step estimate demands inquiry.

We have learned that blatantly adaptive estimates, like TAK, HGL, and JOH seem to come into their own for fairly large samples, probably beyond n = 50. There is clearly place for much further inquiry.

We have learned that one-step versions of iteratively defined estimates do very much better than many supposed. Investigation of one-step versions of hampel-type estimates (plain hampels, rescaled AMT, rescaled OLS) deserves high priority.

We have learned that, with the probable exception of the blatantly adaptive estimates at $n \geq 50$, the estimates developed one by one have not stood up well to a competitive test.

7A2   What Have We Learned About How to Think About Estimates?

We have learned that much more is gained by thinking about families of related estimates than by thinking about individual estimates. So far, the main profit has come from thinking about 1-parameter families, but the case of few-parameter families is open.

We have learned that many estimates can be rescaled, and that we ought to consider the one-parameter families obtained by free rescaling.

We have learned that one-parameter families consisting of unbiased linear combinations of two estimates (i) can be quite profitable, and (ii) have important advantages in use.

7A3   What Have We Learned About How to Look (Visually and Quantitatively) at Estimates?

We have learned that plots in dual-variability planes tell us very much more than any previously used ways of displaying quantitative and semiquantitative information about robust behavior.

We have learned that, when we wish to base comparisons on variance, that there is considerable advantage to be gained from the use of deficiency.

We have learned some of the advantages of assessing estimate variability by the use of (exact or approximate) Pearsonian pseudo-variances -- or by the corresponding Pearsonian pseudo-deficiencies. We have also gained some idea of what such measures miss.

We have learned to begin to face some aspects of the question of the shape (non-Gaussian) of estimate distributions.

We have learned that there are describable techniques to find both rough and sharp bounding curves in dual-variability planes, curves below which invariant estimates cannot exist.

7A4   What Have We Learned About Asymptotics?

We have made some extensions of classical asymptotic techniques.

We have learned how one can go about the asymptotics of iterative skipping procedures.

We have discovered another member, SHO, of the short list of statistics that converge like $n^{-1/3}$ rather than like $n^{-1/2}$.

We have learned that such changes as rescaling dependent on n may have an important role to play in making asymptotic results apply more usefully to smaller samples. As a result, we have begun to realize (see 7A5, below) the opportunities in rethinking such things as the definition of the empirical cumulative.

7A5   Empirical Cumulatives and Formal Simplicity

The early chapters, especially Chapter 3, reflect the modern trend in estimation theory in formulation. To express estimates as functionals on conventional empirical cumulatives, and then to ask for the limits of these functionals as sample size increases is a neat and tidy way to be led to the formulation of asymptotic forms. The process is clearly rewarding, promoting both clarity and facility. We ought to expect to continue to use it and its congeners.

But we need to avoid certain minor traps. Specifically:

- We ought not to close our eyes to other definitions of empirical cumulative.

- We ought not consider estimators defined for all n by the same functional applied to some kind of empirical cumulative as thereby nicer and neater.

Some may feel that there ought to be but one kind of empirical cumulative, the classical sort in which $F_n(x)$ jumps $1/n$ at each observation: Let us note some others:

- $G_n(x)$, say, jumps $1/2n$ at each observation and $1/2(n-1)$ at each midpoint between adjacent observations.

- $H_n(x)$, say, has a density constant between each pair of such midpoints, of area $1/n$ and an exponentially falling density outside of the range of the actual observations.

- $J_n(x)$, say, has a smooth density obtained by spline interpolation among the values of $H_n(x)$.

As $n \rightarrow \infty$, each of these converges to $F(x)$ in the same sense as $F_n(x)$ does. Indeed, the last two converge more smoothly whenever $F(x)$ has a density or a smooth density. The median is defined, for all n, as the solution of $G_n(x) = 1/2$. (We are obliged to take this fact as only a neatening convenience, and not an essential advantage of the median.)

The virtues of asymptotic theory, as distinguished from its beauties, are associated with its performance for finite n, very preferably for small n (if not for very small n). If by redefining the notion of empirical cumulative, or by learning how the apparent forms of functionals on specific kinds of empirical cumulative ought to change with n, we can be led to expressions whose <u>behavior</u> depends less on n, we will have done much to make <u>asymptotic</u> theory more useful.

7A6 What Have We Learned About What Situations to Consider?

We have learned that one may do quite well by looking at as many distributions as possible as being in the form Gaussian/independent.

We have learned that, when we take that attitude, the Cauchy distribution seems unreasonable. (Pure 1/U, the distribution of Gaussian/rectangular, serves most of the purposes to which the Cauchy may be put having a density falling proportional to $1/x^2$, without being, in this sense, unreasonable.)

We have begun to experiment seriously with, and exploit, situations where our data is not a simple random sample. (Thus, for example, our inquiry into Gaussian contamination has been based on fixed numbers of observations randomly sampled from each of two distributions.)

7A7 What Have We Learned About How to Do Robustness Studies?

We have learned that there is no substitute for a comparative study, in which a reasonable number of estimates are <u>compared</u>. (The use of only 10 estimates at a time seems unlikely to be efficient, especially in view of the importance of families. The range from 20 to 100 can hardly fail to be rewarding. Skill and wisdom in the choice of those to be compared is just as essential for many as for few.)

We have learned that there is no substitute for a multistep study, in which results of earlier steps influence the choices of estimates and situations used in later steps.

We have learned that to provide numbers for asymptotic results requires careful planning, of the character if not the amount required for Monte Carlo results. (*Si monumentum dis videre, circumspice.*)

7A8 The Future

As good research should most frequently, almost nothing in this study has brought investigation to its natural limit. We know new areas to explore; we have now techniques to use in this exploration, we know better how to outfit and schedule our exploring parties.

Two things that we have not done need to be put on the list of important jobs for the future:

- We did a little about unsymmetric situations, but we were not able to agree, either between or within individuals, as to the criteria to be used.
- We dealt only with location in the sense of point estimation. Extensions to (1) interval estimation and (2) more general problems are of high importance.

(Separate work by Bickel (Bickel 1971*) on order-statistic related techniques for general linear fitting may point part of the way toward the latter.)

Future work on simple location will have to be much more careful than that of the past. We have narrowed the sizes of the gaps appreciably.

*BICKEL, P. J. (1971). Analogues of linear combinations of order statistics in the general linear model. Symposium on Decision Theory and Related Topics. S. Gupta and J. Yackel (eds.), Academic Press.

7B COMPARING ACROSS SITUATIONS - A PRACTICAL VIEW

7B1 The Need for Such Comparison

There are many instances when a single estimate is to be chosen. For example, suppose location estimation is to be part of a larger, general purpose computer program to be applied to a wide variety of data by a wide variety of users. A program must be written and it must be adequate for almost all situations. Or consider the first time an analyst sees a complex set of data. Initially he will try to summarize the data by taking "robust means" in various ways. At this stage he may not know the distribution's shape; or, if he does, he may not know the model and so the shape information is not yet of use. At this stage the analyst requires a safe procedure that can be applied routinely while his attention, focused on the general nature of the data, is not distracted by details.

In both these instances tolerable overall variance and high breakdown points are more important than high efficiency for large "Gaussian" samples.

Exhibit 7-1 is a table of coded deficiencies of estimators. Only variances greater than 1.5 times the variance of the optimal trimmed mean appear as digits or x's. Clearly many "robust" estimators are very inefficient in some situations, where very inefficient is not measured in % but in hundreds of %! (Many others stay safely below the magic 1.5.) Of course, in rare circumstances, a user may have strong grounds to doubt that his data come from a distribution like the Cauchy and may ignore this table.

Exhibit 7-1

INEFFICIENCIES OF ESTIMATES

Variances divided by variance of trimmed mean optimal for distribution considered.

Variances   (i) rounded to nearest integer
           (ii) 1's are suppressed
          (iii) numbers > 9 are coded x

Normal distribution N
Cauchy distribution C

	40 N	40 C	20 N	20 C	10 N	10 C	5 N	5 C
M	•	x	•	x	•	x	•	x
5%	•	7	•	9	•	x	•	x
10%	•	2	•	3	•	4	•	x
15%	•	2	•	2	•	3	•	x
25%	•	•	•	•	•	•	•	•
50%	2	•	•	•	•	•	•	•
GAS	•	•	•	•	•	•	•	•
TRI	•	•	•	•	•	2	•	•
JAE	•	•	•	•	•	2	•	•
BIC	•	•	•	6	•	3	•	2
SJA	•	•	•	x	•	2	•	2
JBT	•	•	•	•	•	2	•	•
JLJ	−	−	•	•	−	−	−	−
H20	•	3	•	3	•	6	•	x
H17	•	2	•	2	•	3	•	2
H15	•	2	•	2	•	2	•	2
H12	•	2	•	2	•	2	•	2
H10	•	•	•	•	•	2	•	•
H07	•	•	•	•	•	•	•	•
M15	•	x	•	x	•	x	•	2
D20	•	2	•	2	•	3	•	2
D15	•	2	•	2	•	3	•	2
D10	•	•	•	•	•	2	•	2
D07	•	•	•	•	•	2	•	•
A20	•	2	•	2	•	2	•	2
A15	•	2	•	2	•	2	•	•
P15	•	2	•	2	•	2	•	•
HMD	•	•	•	•	•	•	•	•
25A	•	•	•	•	•	•	•	•
22A	•	•	•	•	•	•	•	•
21A	•	•	•	•	•	•	•	•
17A	•	•	•	•	•	•	•	•
12A	•	•	•	•	•	•	•	•

Exhibit 7-1 (cont.)

INEFFICIENCIES OF ESTIMATES

	40 N	40 C	20 N	20 C	10 N	10 C	5 N	5 C
ADA	•	•	•	•	•	•	•	•
AMT	•	•	•	•	•	•	•	•
OLS	•	•	•	•	•	•	•	2
TOL	•	•	•	•	•	•	•	•
MEL	•	2	2	4	•	2	•	2
CML	2	•	2	•	2	•	2	•
LJS	•	•	•	2	•	3	•	2
SST	•	•	•	•	•	•	•	•
CST	•	•	•	•	•	•	•	•
33T	•	2	•	•	•	2	•	•
3T0	•	•	•	•	•	2	•	•
3T1	•	•	•	•	•	•	•	•
5T1	•	•	•	•	•	•	•	•
5T4	•	•	•	•	•	•	•	•
CTS	•	•	•	2	•	•	2	2
TAK	•	•	•	2	•	3	•	•
DFA	•	•	•	•	•	•	•	•
JOH	•	•	•	•	•	•	•	•
HGP	•	•	•	2	•	5	•	x
HGL	•	•	•	•	•	3	•	x
THP	2	2	•	2	•	3	•	2
THL	•	•	•	•	•	2	•	2
JWT	•	•	•	•	•	•	•	•
H/L	•	•	•	2	•	2	•	•
BH1	•	2	•	2	•	3	•	2
2RM	•	5	•	8	•	7	•	x
3RM	•	x	•	x	•	x	•	x
4RM	•	x	•	x	•	x	•	x
3R1	•	3	•	3	•	4	•	•
3R2	2	2	2	•	•	•	•	•
CPL	-	-	-	-	-	-	-	-
SHO	5	2	5	•	4	•	2	•

(i)   JLJ is calculated only for n = 20.

(ii) CPL is not scale invariant.

7B2 Tabular Values Can Be Misleading

In Chapter 5, we tabulated properties of various estimators. These summarize the results of the blind and
repeated use of these estimators under various sampling
situations. Some estimators have very high efficiency at
one particular distribution, say, for example, the Normal,
and very low efficiency at some other distributions. The
high efficiencies are usually not relevant for any real
problem. External evidence to support any distributional
assumptions is weak at best. With little external evidence to support the hypothesis of one distribution, an
estimate efficient for that distribution but inefficient
for another would not be used when the data are much more
plausible under the latter distribution. An alternative
estimate would be used. The result of this procedure is
to reduce the working efficiency of the "efficient"
estimates.
Consider the following idealized situation in which
the location of a sample of n numbers is to be estimated. It is hoped that the data come from a Gaussian
distribution with unit variance and an estimate of location T_1 is required which is relatively efficient for
this distribution. In addition if the data appear very
non-Gaussian another estimate T_2 will be adopted which
is not terribly inefficient for long-tailed distributions.
We believe that this is a simplification of the procedures actually used in the analysis of data but is in
much the same spirit as the development of estimates
which are efficient in a neighborhood of the Gaussian
distribution.
Let T_1 be the sample mean and T_2 the median.
(These are chosen only for purposes of exposition.) The
estimate T_2 will be used only if there is evidence that
the mean is being markedly affected by very long tails.
Thus T_2 will be used rarely -- only if $d = (T_2 - T_1)$
is very large -- say more than 2 standard deviations
away from its mean. T_2 is used only about 5% of the
time. For Gaussian samples of size 20, we wish to calculate the variance of this estimation procedure T.
Without loss of generality assume the true mean is
0. Then

$$Var(T) = \int_{d<2\sigma} T_1^2 f(x) \; dx + \int_{d>2\sigma} T_2^2 f(x) \; dx \; .$$

Writing $T_2 = T_1 + d$ and noting that d and T_1 are independent for the Gaussian distribution

$$\text{Var}(T) = \int_{|d|<2\sigma} T_1^2 f(x)\, \delta x + \int_{T_1}\int_{|d|>2\sigma} [(T_1+d)^2 f_d(d)\delta d] f_{T_1} \delta T_1.$$

From the symmetry of the distribution of d

$$n \cdot \text{var}(T) = n \cdot \text{var}(T_1) + n\int_{T_1}\int_{|d|>2\sigma} d^2 f_{|d|}(d)\delta d\, f_{T_1}\delta T_1$$

$$> n \cdot \text{var}(T_1) + (2\sigma)^2 \cdot n \cdot P(d > 2\sigma)$$

$$= 1 + n \cdot (2\sigma)^2 \cdot .05 .$$

Now

$$\text{var}(d) = \text{var}(T_2 - T_1) = (1 + \pi/2 - 2)/n = .5/n$$

and hence

$$n\, \text{var}(T) > 1.05 .$$

Thus one conservative step has added 5% to the variance of an otherwise efficient estimate. For smaller sample sizes it may be more reasonable to use the median 10% of the time in which case 11% is added to the variance of the mean. In applications many more alternatives are considered and even greater increases in variances may result!

As a consequence, estimates with efficiencies of 90% or more for the Normal distribution are likely to do as well in analyzing Gaussian data as does the mean. A 10% "insurance premium" may cost nothing at all.

7B3 The Need for Alternatives to Efficiencies

Much of the analysis in this and in other studies rests heavily on the efficiency of estimates. This measure is useful when comparing estimates under constant conditions including sample size and distribution shape. However the implications of, say, 70% efficiency depend heavily on both sample size and distribution shape. For example, the following table gives the increase in variance resulting from use of only a 70% efficient estimate

where efficiency is measured relative to the minimum
variance estimate, for four sample sizes and two distributions.

	Normal	Cauchy rescaled (with interquartile range the same as for the normal distribution)
n = 5	.086	.24
10	.043	.060
20	.022	.022
40	.011	.010

The variation in the above table suggests that constant efficiency means different things in different
situations. Alternative measures are required for comparisons across situations.

7B4 An Alternative Measure

In an application a quantity is to be estimated so
that the error in estimation is relatively small. In
certain restricted instances where a loss function can
be prescribed, one may choose the estimate to minimize
in some sense the expected loss. In other cases we may
wish to minimize the mean square error or the average
length of a confidence interval. In all these cases,
the measure to be minimized is *independent of the sample
size* -- we are interested in the number of correct digits
in the answer, not this number decreased by $\frac{1}{2}$ log n.

The variance of an estimate is a possible candidate
for our analysis. However, it may be unduly inflated by
events with very small probabilities. Since these events
do not matter in practice, undue inflation is misleading.

A quantile not too far out in the tail of the distribution of the estimate is not affected by these improbable
events. We will use the 97 1/2% quantile partly because
of the widespread acceptance of 95% confidence intervals.

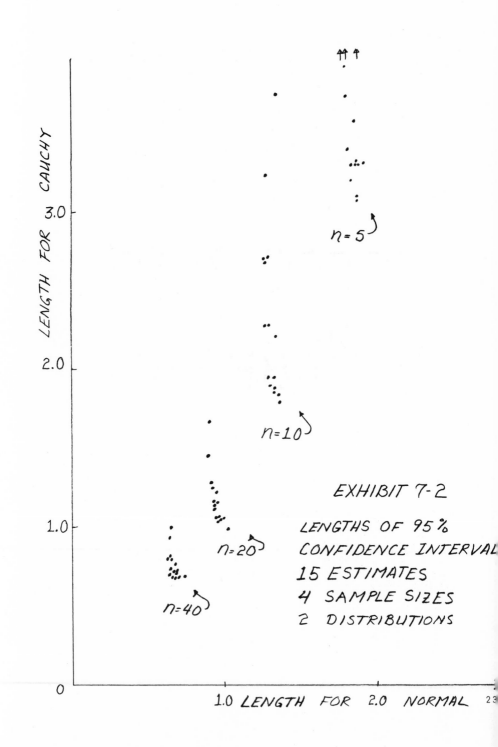

EXHIBIT 7-2

LENGTHS OF 95%
CONFIDENCE INTERVAL
15 ESTIMATES
4 SAMPLE SIZES
2 DISTRIBUTIONS

LENGTH FOR CAUCHY

3.0

2.0

1.0

0

$n=5$

$n=10$

$n=20$

$n=40$

1.0 LENGTH FOR 2.0 NORMAL

23

7B5 The Importance of Small Samples

The effect of this measure is a fact well known in practice but often lost in theory: estimates are asymptotically equivalent. In any problem there is a fixed practical precision beyond which no useful information can be gained. (In estimating average annual income of a population precision beyond the nearest dollar is useless and to estimate this quantity in cents or mills is senseless.) For large enough sample sizes, almost all the estimates will be equivalent to any fixed precision. Thus practical differences, if they exist, will be exhibited by small sample size.

7B6 The Importance of Non-normality

Distributions cannot reliably be distinguished on the basis of small samples. We must therefore consider carefully the properties of estimates under a wide variety of situations.

Exhibit 7-2 shows the length of 95% confidence intervals of a subset of the estimates for two distributions (normal horizontally and Cauchy vertically) and four sample sizes.

From this exhibit we note:

(i) The differences between the estimates are most marked for sample sizes 5 and 10.

(ii) The differences between the estimates are greater in the Cauchy case.

(iii) Corollary: for large sample sizes and particularly for the Gaussian distribution, the estimates are, for all practical purposes, equivalent.

7B7 Asymptotic Feasibility

It is difficult to imagine instances in which it is useful to consider a large body of data as a random sample. It is always useful to partition such data to investigate lack of stationarity or for dependence on exogenous variables. Nevertheless it is interesting to speculate on the effect of various estimators in such large sample situations.

The resources available for analysis increase with n -- sometimes -- but usually with only a power of n less than 1. Some of the estimates may be computed using only a partial sort (see Chambers 1971) of the data, a procedure involving only the order of n operations. The trimmed means, GAS, TRI, 50% (median) are the only estimates which formally require less than order $n \log n$ operations. For very large sample sizes $(>> 10^3)$ these are the only feasible estimates barring modification.

* Modification for very large sample sizes *

For very large samples the following may be used.
(i)   Perform a partial sort (order n operations) to get quartiles, eighths, perhaps a few more.
(ii)   Use these to define grid points and group data to these locations (order n operations).
(iii)   Evaluate the estimate on this grouped data.

7B8   Some Useful Estimates

The following estimates have high breakdown bounds and yield relatively short confidence sets over all distributions

GAS	AMT*	DFA*
H07	TOL	JWT
12A-25A*		5T4
ADA*	CST	

An * denotes breakdown bound as close to 50% as possible over all sample sizes.
These estimates have many different properties and choice is difficult. This author will <u>tend</u> to use the sine M-estimate AMT. The estimate had minimum or near minimum variance in several situations. It was never markedly inefficient. It is an M-estimate with readily computable asymptotic properties. The one-step version may be calculated by hand using tables of trigonometric functions.

7C SOME QUESTIONS AND ANSWERS (FOR HURRIED READERS)

7C1 Organization and Scope

1. What is the Princeton Monte Carlo Project?

The Princeton Monte Carlo study of robust estimators of location is an extensive study of the finite-sample behavior of about 70 estimators, many of them new, at about 2 dozen distributions in the range from the Normal to the Cauchy distribution. The sample sizes were 5, 10, 20 and 40, with main emphasis on 20. The main output were rather accurate variances and percentage points of the distributions of the estimators and correlations between them. Side studies (partly still incomplete) include the effects of asymmetric contamination, finite-sample "influence curves" and "breakdown points" of the estimators, the (lack of) normality of their distributions, formulas for asymptotic variances and, in a different field, various Monte Carlo techniques.

2. What was the main purpose of the study?

This depends on one's personal viewpoint. A partial answer may be: to gather and interpret a large number of rather accurate finite-sample results, to extend the scope of previous Monte Carlo studies in this area and to serve as a stepping-stone for future studies of robust regression, and for this to utilize the abilities and great diversity of interests of the residents and visitors that had come together for the year 1970/71 at Princeton.

3. Why was this another study of robust location estimators and not of, say, robust regression?

Critics from outside and also from inside have maintained that a study of robust regression would be much more important at present, both because of practical needs and because of the greater lack of knowledge in this area. There are several answers. If attempted, such a program probably would have surpassed the limits of time and energy available by far. On the other hand, the results of the location estimator study showed a posteriori that we still had (and have) to learn basic things there, things which are also fundamental to any robust regression study but which are easier to obtain and interpret in the location case. Furthermore, the empirical facts gathered seem to be of ample interest also for other reasons, e.g., as a potential basis for the development of a better finite-sample distribution theory.

4. Why were there no Bayesian estimators included?

    For this we have several answers, each of them suffi-
cient. Some noted Bayesians have been asked for their
proposals, and contrary to expectation they suggested not
Bayesian, but other (maximum likelihood type) estimators
to be included in the study. The program has been
restricted to location and scale invariant estimators for
important technical (Monte Carlo) reasons. Finally, contrary to what some Bayesians seem to believe, the main
features and problems (with regard to robustness, etc.) of
non-Bayesian estimators still persist almost unabated
when these estimators are mixed up with an _apriori_ distribution. The   new robustness problems added are
those connected with the choice of the _apriori_ distribution.

7C2 Results

5. Which was the best estimator?

    This question, though asked frequently, does not and
will never admit a unique general answer, since the great
variety of situations in applied statistics and real life
will always demand a variety of tools. We can say, however, which classes or groups of estimators turned out to
be most successful or promising under the aspects of the
study. These were, in the first place, the three-part
descending M-estimators, like 17A or 21A, including the
closely related sine-type M-estimator AMT. In some sense,
they form an almost admissible class of rather "stable"
estimators within which one can trade off better behavior
near (and at) the Normal distribution against worse
behavior for longtailed distributions and vice versa.
Beyond these estimates apparently one cannot improve very
much over a wide range of contamination with non-adaptive
estimates. Another very good group are three estimates
5T4, 5Tl and 3Tl which combine skipping (a rejection
procedure) with moderately adaptive trimming and which are
also suitable for hand-calculations. A third successful
group are adaptive estimators like JOH which try to
estimate more characteristics of the actual underlying
distribution. They lose for intermediate contamination,
but gain at the extremes (near-Normal resp. long-tailed
distributions), reaching a more uniform overall-behavior.
That such gains are possible for sample size 20 is
remarkable and indicates the possibility of further
improvement in that direction.

6. Why are the three-part descending M-estimators so good?

They fulfill the basic intuitive requirements for robust estimators: to restrict the influence of any fixed fraction (< 1/2) of wrong observations on the value of the estimator, to reject outliers which are too far away from the bulk of the data (leaving them for separate treatment), to be little affected by rounding, grouping and other local inaccuracies (in particular, to reject "smoothly"), and under these side conditions to be as good as possible under the "ideal" parametric distribution (here: the Normal distribution). For the scaling necessary in any proper rejection procedure, the "most robust" scale estimator is used, which is the counterpart to the median in the location case. (This choice, by the way, (with some more side points) also ensures consistency and uniqueness which in general can cause problems for estimators with non-monotone influence curves such as estimators coupled with rejection procedures or based on non-monotone likelihood ratios.) The three-part M-estimators are based upon and are a step or two beyond the now classical Huber-estimators. They differ from these by using the most robust scale estimator. This change alone, as the study showed, yields on the whole a clear improvement over the specific "proposal 2" with large k. They also differ by using a defining ψ-function which, instead of staying constant, slides to zero in the tails. The estimators are defined by one implicit equation (to be solved iteratively in one or a few steps), with one more formula (for the scale estimator) and a trivial side condition (to select the root closest to the median).

7. Which was the worst estimator in the study?

If there is any clear candidate for such an overall statement, it is the arithmetic mean, long celebrated because of its many "optimality properties" and its revered use in applications. There is no contradiction: the optimality questions of mathematical theory are important for that theory and very useful, as orientation points, for applicable procedures. If taken as anything more than that, they are completely irrelevant and misleading for the broad requirements of practice. Good applied statisticians will either look at the data and set aside (reject) any clear outliers before using the "mean" (which, as the study shows, will prevent the worst), or they will switch to taking the median if the distribution looks heavy-tailed. As Hogg's proposals show, the latter may be rather satisfactory for moderate standards.

239

8. Which results appear to be of special interest for
   applied statisticians?

   The arithmetic mean, in its strict mathematical
sense, is "out". The mean combined with any reasonable
rejection procedure, however, can survive, though not
very well. "Throwing out" observations, for a long time
frowned upon, is thus justified. We now even have computer methods (the three-part descending M-estimators and
related estimators) which do this rejection in a smooth,
objective, conceptually simple and very efficient way.
(For practical applications, the programs should be
supplemented by having them tell the values of all "outliers" and "partly rejected" observations.) We also have
simple yet very efficient methods for hand-calculations,
like 5T4 and CST. Adaptive estimators, which in a sense
(asymptotically) aim at estimating the full underlying
distribution in order to get nearly fully efficient estimates of location, do work well already for n = 20,
and quite likely they will be further improved.
However, there is, apart from the amount of computation,
some question about what we want to estimate, and what
the estimator actually does in the presence of a bunch of
outliers.
   Of the "classical" three classes of estimators:
linear functions of the order statistics, M-estimators
in the sense of Huber and estimators derived from
rank tests, the second class does best, since it is the
only one which permits genuine rejection of outliers
(i.e., rejection based on the distance from the bulk of
the data). But trimmed means which don't trim too little
on either end, e.g., the 25%-trimmed mean ("midmean"),
are also fairly safe and still quite good (the efficiency
loss for n = 20, compared with the best available estimator out of 65, is "typically" between 0 and 25%). The
median, of course, is still a reliable stand-by, being
simple and even "most robust" both with respect to
"breakdown point" and to "gross-error-sensitivity".
   Another interesting feature of the study is the computation of a number of finite-sample "breakdown points"
("how many extreme outliers can an estimator tolerate
until it breaks down completely?"). This simple and useful concept tells us something about the "global" aspect
of robustness, about the boundary beyond which we may end
up with sheer nonsense. The "local" behavior of an estimator, the way in which it reacts to small changes of the
sample (by adding, deleting or slightly changing a few
observations) is often approximately described by the
"influence curve", in the same way in which a function
near one point can often roughly be described by its de-

rivative in this point. Many graphs of "finite-sample influence curves" have been computed and included in the study. The influence curve (a function on the sample space) can often give a good intuition about the properties of an estimator. It has already served in clarifying the relations between trimming (which is not a rejection procedure!), Winsorizing, Huberizing and rejecting. It has supplied a technology for constructing new estimators with certain prescribed properties, for robustifying estimators, for computing asymptotic variances, etc. The first function of the computed graphs in the study, quite unexpectedly, was to weed out errors in the computer programs of some estimators.

9. Which results appear to be of special interest for mathematical statisticians?

The study contains finite-sample distributions of several peculiar estimators about whose theory little is known, like the Bickel-Hodges estimator which is not asymptotically normal, and the shorth (mean of the shortest half) whose rate of convergence is not even the usual one. Another challenging area is the study of adaptive (and, more generally, multi-parameter) estimators. A broad outstanding field is the development of a better finite-sample theory for which the study provides a great wealth of empirical facts which wait to be properly interpreted and described. - The study shows the success of a good, reality-oriented theory of robust estimation. It also contains a large number of formulas for asymptotic variances derived by the unifying tool of the "influence curve". Specialists for Monte Carlo methods may want to look at the methods discussed and used in the project.

7C3 Where Do We Stand?

10. Where do we stand now with regard to robust estimation of a location parameter?

It seems to this writer that we now have a fairly comprehensive framework of a useful (though still partly incomplete) theory for robust estimation of a location parameter, and a good intuitive understanding of most non-adaptive estimators. This is reflected in the study by the several groups of new robust estimators which have set new standards of quality. We still have to learn much more about adaptive estimators; the few in the study can only be viewed as pioneer examples. Further progress is to be expected. Regarding the effect of sample size, the study seems to indicate that for reasonable non-adaptive estimators asymptotic results are a good first approximation even for small sample sizes. However, for finer effects even the four sample sizes 5, 10, 20 and 40 may be barely enough to give a complete picture and perhaps may have to be supplemented by more sample sizes, especially between 3 and 20. There remain theoretical questions regarding the proper choice of different estimators for different sample sizes (e.g., of whether and how to choose more robust estimators for larger sample sizes).

Perhaps the most immediate problem now is the exploration (theoretical and empirical) of Studentizing procedures. This should then satisfy the most important practical needs with regard to a location parameter.

7C4 An Incomplete Analysis of the "Bad Cases", Including "Catastrophes"

Perhaps even more important than to say which estimators should be used, is to say which shouldn't. An important reason for not using an estimator is that it may behave catastrophically under some circumstances. Opinions may differ about what "catastrophically" means, and accordingly we shall distinguish several classes of "bad cases". We take as a basis of comparison the variance of the best available estimator (out of 65). An estimator with twice that variance already looks quite bad, but the length of the corresponding confidence interval is only about 1.4 times the shortest one, and hence the estimator may still be considered tolerable especially if the distribution at which this occurs is judged to be rather unlikely. An estimator with four times the minimal available variance (twice the length of the confidence interval) or more may be considered as "seriously bad", 10 times the variance may be regarded as "catastrophic" and 100 times the minimal variance as "completely disastrous". Between 2 and 4 times the minimal variance will be called "mildly bad".

The following discussion is based on 65 estimators and on 17 underlying distributions (resp. contamponents), 11 of which are for sample size 20. These distributions are: 18 N(0,1) & 2 N(0,9), 18 N(0,1) & 2 N(0,100), .9 N(0,1) + .1 N/U(0,1/3) (20 obs.), 15 N(0,1) & 5 N(0,9), Student's t with 3 d.f. (20 obs.), N/U (20 obs.), Cauchy (20 obs.), 5 N(0,1) & 15 N(0,9), 10 N(0,1) & 10 N(0,9), Cauchy (5 obs.), Cauchy (10), Cauchy (40), Normal (40), 9 N(0,1) & 1 N(0,9), .75 N(0,1) + .25 N/U (5 obs.).

Of course, part of the reasonableness of this discussion depends on the underlying distributions: if one would include the rectangular distribution
and the midrange, this would be very unfair against the
other estimators. Nevertheless, with the distributions
at hand, the summary result is the following: out of 65
(mostly) "robust" estimators, 42 are "bad", and 25 of
those are "seriously bad". There are 11 "complete
disasters", and 4 more "catastrophes". Only 23 estimators survive as never bad for the distributions considered,
some of them in dull mediocrity: they are never very good
either. Among the "bad" estimators are to be found LJS,
TAK, JAE, H/L, OLS, all Hogg's, and all but one Huber's.
The situation looks somewhat better if one regards only
sample size n = 20, as many "mildly bad" cases are found
only under Cauchy 10 (but who knows what happens at
other sample sizes?). For n = 20, we still have 23 "bad"
estimators, 11 of them "seriously bad". Among these are
still 4 "complete disasters" and 3 more "catastrophes".

The "complete disasters" for n = 20 are M (in 4
cases; in 7 out of 11 cases bad), M15 (in 4 cases; bad in
6 cases), 4RM (in 4 cases; bad in 6 cases) and SJA (in 2
cases; only in these 2 cases bad). The further catastrophes are 3RM (4 cases; 6 cases bad), 5% (2 cases; 6
cases bad) and 2RM (3 cases; 5 cases bad). Seriously bad
are furthermore H20, 3R1, 10%, and MEL. The "mildly bad"
estimators are SHO, BIC, HGP, H17, BH, H15, D20, A20,
15%, CTS, D15, and LJS.

The mean M deserves some special remarks. As everybody familiar with robust estimation knows, the mean is a
horrible estimator, except under the strict normal distribution; it gets rapidly worse even under very mild
deviations from normality, and at some distance from the
Gaussian distribution, it is totally disastrous. However,
as everybody familiar with practical statistics (not to
mention a number of theories) knows, the mean is the
estimator of location, to be applied under almost all
circumstances (especially in more complex models); and
it is even used by some very good statisticians, or so it
seems at first glance.

There are some things to be clarified. Our analysis
of bad cases shows that the mean is not even mildly bad
if the contamination is not too far away, even if it is
as heavy as in 15 N(0,1) & 5 N(0,9). It is true that
already for 18 N(0,1) & 2 N(0,9) the loss of efficiency
is about 30%, but even a loss of 50% or more can easily
be compensated by other features of a statistical analysis, like choice of a good model, discovery of unsus-
pected effects and structures, etc. On the other hand,

243

in the presence of distant contamination, the mean is disastrous (and, as a side remark, the case for the sample variance would be even much worse).

A good statistician (or research worker) never uses the mean, even though he may honestly claim to do so. In reality, he will look at the data and set aside outliers and suspected outliers in order to treat them separately; then he will take the mean of the rest. But this amounts to using the mean with a (possibly very vague and subjective) rejection procedure, not "the mean" in its strict mathematical sense. And since such a rejection procedure will get rid of distant contamination, any such method will work in that it prevents the worst. (And from a practical viewpoint, it is much more important to prevent the worst in every conceivable aspect, including the ones which are not covered by any statistical theory, than to optimize in one or a few directions, which deeper examination might have shown to be totally irrelevant and misleading.)

Of course, we can do better now (with regard to controlling contamination, not interpreting it), even with objective routine methods; and this seems the more important as more and more data are processed on the computer without being looked at thoroughly. But it is to be remembered that good routine methods (even methods for making the data more transparent) are an aid, but in no way a replacement for detailed and flexible thinking about the results.

Returning to the other bad cases, we see that a good l-step (M15) starting from a bad estimate (M) is not much better than this estimate. Furthermore, refolded medians (including 3R1, but not 3R2) are also bad for longtailed contamination, as to be expected; the more foldings, the worse (the breakdown point goes down geometrically with the number of foldings). The erratic mean likelihood is seriously bad for two longtails. Light trims and Hubers with large k are also bad at longtails, the direction of increasing resistance being typified by 5%, H20, D20, A20. SJA breaks down completely in the two cases with N/U-tails and a near-normal center, but not otherwise. BIC is only mildly bad in three cases (the same two and another one). The shorth is uniformly "mildly bad" (except that for longtails it may even fall below the double-variance line). HGP and LJS are not quite sufficiently adaptive, and also CTS fails once.

If we now include the other sample sizes, we find more disasters mainly at Cauchy 5 (in much milder form at 25% l/U)and more bad cases (often only mildly bad) mainly at Cauchy 10. In addition to those mentioned above, 3RM,

2RM, 5%, 10%, H20, HGP, and HGL are completely disastrous.
The failure of both Hogg's for smaller sample sizes is
especially noteworthy. Catastrophic are now BIC, 15%,
CTS, and - another surprize - OLS (all for n = 5); for
larger n OLS looks like a strongly robust estimator with
a broad resistance spectrum. Among the seriously bad
cases we find also SHO, H17, D20, D15 (but neither H15
nor A20), BH, and of the more adaptive procedures LJS,
THP, and TAK. THL, JAE and JBT (among others) are only
mildly bad.

The list of remaining estimators (those which are
never bad at the distributions considered) is instructive.
They are either of the heavily trimmed type (25%, 50%,
H07, GAS) or specifically for longtailed distributions
(CML, CPL, TOL), or adaptive with the stress on longtails
(JWT, DFA, JOH, JLJ), or they explicitly include a rejection procedure (SST, CST, 5T1, 5T4, HMD, AMT, 25A, 22A,
21A, 17A, 12A, ADA). (Of these, only 25A, AMT, ADA, 21A,
and 22A have less than 10% excess variance at the Normal
40; in addition, 25%, H07, JWT, JOH, CST, 5T1, 5T4, HMD,
17A, and 12A have less than 20% excess variance.) This
shows clearly the importance of having a (reasonable)
rejection procedure (or at least an influence curve close
to zero in the extreme tails) combined with any (also
otherwise reasonable) estimator.

7C5 Some Crude Ways of Screening Out Good Estimators

We are still working with the following 11 distributions (resp. contamponents) for sample size 20: 18 &
2 N(0,9), 18 & 2 N(0,100), .9 + .1 N/U(0,1/3), 15 &
5 N(0,9), t with 3 d.f., 15 & 5 N(0,100), .75 + .25 N/U
(0,1/3), N/U, Cauchy, 5 & 15 N(0,9), 10 & 10 N(0,9).
Of these, the last two can be ignored, as they are neither
very realistic nor very difficult. The others are roughly
ordered according to longtailedness. We may ask for those
estimators which for some range of distributions are never
much worse than the best available estimator (say, by not
more than 10% excess variance). The estimates may be
ordered according to the number of situations for which
this condition is satisfied. This ordering selects estimator 17A; then 12A; then CPL, 22A, ADA, 5T4; then JOH,
AMT, 25A, 21A, JLJ; and finally 3T1 and 5T1. Furthermore,
OLS, TOL and CST only barely missed being on the list.

Against this procedure one may object that the Normal
20 is missing (so that estimators which are not very good
at the Normal can appear) and that the standardization by
the best estimator that happens to be around is somewhat

245

too haphazard (e.g., the standard for the Cauchy, with
CML and CPL as competitors, would be too strict, even
though this has been partly balanced by putting a higher
bound on the excess variance). The first objection can
be partly met by including Normal 40 (as long as Normal 20
is not at hand)(only for JLJ the value for Normal 20 has
been used); the second objection can be partly met by
taking, e.g., the 10-th best estimator as standard
(because then, but not yet with the 5-th best, adding or
taking out a few estimators makes hardly any difference).
Furthermore, we may introduce loss units (negative scores)
as follows: no loss, if the estimator is among the 10
best ones, 1 unit if its variance is less than 10% higher
than that of the 10-th best, and 4 units if it is less
than 20% higher. The other estimators drop out. Since t
with 3 d.f. is almost identical with 15 & 5 N(0,9), we may
drop it here. The distributions are grouped in Normal 40,
18 & 2 N(0,9), 18 & 2 N(0,100), .9 + .1 N/U(0,1/3); 15 &
5 N(0,9) added; 15 & 5 N(0,100) and .75 + .25 N/U(0,1/3)
added; N/U (20) and Cauchy (20) added. Finally, a special
group may include all but the worst run, giving a chance
also to those estimators which only once didn't make it
at all.

The first two groups (distributions near the Normal)
include 34 estimators which never go beyond 1.2 times the
variance of the 10-th best one. Including heavy contamination in the tails makes the number shrink to 17, and
only 14 survive everything up to the Cauchy. These 14
estimators can be ordered in different ways. If one uses
basically the overall loss units and looks at the behavior
near the Normal only to decide between ties and close
groups, the order is 17A (4 loss units), ADA (4), 5T4 (6),
12A (6), 5T1 (8), JOH (8), 21A (12), 22A (12), 3T1 (13),
CST (11), JLJ (12), JBT (21), H07 (19) and 25% (19). The
three estimators that make it up to 25% contamination
follow as AMT, 25A, and THL. - If we add up the loss
units in each of the four groups, thus giving a higher
weight to near-normal distributions, the sequence (with
the number of units in parentheses) becomes: 17A (10),
ADA (12), 5T4 (17), 21A (19), 5T1 and 22A (21), 3T1 (23),
12A (24), JOH (31), CST (35), JLJ (44), JBT (52), H07 and
25% (57). If we add up the loss units only up to the
third group (for the 17 surviving estimators), ignoring
the behavior at N/U and Cauchy, the order is 17A (6), 21A
and AMT (7), ADA (8), 22A (9), 3T1 (10), 5T4 (11), 5T1
and 25A (13), 12A (18), THL (22), JOH (23), CST (24),
JBT (31), JLJ (32), H07 and 25% (38).

Thus we find roughly the following structure: a
group of 9 best estimators, followed by CST and then JLJ
and, in some distance, JBT, H07, and 25%, with up to 3

more estimators which may be included near the Normal.
Rather uniquely (and perhaps a bit surprisingly) 17A
comes out best, followed roughly by ADA and (remarkably)
5T4, then perhaps 5T1. This group is joined for overall-
goodness by 12A and then JOH, for near-Normal good
behavior by 21A and then 22A and 3T1, and in addition, if
the condition on the 2 longtails is omitted, by AMT (near
the top) and by 25A (near the bottom of this group). THL
would follow this last group in some distance, while CST
follows the other ("strongly robust") group fairly closely
behind JOH.

Of these 17 estimators, only 3 have been mildly "bad
cases" for other sample sizes: 3T1 and JBT only barely
once at Cauchy 10, THL somewhat worse at the same distribution and in addition twice for sample size 5. The
others (disregarding JLJ, which is properly defined only
for n = 20) can be judged fairly safe for other reasonable
distributions not included in the study also.

This leaves us with the following choices from the
estimators in the study: for hand-calculations, depending
on accuracy and simplicity required, 5T4, CST or 25% (for
near-Normal data perhaps also 3T1), for the computer and
general purposes 17A, for near-Normal data 21A or AMT,
for·very broad resistance 12A; for larger sample sizes
perhaps also ADA or, for heavier tails, JOH.

We now have to look back at the estimators that
survived only the first two groups of distributions (near-
Normal), and at those estimators that survived all but
one distribution to see whether they could alter our
interpretation materially. We see that the top (10 resp.
9) places in the near-Normal distribution groups are
occupied by distributions which are among the 17 overall-
best ones. Only then follow TAK, THP, 15%, 3R2, H12, H10,
D10, JAE, and (after a small gap) the others. There may,
however, be some merit in looking for estimators which
are good for very nearly Normal distributions (and still
bearable for heavier contamination). First, we realize
that none of the 10 best estimators at Normal 40, except
possibly LJS, seems to have any fighting chance. Next,
we find that with our coarse screening perhaps 15%, H15,
A15, P15, 3R2 and H12 may merit some consideration,
besides 25A, AMT and 21A. Of these, 15% may have to be
excluded because of its bad behavior elsewhere and its
fairly low breakdown point. The others would have to be
subject to a more detailed analysis. We note that all 17
estimators which survive only in the first two groups show
bad (even catastrophic) behavior somewhere else, though
mostly only at Cauchy 10.

Finally, we may look at the loss units if the worst distribution in each case (for which the estimator may not even fall below the bound on the variance) is omitted. Only 7 new estimators are added, plus one (D07) that appeared in the near-Normal groups. The 3 first places are still occupied by 12A, 17A and ADA, and only two new estimators, namely CPL (slightly improper, because not scale-invariant) and HMD, manage to get into the top group of 14 estimators which otherwise includes the 9 best of the overall analysis, CST, AMT and JLJ. The other new ones (TOL, OLS, 30%, GAS, finally JWT) follow in some distance. Most of them suffer too much at the Normal, while they are good at longtails. But since we have already other good estimators for this situation, there is no need for them to be screened out as good competitors unless special emphasis is put on special longtails.

7C6 Notes on the Standardized Variances, Sample Size 20

The following lines give a description and interpretation of graphs which show the variances (always divided by the variance of the best available estimator) against the underlying distributions (roughly ordered according to tail length), for each estimator. The graphs themselves are not included. However, they can be recovered from the tables.

The distributions were ordered, as well as possible, on the horizontal axis and spaced in a semi-intuitive and semi-empirical way. The distributions included (in parentheses the relative position) are Normal 20 (0), 18 N(0,1) & 2 N(0,9) (3), 18 N(0,1) & 2 N(0,100) (3.5), .9 N(0,1) + .1 N/U(0,1/3) (4), 15 N(0,1) & 5 N(0,9) (6), t with 3 d.f. (6.5), 15 N(0,1) & 5 N(0,100) (8), .75 N(0,1) + .25 N/U(0,1/3) (8.5), 5 N(0,1) & 15 N(0,9) (0), 10 N(0,1) & 10 N(0,9) (9.5), N/U 20 (10), Cauchy 20 (11.5), = all for sample size 20. Other distributions were not available at the time of writing. Different types of distributions were marked using different symbols. The vertical axis described the standardized variance (variance divided by variance of best available estimator, for each distribution). Other scales (e.g. standardization by best trimmed mean, for each distribution) would have been possibl and would have yielded qualitatively different pictures in general. The choice described here yields graphs where most symbols lie roughly on either one or on two U-shaped (resp. J-shaped) curves (with the exception of some bad estimators). Position and height of the minimum of the (lower) U-shaped curve, its flatness, and whether there are one or two branches (one for "close", one for "distant"(N(0,100), N/U(0,1/3)) contamination); these "dimensions" describe fairly well the behavior of most estimators. Their meaning, and their relation to the "influence curves" of the estimators, are now analyzed briefly.

The estimators can essentially be described by two dimensions. The first one can be characterized by the sort of distribution at which the estimator does best: short-tailed or long-tailed. There is a continuous range, e.g. from the mean to the maximum likelihood estimator for the Cauchy distribution, and in some cases a broad part is covered by a single class of estimators (e.g., trimmed means, Huber-estimators). All estimators though sometimes only crudely, find their place in this spectrum. This place can be described, e.g. by the distributions at which they do best, or by the trimmed means or the Huber-estimators which correspond most closely to them near the optimum or in overall-shape of their graphs.

The second dimension, which has almost the form of a qualitative dichotomy for the estimators in the study, is determined by the treatment of rather distant con- tamination: essentially, whether it is rejected or not. If not, the graphs split into two curves: one for close, and one for distant contamination. If outliers are rejected, the representation chosen yields roughly a single curve. In general (excluding cases like M which perhaps should be put into a separate group), not rejecting means some sort of trimming (keeping down the influence of outliers, but not bringing it to zero). This class "t" comprises about 38 estimators, about 26 of them clearly. The class "r" of rejecting estimators consists of about 20 estimators (13 of them clearly, 7 roughly). 7 estimators remain which don't quite fit into either group (see below).

The class "t" contains all trims, Hubers with modi- fications, H/L, most three-parameter estimators (unskipped Hogg's, Jaeckel's, LJS), GAS and TRI as "abbreviated trims", the refolded medians, and (to some extent) OLS (even less so CPL). If one plots the increased loss in efficiency at, say, 15 & 5 N(0,100) against the variance at the Normal, most estimators of "t" lie roughly on a

curve, showing the dichotomic character of this dimension
for the estimators under study. (More precisely, the
curve splits into several very similar curves in the
familiar way, for high efficiency at the Normal described
in the direction of increasing loss by distant outliers
by: A15, D15, H15, and, in some more distance, 10%. For
lower efficiencies at the Normal, crossovers can occur.)
This band, determined by trims and Hubers, also contains
H/L and 3R2, for example. TRI and GAS lie slightly
above. Three-parameter-estimators, in general, lie below
the curve (LJS, HGL, JAE, JBT), as it should be, since
they adapt better. However, SJA(?), BIC and HGP show high
losses. Finally, OLS (in the neighborhood of heavily
trimmed means) shows only about half the loss of these
trims and thus is standing between the groups.

The class "r" contains the three-part descending
M-estimators, most skipped estimators and the adaptive
estimators (TAK, JOH, JLJ, DFA) as well as ADA, CML,
SHO and (pretty much) TOL. That the adaptive estimators
belong to this group (though they all have about one
exceptional point) is a revealing fact. The word which
once fell in a very early stage of the project (when the
first results came in) that TAK seemed to be "the fastest
moving estimator", can now be seen under new light; in a
way, it is the trims and their allies which move fast
by reacting unnecessarily strongly to distant outliers.
(There are, however, exceptions: JOH and JLJ react too
much to 2 N(0,100), and TAK reacts too much to 5 N(0,100).
We leave these details to more refined studies.) Of the
7 estimators not quite fitting into either group, 3
behave like "super r", namely 33T and (much less) THP and
THL. Especially 33T is even relatively better for
distant than close contamination. Two are like "r" (HMD,
CTS), but react somewhat too much to mixtures with
N/U(0,1/3). Finally, MEL and JWT are not readily classified, but are perhaps both somewhat close to "t". Thus,
all estimators (with the main exception of 33T as clear
"super r") can more or less be put into one of the two
groups.

It is remarkable that these two dimensions can in most
cases be described by just 2 points of the influence curve.
Unfortunately, the finite-sample influence curves are not
at hand to check this statement in more detail, but in
general the following can be said: the first dimension
corresponds to the maximum of the influence curve, the
"gross-error-sensitivity", and the second dimension

corresponds to the height of the IC in some point farther
out in the tail -- or else to something similar to the
"rejection point" (for which a satisfactory general
definition may still have to be found, since estimators
with very low IC in the tails, like CML, behave almost
as if they had a low rejection point - the point at which
the IC becomes exactly zero). It fits into this picture
that the loss by distant outliers for most "t"-estimators
lies on a curve; for them the IC is (practically) constant
in the tail and equal to the gross-error-sensitivity. Good
three-parameter estimators are able to decrease the IC
somewhat. OLS has a fairly slowly decreasing IC, which
results in its intermediate position. The "r"-group contains
mainly estimators with fast decreasing IC or with IC
jumping to zero.

A third dimension, which in a few cases is about as
important as the first two, is the shift of the whole
curve upwards. This shift is very large for SHO, large
for 33T and DFA, still fairly large for 50%, SST and JWT.
(In other words, these estimators are not very good for
sample size 20.) Next come CTS, THP, THL, JLJ and finally
JOH; even less shift show OLS, TOL, CST, JBT, HGP, HGL,
TAK, JAE, BIC, SJA, ADA and 5Tl. Traces of a shift may be
found in GAS, TRI, 5T4, perhaps some one-steps, and LJS.
We see that, ignoring a few bad or otherwise not quite
efficient estimators, this shift is a typical feature of
adaptive and three-parameter estimators, and more pronounced
for those which are better for longtails. They lose somewhere in the middle in order to gain at the ends.

This gain at the extremes constitutes a fourth dimension, which for more adaptive estimators is closely
related with the loss in the middle and the resulting
shift; however, there are also differences, e.g. between
Huber-estimators and 3-part M-estimators, the latter
gaining at the extreme ends. Another, though more
complicated, assignment of dimensions would be the "flat-
tening" with respect to some "center of gravity" and the
remaining shift of this center (rather than the minimum).
This latter shift would correspond to an inappropriate IC
to start with; the gain by more adaptive estimators could
be described by the change of the IC (e.g. in the two
points mentioned, or just mainly the gross-error-sensitivity)
from short-tailed to long-tailed distributions (containing,
e.g., the shift of the "corresponding trimmed mean").

There are still many finer effects left which deserve
study, but the four dimensions mentioned seem to give a
fairly good crude first description of the situation.

251

7D PRELIMINARY IMPRESSIONS AND TENTATIVE CONCLUSIONS

In a study of this size and complexity, striking and obvious features may get buried under a mass of interesting but bewildering details. As a partial remedy, I would like to point out a few simple observations which struck me most, in an arbitrary order.

7D1 For a normal parent distribution, most estimates approach their asymptotic behavior very quickly

The numerical results for the asymptotic variances were not available in time. But, a glance at exhibit 5-14 shows that for a normal parent distribution:

(i) most estimates are very nearly normal even for sample size n=5;

(ii) for nonadaptive estimates, the normalized variances change little from n=10 on (sometimes even from n=5 on) and seem to have stabilized near their asymptotic values.

As expected, the shorth (SHO) does not share this property, but, surprisingly, the Hodges estimate (BH) is almost perfectly normal.
For longer tailed parent distributions, the convergence is much slower. Still, asymptotics may give useful approximations for sample size 20 and not too extreme percentage points, as a few spot checks show.

7D2 One-step M-estimates behave almost like the iterative ones already for small samples

Only the pair A15, P15 allows an exact comparison. Both use the symmetrized interquartile range as their estimate of scale, but A15 is iterative and P15 is one-step, starting from the median.
P15 ranks almost consistently just behind A15 (with the N(0,100) contamponents it seems to be a shade better), but the difference in efficiency is of the order of 0.5 percent or less, and thus negligible in practice. This holds for all sample sizes and distributions in the study, including the asymmetric ones.

This agreement is much better than one could expect a priori. I conjecture that it is equally good also for other scale estimates and other ψ-functions, but this point remains to be checked.

The mean is an extremely poor starting point for a one-step estimate (M15).

7D3 Symmetrized scale is quite different from ordinary scale

For some M-estimates scale is estimated from a symmetrized version of the empirical distribution function.

The one-step estimates D15 and P15 differ by using the ordinary and the symmetrized interquartile range respectively to estimate scale. The difference in behavior is considerable: for the normal distribution, P15 is 0.5 to 4 percent less efficient, depending on the sample size, but for longer tailed distributions it is sometimes considerably more efficient than D15.

While for P15, A15 and A20 symmetrization seems to have a beneficial overall influence, it has quite undesirable side effects for SJA and BIC (compared to JAE). The explanation seems to be that symmetrizing leads to undertrimming. An asymmetric configuration of outliers (say 3 left, 1 right) obviously needs more symmetric trimming than a symmetric one with the same total number of outliers (2 left, 2 right). This feature is also expressed in the sensitivity curves (exhibit 5-15: curves for JAE and BIC $\sim$ SJA).

7D4 No estimate in this study achieves 95% efficiency at the normal distribution (relative to the mean) and 33% efficiency at the Cauchy (relative to the maximum likelihood estimate) simultaneously for all sample sizes 5, 10, 20, 40.

The nearest misses are H15 (32% at Cauchy, n=10) and A20 (93% at normal, n=5). This sobering result may induce you either to lower the requirements at the normal or to throw out the Cauchy as too unrealistic, or both.

7D5 Subjective preferences (for sample size 20)

The performance of about one-third of the estimates is dominated (or very nearly dominated) by that of others. Each of the following estimates is (for practical purposes) improved by the one in parentheses:

5%(H17), 50%(12A), GAS(H07), M15(A15), D20(A20), D15(A15), HGP(TAK), THP(22A), THL(22A), CTS(OLS), SST(TOL), 33T(H10), JWT(JOH), SHO(25%), DFA(12A), BH(A15), 2RM(H20), 3RM(H20), 4RM(H20), 3R1(H15), 5T1(5T4), CST(17A).

While I do not attach too much significance to this list (the selection of distributions may not be comprehensive enough; moreover the list was prepared from an intermediate and somewhat incomplete stage of the tabulation of the results), it gives some hints as to the selection of robust estimates. Of course, ease of computation may still override inadmissibility. Apart from these admissibility aspects, preferences are largely subjective by necessity, and depend on what kind of data one believes to have.

After gazing at the Monte Carlo results and weighing also conceptual ease and ease of computation, I would favor the following collection of estimates:

- the trimmed means 10%, 15%, 25% (but with the number of trimmed observations an rounded upward to the nearest integer);

- the one-step hubers with Hampel's scale: P12 to P20;

- the hampels, in particular 25A and 21A (the behavior of their one-step versions needs to be checked);

- as a moderately adaptive estimate: JBT.

7E   AN UNSYSTEMATIC OVERVIEW

A body of data as huge as the one produced by this study can obviously be considered under many aspects. What follows is a description of those features (or pseudo features) of the numbers that I found most interesting.

7E1   General qualitative behavior of most estimates

1.   It is well known that the information inequality lower bound is not sharp when a location parameter is estimated save for the normal (and another shape which we did not consider). Thus even in terms of normalized variance the efficient estimate has a harder time for finite n than for n = ∞. This appears to be true for most of the procedures considered. Whether performance is measured by normalized or percent point variances, almost invariably these quantities decrease monotonely as functions of n for both the 25% 1/U and Cauchy distributions.

2.   The stabilization of the variances and convergence of the distribution of the estimate to normality seem to be closely related to the efficiency of the procedure for the given underlying distribution and in some loose sense the heaviness of the "tails" of the underlying shape. Roughly speaking the better the estimate the more it is Gaussian and the less its variance changes as n increases in the range considered. Convergence of the variance is excellent for the Gaussian, mediocre to atrocious for the Cauchy.

3.   The same seems true but less so for convergence to normality.

4.   Typically, in the nonGaussian cases considered, the distributions of the absolute values of the estimates are stochastically larger than those of the absolute value of Gaussian variables with the same variance. This, coupled with the slow convergence from above of the variances to the asymptotic value, suggests that formation of t statistics with denominators based on estimated asymptotic variances may not be satisfactory. Perhaps jackknifing would be better. Of course more Monte Carlo work would have to be done to support or destroy such a conjecture.

These impressions were gained essentially by considering the following eleven estimates chosen rather arbitrarily.

15%	H10	ADA
GAS	A15	TAK
JAE	P15	H/L
JBT	25A	

If we measure closeness of the (normalized) variance to its limit by the ratio of the normalized variance for the sample size in question to its value for n=40 we find that

(i) For the Gaussian and n=5 all but those of ADA and 25A are between .90 and 1.1. These are also the only procedures having efficiencies of less than 80% in this group. By sample size 10 all have ratios between .95 and 1.05. For sample size 20 all are within these limits and percentage differences of only 1 or 2 are common.

(ii) The picture for 25% 1/U is much less rosy. For sample size 5 no ratio falls below 1.5. Stability is more evident for n=10 with most estimates having ratios of the order of 1.1. However the poor procedures 15%, JAE, JBT, H/L and TAK still have ratios of 1.3 or more. By n=20 the asymptotic is almost reached for the entire list with all ratios of the order of 1.05.

(iii) For the Cauchy all ratios are above 1.25 for n=10 and typically above 2. For n=20 only 25A has a ratio below 1.1. For this shape one can make a comparison with the asymptotic values of the normalized variance for selected estimates. The ratios of normalized variances for n=40 to n=∞ seem to be in the range 1.0 to 1.05 save for TAK for which the asymptotic value is questionable.

To measure closeness to normality I used a crude measure: the ratio of the 1% point of the Gaussian distribution with mean 0 and with the variance of the distribution of the estimate to the actual 2.5% percent point of this distribution and in particular noted whether the first fell below the second, i.e., whether a nominal 98% confidence interval had confidence level less than 95%.

For the estimates considered and the following set of
shapes and sample sizes: n=5 (Cauchy), n=10 (1/U Cauchy),
n=20 (1/U, 25% 1/U), n=40 (Cauchy), this happened in only
5 cases. The estimates and sample sizes were:

n=5 (Cauchy)	15%
n=10 (Cauchy)	JAE, JBT, H/L, TAK.

In all of these situations, the estimates in question
were very poor.

7E2 Maximin efficiency

The point of view that I took was, in the terminology
of Chapter 6, either that of a wary classicist or of a
cautious realist. However, rather than attempting a
detailed analysis, I tried a simple minded "shotgun"
approach reminiscent of the two error theory of testing.
That is, for n=20, first attention was restricted to
procedures whose efficiency (in the sense of Chapter 6)
was 90% (95%, 80%) or more at the Gaussian (wary
classicist) or 5% contaminated distribution (cautious
realist). Subject to this requirement I sought estimates
which maximized the minimum efficiency for all other
(all other reasonable) symmetric situations or were near
maximin in some vague way. In the process, Exhibit 7-3
of all (10) estimates which have efficiency 90% for the
Gaussian and at least 70% for all other symmetric shapes
was produced. When the "unreasonable" Cauchy and 50% 3N,
75% 3N distributions were deleted three more procedures
could be added to this list. The following points seemed
noteworthy.

1) Taking a "wary classicist" or "cautiously
realistic" attitude made no difference from this
point of view. Procedures which behaved very well
for the Gaussian and moderately well for the assorted
heavy tails also performed very well for the limited
range of 5% contamination available.

2) Deleting the "unreasonable" situations did
add some estimates, raised performance impressively,
but did not affect the overall minimax choices.

Exhibit 7-3
"Good" estimates for n=20

EST.	Eff. for N	Min. eff. (all)	Min. eff. (reasonable)	Least favorable shape	
JAE	.91	.78	.86	Cauchy	1/U
JBT	.90	.83	.88	Cauchy	1/U
H10	.90	.74	.83	Cauchy	1/U
22A	.90	.79	.84	50% 3N	1/U
25A	.96	.73	.78	Cauchy	1/U
ADA	.91	.86	.94	50% 3N	1/U
AMT	.94	.79	.81	Cauchy	1/U
TAK	.95	.78	.82	Cauchy	1/U
HGL	.94	.73	.86	50% 3N	1/U
D10	.90	.70	.80	Cauchy	1/U, 25% 3/U
A15	.95	.61	.71	Cauchy	1/U
P15	.95	.61	.71	Cauchy	1/U
H/L	.94	.66	.71	Cauchy	25% 3/U

Maximin estimate for all shapes: ADA

Maximin estimate for "reasonable" shapes: ADA

Of course, the optimality of ADA, TAK etc. and
the overall minimum achieved when the unreasonable
distributions are deleted is to be taken with a grain
of salt. It seems to me that one thing these results
call for is a more extensive inquiry into what it means
for one shape (scale family) to be "longer tailed"
(harder to estimate location for) than another. It
might then be easier to see how representative our
selection really was. On the other hand one should
try and see to what extent these "difficult" distributions occur in the real world. Nevertheless it seems
clear that the estimates of Exhibit 7-3 are promising.
Some (JBT, P15, D10, and AMT) are even easy to compute
by hand.

Sample size 40

The small selection of shapes made the minimax
game questionable. However, there was no change in the
list of procedures with efficiency 90% at the Gaussian
and at least 70% overall and the minimax estimate (ADA) was
unchanged if the sample sizes were pooled.

Sample size 10

Only one estimate, ADA, had efficiency with respect
to the best scale invariant estimate in the study of more
than 90% for the Gaussian and 70% elsewhere. With the
Cauchy deleted the list grew to:

21A
ADA
AMT

and again ADA was maximin with minimum efficiency 84%
at 1/U. If we permit efficiencies of less than 90% but
more than 80% for the Gaussian and at least 70% for the
Cauchy we obtain as reasonable estimates:

GAS	17A	AMT
HO7	12A	DFA
HMD	ADA	

Sample size 5

Given the small number of distributions and the large variability of the Monte Carlo variances a maximin comparison seemed inappropriate here.

7E3   Pet estimates

The very close agreement between the variances (and percent point variances) of A15 and P15 is gratifying and gives hope for one step approximations from the median for the excellent ADA and 25A estimates. The one steps are, of course, as easy to compute as trimmed means.

The dangers of a naive faith in asymptotics may also be perceived in the one steps. The estimates D20 and D15 are also equivalent to A20 and A15 respectively for large n. But the asymptotically negligible difference in the scale estimates used cause substantial differences in the small sample behavior of these procedures. (For n=10 (25% 1/U) the efficiency is of the order of 40%.) This effect predicted by F. Hampel came as a surprise to the author.

The simplest one step, M15, unfortunately appears to be sensitive to its nonrobust starting point. This is to be expected for all of the distributions considered in which no second moment exists. However, poor performance was also evident for n=20 in cases such as the t_3 and 5%

10N in which the mean had efficiency of about 50% or less. The picture is better when one considers 2.5% percent point pseudovariances. Efficiencies with respect to D15 change from 61% to 81% for 5% 10N and from 84% to 98% for t_3. However, the general effect does not inspire confidence.

11. PROGRAMS OF THE ESTIMATES

This appendix contains listings of the programs of estimates used in the study. These programs are included as a record of the routines used to produce the numerical part of the study. They may also be useful as source listings for users wanting programs for particular estimates.

Some estimates have been very carefully programmed with both numerical precision and computational efficiency in mind. FORTRAN4 was used throughout. Some programs were supplied by others with the result that the style of programming is not constant nor is the quantity of comment cards. However, all the programs work. Some estimates use several subroutines; some subroutines serve several estimates. The subroutines are listed in alphabetical order, by subroutine name. They are preceded by the calling sequences for each estimate. Exhibit 11-1 gives for each estimate (tabled according to the extended alphabet used in Chapter 2) the FORTRAN statement number of the associated calling sequence. For example, to use the estimate JAE, obtain the FORTRAN statement number 21 from Exhibit 11-1 and find from the list of calling sequences

21 CALL JAECKL(X,N,ESTIM)

The subroutine JAECKL is then found in the following listings.

In the calling sequence,

X   denotes an array of N, ordered, single-precision numbers,

X8   denotes an array of N, ordered, double-precision numbers,

N   denotes the size of the array,

and

ESTIM   denotes the returned value of the estimate.

Note: The programs assume that the data has been ordered.

Exhibit 11-1

INDEX TO CALLING SEQUENCES OF ESTIMATES

TAG	FORTRAN Statement Number	TAG	FORTRAN Statement Number
5%	2	D20	16
10%	3	DFA	36
12A	57	GAS	7
15%	4	HO7	11
17A	56	H10	10
21A	55	H12	63
22A	54	H15	9
25%	5	H17	64
25A	53	H20	8
2RM	44	H/L	38
33T	33	HGL	26
3R1	50	HGP	25
3R2	49	HMD	14
3RM	47	JAE	21
3T0	12	JBT	31
3T1	51	JLJ	59
4RM	48	JOH	40
50%	6	JWT	34
5T1	52	LJS	39
5T4	61	M	1
A15	20	M15	13
A20	65	MEL	37
ADA	58	OLS	24
AMT	46	P15	60
BH	43	SHO	35
BIC	22	SJA	42
CML	15	SST	30
CPL	41	TAK	23
CST	62	THL	28
CTS	29	THP	27
D07	19	TOL	32
D10	18	TRI	45
D15	17		

```
 1 CALL TRMDMN (X8, N, ESTIM, 0.D0)
 2 CALL TRMDMN (X8, N, ESTIM, 0.05D0)
 3 CALL TRMDMN (X8, N, ESTIM, 0.10D0)
 4 CALL TRMDMN (X8, N, ESTIM, 0.15D0)
 5 CALL TRMDMN (X8, N, ESTIM, 0.25D0)
 6 ML = (N+1) / 2
   ESTIM = (X8(ML) + X8(N-ML+1)) / 2.D0
 7 CALL GASTWT (X8, N, ESTIM)
 8 CALL H2 (X8, N, ESTIM, 2.0)
 9 CALL H2 (X8, N, ESTIM, 1.5)
10 CALL H2 (X8, N, ESTIM, 1.0)
11 CALL H2 (X8, N, ESTIM, 0.7)
12 CALL MSKIP0 (X, N, ESTIM, 1.0, 1.5)
13 CALL TRMDMN (X8, N, ESTIM, 0.D0)
   EST1 = ESTIM
   ESTIM = PB2 (X, N, EST1, 1.5)
14 CALL HAMPEL (X8, N, ESTIM)
15 ESTIM = CMLE (X8, N)
16 ML = (N+1)/2
   XM = (X8(ML)+X8(N-ML+1))*0.5D0
   ESTIM = PB2 (X, N, XM, 2.0)
17 ML = (N+1)/2
   XM = (X8(ML)+X8(N-ML+1))*0.5D0
   ESTIM = PB2 (X, N, XM, 1.5)
18 ML = (N+1)/2
   XM = (X8(ML)+X8(N-ML+1)) / 2.D0
   ESTIM = PB2 (X, N, XM, 1.0)
19 ML = (N+1)/2
   XM = (X8(ML)+X8(N-ML+1))*0.5D0
   ESTIM = PB2 (X, N, XM, 0.7)
20 CALL HH (X8, N, ESTIM, 1.5)
21 CALL JAECKL (X, N, ESTIM)
22 CALL BJTM (X8, N, ESTIM)
23 ESTIM = TKUCHI (X, N)
24 CALL OLSHEN(X, N, EST1)
   ESTIM = EST1
25 CALL HOGG2 (X8, N, ESTIM)
26 CALL HOGG1 (X8, N, ESTIM)
27 CALL OSKIP (X, N, 1.5, INIT, LEN)
   CALL HOGG2 (X8(INIT), LEN, ESTIM)
28 CALL OSKIP (X, N, 1.5, INIT, LEN)
   CALL HOGG1 (X8(INIT), LEN, ESTIM)
29 CALL OSKIP (X, N, 2.0, INIT, LEN)
   CALL OSKIP (X(INIT), LEN, 1.5, INIT1, LEN1)
   INIT = INIT1 + INIT - 1
   CALL OSKIP (X(INIT), LEN1, 1.0, INIT1, LEN)
   INIT = INIT1 + INIT - 1
   ESTIM = TRIMEN (X(INIT), LEN)
30 CALL SKIP (X, N, 1.0, INIT, LEN)
   ESTIM = TRIMEN (X(INIT), LEN)
31 CALL JBT (X, N, ESTIM)
32 CALL OSKIP (X, N, 1.5, INIT, LEN)
   CALL OLSHEN (X(INIT), LEN, EST1)
   ESTIM = EST1
33 CALL MSKIPS (X, N, ESTIM, 1.0, 1.5)
34 CALL JWTC2 (X, N, EST1)
   ESTIM = EST1
35 CALL SHORTH (X8, N, ESTIM)
36 CALL DFA2 (X, N, EST1)
```

```
      ESTIM = EST1
37 CALL MELE (N, X, EST1)
      ESTIM = EST1
38 ESTIM = HDGSL1 (X8, N)
39 CALL LFD (X8, N, ESTIM)
40 CALL JOHNS (X, N, ESTIM)
41 ESTIM = CPL (X8, N)
42 CALL SYMJAE (X8, N, ESTIM)
43 CALL FT (X8, N, ESTIM, 0, 1)
44 CALL FT (X8, N, ESTIM, 0, 2)
45 ESTIM = TRIMEN (X, N)
46 CALL SH (X8, N, ESTIM)
47 CALL FT (X8, N, ESTIM, 0, 3)
48 CALL FT (X8, N, ESTIM, 0, 4)
49 CALL FT (X8, N, ESTIM, 2, 3)
50 CALL FT (X8, N, ESTIM, 1, 3)
51 CALL MSKIP (X, N, ESTIM, 1.0, 1.5)
52 CALL MSKIP (X, N, ESTIM, 2.0, 1.5)
53 CALL HAM3MD (X8, N, ESTIM, 2.5D0, 4.5D0, 9.5D0)
54 CALL HAM3MD (X8, N, ESTIM, 2.22D0, 3.71D0, 5.93D0)
55 CALL HAM3MD (X8, N, ESTIM, 2.1D0, 4.0D0, 8.2D0)
56 CALL HAM3MD (X8, N, ESTIM, 1.7D0, 3.4D0, 8.5D0)
57 CALL HAM3MD (X8, N, ESTIM, 1.25D0, 3.5D0, 8.0D0)
58 CALL HAMABC (X8, N, A, B, C)
      CALL HAM3MD (X8, N, ESTIM, A, B, C)
59 CALL JKL (X, N, ESTIM)
60 ESTIM = PB3 (X, N, 1.5)
61 CALL MSKIP4 (X, N, ESTIM, 2.0, 1.5)
62 CALL SKIP (X, N, 2.0, INIT, LEN)
      ESTIM = TRIMEN (X(INIT), LEN)
63 CALL H2 (X8, N, ESTIM, 1.2)
64 CALL H2 (X8, N, ESTIM, 1.7)
65 CALL HH (X8, N, ESTIM, 2.0)
      RETURN
      END
```

```
      FUNCTION ALGMMA(Z)
         GIVES LN GAMMA(Z)
      IMPLICIT REAL*8(A-H,O-Z)
      ALG=0.DO
      X=Z
      IF(X.GT.7.DO)GOTO 10
      M=8.DO-X
      DO 1 I=1,M
      ALG=-DLOG(X+I-1.DO)+ALG
      X=X+M
      ALGMMA=(X-.5DO)*DLOG(X)-X+.91893853320467+1.DO/(12.DO*X)-
     * 1.DO/(360.DO*X**3)+1.DO/(1260.DO*X**5)-1.DC/(1680.DO*X**7)+ALG
      RETURN
      END




      SUBROUTINE BJTM (X, N, EST)
         **    PETER BICKEL REFINED JAECKEL ESTIMATE    **
         LAST CHANGES ON 27 FEB. 1971
      IMPLICIT REAL*8 (A-H,O-Z)
      REAL*8 X(N), EST, GN1(50)
      REAL*8 TN(26)
      DATA PI /3.14159265357828/
      FNI = 1. / DFLOAT(N)
      SQRTFN = DSQRT (DFLOAT(N))
         **    SYMMETRIZE ON MEDIAN    **
      NL = (N+1)/2
      NU = N+1-NL
      XM = (X(NL)+X(NU))*0.5DO
      DO 1 I = 1, NL
      GN1 (2*I-1) = XM - X(I)
    1 GN1(2*I) = X(N-I+1) - XM
      CALL SORT8 (GN1, N)
         **    COMPUTE FUNCTIONS FOR TESTING    **
         TN -- JAECKEL'S ORIGINAL FUNCTION
         QPB -- PETER BICKEL'S ALLOWABILITY CRITERION
      C1 = 0.DO
      NLM1 = NL - 1
      DA = 0.5 * FNI
      DO 2 I = 1, NLM1
      CO = GN1(I)**2
      C1 = C1 + DA * CO
    2 CONTINUE
      DO 3 J = NL, N
      J1 = N - J + 1
      ALPHA = DFLOAT(N-J) *DA
      CO = GN1(J)**2
      C1 = C1 + DA * CO
      C2 = ALPHA * CO
      C6 = 2.DO / (1.DO - 2.DO * ALPHA)**2
      TN(J1) = (C1+C2) * C6
    3 CONTINUE
      NJALW = N/2 + 1
         **    SELECT CHOICE    **
      DO 6 I = 1, 3
```

```
      TNMIN = 1.D75
      DO 7 J = 1, NJALW
      IF (TN(J) .GT. TNMIN) GOTO 7
      TNMIN = TN(J)
      JMIN = J
    7 CONTINUE
      ALPHA = (DFLOAT(JMIN-1)) * DA
      IF (JMIN .LE. 2) GOTO 8
      ALPHA1 = ALPHA - DA
      ALPHA2 = ALPHA1 - DA
      TNTAO = 2.DO * ALPHA * (GN1(N-JMIN+1)/(1.DO-2.DO*ALPHA))**2
      TNTA1 = 2.DO * ALPHA1 * (GN1(N-JMIN+2)/(1.DO-2.DO*ALPHA1))**2
      TNTA2 = 2.DO * ALPHA2 * (GN1(N-JMIN+3)/(1.DO-2.DO*ALPHA2))**2
      IF(TNTAO.EQ.TNTA2) GOTO 8
      QPB = (TNTAO - TNTA1) / (TNTAO - TNTA2)
      IF (QPB .LE. 1.DO) GOTO 8
      NJALW = JMIN - 1
    6 CONTINUE
    8 CALL TRMDMN (X, N, EST, ALPHA)
      RETURN
      END



      FUNCTION CHECK(X)
      REAL*8 X(4)
      LOGICAL CHECK
      CHECK=.FALSE.
      IF(X(2).LE.0.DO)CHECK=.TRUE.
      RETURN
      END



      FUNCTION CMLE(X, N)
      IMPLICIT REAL*8 (A-H,O-Z)
      REAL*8 X(N)
      REAL X1(40)
      DATA IMAX /25/
      ITER = 0
      MH = (N+2)/2
      T = 0.5DO * (X(N-MH+1)+X(MH))
      SIGMA = 0.DO
      IH = N
      IL = 1
      DO 1 I = 1, MH
      D1 = T-X(IL)
      D2 = X(IH)-T
      SIGOLD = SIGMA
      IF (D1 .GT. D2) GOTO 2
      IH = IH - 1
      SIGMA = D2
      GOTO 1
    2 IL = IL + 1
      SIGMA = D1
    1 CONTINUE
      IF (MOD(N,2) .EQ. 0) SIGMA = 0.5DO * (SIGMA + SIGOLD)
   20 TOLD = T
      SIGOLD = SIGMA
```

```fortran
      SUMO = 0.D0
      SUM1 = 0.D0
      SIGSQ = SIGMA**2
      DO 10 I = 1, N
      Z = SIGSQ + (X(I)-T)**2
      SUMO = SUMO + 1.D0/Z
 10   SUM1 = SUM1 + X(I)/Z
      TSIGMA = DFLOAT(N)/2.D0/SUMO/SIGMA**(1.5)
      SIGMA = TSIGMA**2
      T = SUM1/SUMO
      ITER = ITER + 1
      IF (ITER .GT. IMAX) GOTO 99
      IF ( DABS(T-TOLD) .GT. 1.D-3*SIGMA          .OR.
     .     DABS(SIGMA-SIGOLD) .GT. 0.05*SIGMA) GOTO 20
      CMLE = T
      RETURN
 99   DO 990 I = 1, N
990   X1(I) = (X(I)-T)/SIGMA
      CMLE = T
      WRITE (6, 2000) (X1(I), I=1, N)
000   FORMAT ('0CAUCHY MAX. LIKE DOES NOT CONVERGE.  STANDARDIZED ',
     .  'SAMPLE IS:'/(16F8.3))
      RETURN
      END



      FUNCTION CPL (X, N)
      IMPLICIT REAL*8 (A-H,O-Z)
      REAL*8 X(N)
      REAL*8 X40A(20), W40A(20), XI(40), FI(40)
      LOGICAL FIRST
      DATA FIRST /.TRUE./
      DATA NPT /40/
      DATA X40A /
     1    3.877 24175 06050 821933 D-2, 1.1608 40706 75255 208483 D-1,
     2    1.9269 75807 01371 099716 D-1, 2.6815 21850 07253 681141 D-1,
     3    3.4199 40908 25758 473007 D-1, 4.1377 92043 71605 001525 D-1,
     4    4.8307 58016 86178 712909 D-1, 5.4946 71250 95128 202076 D-1,
     5    6.1255 38896 67980 237953 D-1, 6.7195 66846 14179 548379 D-1,
     6    7.2731 82551 89927 103281 D-1, 7.7830 56514 26519 387695 D-1,
     7    8.2461 22308 33311 663196 D-1, 8.6595 95032 12259 503821 D-1,
     8    9.0209 88069 68874 296728 D-1, 9.3281 28082 78676 533361 D-1,
     9    9.5791 68192 13791 655805 D-1, 9.7725 99499 83774 262663 D-1,
     A    9.9072 62386 99457 006453 D-1, 9.9823 77097 10559 200350 D-1/
      DATA W40A /
     1    7.750 59479 78424 811264 D-2, 7.703 98181 64247 965588 D-2,
     2    7.611 03619 00626 242372 D-2, 7.472 31691 57968 264200 D-2,
     3    7.288 65823 95804 059061 D-2, 7.061 16473 91286 779695 D-2,
     4    6.791 20458 15233 903826 D-2, 6.480 40134 56601 038705 D-2,
     5    6.130 62424 92928 939167 D-2, 5.743 97693 99391 551367 D-2,
     6    5.322 78469 83936 824355 D-2, 4.869 58076 35072 232061 D-2,
     7    4.387 09081 85673 271992 D-2, 3.878 21679 74472 017640 D-2,
     8    3.346 01952 82547 847393 D-2, 2.793 70069 80023 401098 D-2,
     9    2.224 58491 94166 657262 D-2, 1.642 10563 81907 888713 D-2,
     A    1.049 82845 31152 813615 D-2, 4.52 12770 98523 191258 D-3/
      IF (FIRST) GOTO 100
200   ML = (N+1)/2
      XMED = 0.5D0 * (X(ML)+X(N-ML+1))
```

```
      XSCAL = DSQRT(1.D0/DFLOAT(N))
      SUM0 = 0.D0
      SUM1 = 0.D0
      DO 1 I = 1, NPT
      XLIKE = 1.D0
      T = XMED + XSCAL*XI(I)
      DO 2 J = 1, N
    2 XLIKE = XLIKE / (1.D0+(X(J)-T)**2)
      XLIKE = XLIKE / FI(I)
      SUM1 = SUM1 + (T-XMED)*XLIKE
      SUM0 = SUM0 + XLIKE
    1 CONTINUE
      CPL = XMED + SUM1/SUM0
      RETURN
  100 NPD2 = NPT/2
      SQRTPI = DSQRT (2.D0 * 3.141592653578)
      DO 110 I = 1, NPD2
  110 X4OA(I) = 0.5D0 + X4OA(I)/2.D0
      XK = 10.D0 / PHIINV(X4OA(NPD2))
      DO 101 I = 1, NPD2
      XI(I) = XK * PHIINV (X4OA(I))
      FI(I) = W4OA(I)/DEXP(-X4OA(I)**2/2.D0)*SQRTPI
      XI(NPD2+I) = -XI(I)
  101 FI(NPD2+I) = FI(I)
      FIRST = .FALSE.
      GOTO 200
      END




      SUBROUTINE DAH2 (X, N, T, S, HUBERK, II)
C         MODIFIED HUBER 2 ESTIMATOR FOR LEAST FAVORABLE DISTRIBUTIO
C         INTENDED AS A SUBROUTINE FOR LFD.
C         VECTOR X (LENGTH N) IS INPUT, LOCATION T IS OUTPUT.
C         ---> X IS ASSUMED TO BE SORTED <---
C         ASSOCIATED SCALE S IS COMPUTED INTERNALLY.
      IMPLICIT REAL*8 (A-H,O-Z)
      REAL*8 X(N)
C
      DATA PI /3.14159265357828/
      DATA ISTOP /20/
      INTEGER ITER(20)
      DATA ITER /20*0/
C         **    COMPUTE SOME NECESSARY CONSTANTS    **
      BETA=(N-II)/DFLOAT(N)
C         **    COMPUTE INITIAL ESTIMATES S AND T    **
      ML = (N+1)/2
      T = (X(ML)+X(N-ML+1))*0.5D0
      ND4 = N/4
      F1 = DFLOAT (4*ND4 + 3 - N) / 4.D0
      RIQ = (1.D0-F1)*(X(N-ND4)-X(ND4+1)) + F1*(X(N-ND4+1)-X(ND4))
      S = RIQ / 1.34D0
C         **    COMPUTE WINSORIZING COUNTS AND DECIDE ON STOPPING    *
      I = 0
      M1OLD = -1
      M3OLD = -1
   10 CONTINUE
      XL = T - S*HUBERK
      XU = T + S*HUBERK
```

```
      DO 11 J = 1, N
      IF (X(J) .GT. XL) GOTO 12
   11 CONTINUE
      J = N + 1
   12 M1 = J - 1
      DO 13 J = 1, N
      IF (X(N-J+1) .LT. XU) GOTO 14
   13 CONTINUE
      J = N + 1
   14 M3 = J - 1
      IF (M1 .EQ. M1OLD .AND. M3 .EQ. M3OLD) GOTO 50
   16 I = I + 1
      IF (I .GE. 20) GOTO 60
      FM1 = M1
      FM3 = M3
      FM2 = N - M3 - M1
      IF (FM2 .LT. 0.5D0) GOTO 60
      M1OLD = M1
      M3OLD = M3
C        **   UPDATE ESTIMATE SET    **
      XBAR = 0.D0
      M1P1 = M1 + 1
      NMM3 = N - M3
      DO 20 J = M1P1, NMM3
   20 XBAR = XBAR + X(J)
      XBAR = XBAR / FM2
      S = 0.D0
      DO 21 J = M1P1, NMM3
   21 S = S + (X(J)-XBAR)**2
      SD1 = 0.D0
      IF (M1 .NE. M3) SD1 = (FM3-FM1)**2/FM2
      S = S / (DFLOAT(N)*BETA - (FM1+FM3+SD1)*HUBERK**2)
      IF (S .LE. 0.D0) GOTO 22
      S = DSQRT (S)
      T = XBAR + HUBERK*S*(FM3-FM1)/FM2
      GOTO 10
   22 IF (M1 .GT. 0) M1 = M1 - 1
      IF (M3 .GT. 0) M3 = M3 - 1
      GOTO 16
   50 ITER(I) = ITER(I) + 1
      RETURN
   60 WRITE (6,1000) I, M1, M3, X
 1000 FORMAT ('0ABORTING HUBER ESTIMATE EVALUATION ON TRIAL',I4,
     .  '. M1 AND M3 ARE', 2I5,', AND THE SAMPLE IS:' / (8D16.7))
      RETURN
      END



      SUBROUTINE DFA2 (X, N, T)
C         DF ANDREWS MAXIMUM LIKELIHOOD OVER
C         DISTRIBUTION AND PARAMETER
      DIMENSION X(N), XC(1000), D(100)
     1,XLS(1000),P(100),DZ(100),DMS(100),DP(100)
      DATA NOLD/0/
      IF(N.EQ.NOLD) GO TO 20
      C=0.1
      FN=N
      FNB2=FN/2.0
```

```
              NB3=SQRT(FN)
              NC2=(N*N+N)/2
              IF(N.GT.30) NC2=N-1
              NP1=N+1
              NOLD=N
              FNC=0.0
              DO 21 I21=1,NP1
              FNC=FNC+1.0
              Z=((FNC-FNB2)**2)*2.0 /FN
              Z=-1.0*Z
              P(I21)=EXP(Z)
21            CONTINUE
20            CONTINUE
              T=0.0
              I1I2=0.0
              DO 1 I1=1,N
              I21=1
              IF(N.GT.30) I21=I1
              DO 1 I2=I21,I1
              I1I2=I1I2+1
              XC(I1I2)=(X(I1)+X(I2))/2.0
1             CONTINUE
              XL4=0.0
              N1=(NC2+3)/4
              N2=NC2+1-N1
              L=1
              CALL FSORT(XC,L,NC2)
              N11=(FN+1.0)/2.0-SQRT(FN)+1.0
              N22=N+1-N11
              T1=X(N11)
              T2=X(N22)
              N1C=N1
              N2C=N2
              DO 3 I3=N1,N2
              TH=XC(I3)
              IF(TH.LT.T1) GO TO 8
              IF(TH.GT.T2) GO TO 9
              FNC=0.0
              NP=1
              NM=0
              DP(1)=0.0
              DO 4 I4=1,N
              DT=X(I4)-TH
              IF(DT.LT.0.0) GO TO 41
              NP=NP+1
              DP(NP)=DT
              GO TO 4
41            NM=NM+1
              DZ(NM)=ABS(DT)
4             CONTINUE
              DO 42 I42=1,NM
              DMS(I42)=DZ(NM+1-I42)
42            CONTINUE
              CALL FMERGE(DP,NP,DMS,NM,D)
              XL=1.0
              I5=1
5             IS=I5+1
              IF(IS.GT.NP1) IS=NP1
              IF(IS.LT.NB3) IS=NB3+1
```

```
      F=0.0
      DO 6 I6=IS,NP1
      DN=I6-I5
      DD=D(I6)-D(I5)
      IF(DD.LT.0.000001) GO TO 6
      FT=DN/DD
      IF(FT.LT.F) GO TO 6
      F=FT
      DM=DN
6     CONTINUE
      XL=XL*F**DM
      I5=I5+DM
      IF(I5.LT.NP1) GO TO 5
      XL=XL*P(NP-1)
      XLS(I3)=XL
      IF(XL.LT.XLM) GO TO 3
      XLM=XL
      T=TH
      GO TO 3
8     N1C=N1C+1
      GO TO 3
9     N2C=N2C-1
3      CONTINUE
      TMIN=T
      TMAX=T
      DO 7 I7=N1C,N2C
      R=XLS(I7)/XLM
      IF(R.LT.C) GO TO 7
      TH=XC(I7)
      IF(TH.LT.TMIN)  TMIN=TH
      IF(TH.GT.TMAX) TMAX=TH
7     CONTINUE
      T=(TMAX+TMIN)/2.0
      RETURN
      END




      SUBROUTINE DIMEN1 (A,NROW,NCOL)
      DOUBLE PRECISION A(1,1)
      COMMON /DIM/ NROW1, NCOL1
      NROW1 = NROW
      NCOL1 = NCOL
      RETURN
      END




      SUBROUTINE FMERGE(X,NP,Y,NM,Z)
      DIMENSION X(100),Y(100),Z(100)
      DATA XBIG/9.9E09/
      X(NP+1)=XBIG
      Y(NM+1)=XBIG
      NEND=NP+NM+1
      N=1
      NX=1
      NY=1
      XN=X(1)
      YN=Y(1)
```

```
  1     IF(XN.LT.YN) GO TO 2
        Z(N)=YN
        NY=NY+1
        YN=Y(NY)
        GO TO 3
  2     Z(N)=XN
        NX=NX+1
        XN=X(NX)
  3     N=N+1
        IF(N.LT.NEND) GO TO 1
        RETURN
        END



        SUBROUTINE FOLD(X,N)
        IMPLICIT REAL*8(A-H,O-Z)
        REAL*8 X(N)
        NP1B2=(N+1)/2
        DO 1 I=1,NP1B2
  1     X(I)=(X(I)+X(N+1-I))/2.CD0
        N=NP1B2
        CALL SORT8(X,N)
        RETURN
        END



        SUBROUTINE FSORT(A,II,JJ)
C  THIS IS A MINOR MODIFICATION OF ALGORITHM 347,  AN EFFICIENT
C  ALGORITHM FOR SORTING WITH MINIMAL STORAGE, RICHARD C. SINGLE
C  COMMUNICATIONS OF THE ASSOCIATION FOR COMPUTING MACHINERY  12
C  (MARCH 1969), 185-187.
C  SORTS ARRAY A INTO INCREASING ORDER FROM A(II) TO A(JJ)
C  ARRAYS IU(K) AND IL(K) PERMIT SORTING UP TO 2**(K+1)-1 ELEMEN
        DIMENSION A(JJ), IU(16), IL(16)
        M = 1
        I = II
        J = JJ
  5     IF(I .GE. J) GO TO 70
 10     K = I
        IJ = (J + I) / 2
        T = A(IJ)
        IF(A(I) .LE. T) GO TO 2C
        A(IJ) = A(I)
        A(I) = T
        T = A(IJ)
 20     L = J
        IF(A(J) .GE. T) GO TO 4C
        A(IJ) = A(J)
        A(J) = T
        T = A(IJ)
        IF(A(I) .LE. T) GO TO 4C
        A(IJ) = A(I)
        A(I) = T
        T = A(IJ)
        GO TO 40
 3C     A(L) = A(K)
        A(K) = TT
```

```
   40 L = L - 1
      IF(A(L) .GT. T) GO TO 40
      TT = A(L)
   50 K = K + 1
      IF(A(K) .LT. T) GO TO 50
      IF(K .LE. L) GO TO 30
      IF(L-I .LE. J-K) GO TO 60
      IL(M) = I
      IU(M) = L
      I = K
      M = M + 1
      GO TO 80
   60 IL(M) = K
      IU(M) = J
      J = L
      M = M + 1
      GO TO 80
   70 M = M - 1
      IF(M .EQ. 0) RETURN
      I = IL(M)
      J = IU(M)
   80 IF(J-I .GE. 11) GO TO 10
      IF(I .EQ. II) GO TO 5
      I = I - 1
   90 I = I + 1
      IF(I .EQ. J) GO TO 70
      T = A(I+1)
      IF(A(I) .LE. T) GO TO 90
      K = I
  100 A(K+1) = A(K)
      K = K - 1
      IF(T .LT. A(K)) GO TO 100
      A(K+1) = T
      GO TO 90
      END



      SUBROUTINE FT(XG,NG,T,ISHORT,NFOLD)
C FOLDED TRIMMED MEANS
CFOLD NFOLD TIMES
C  SHORTEN STRING IF ISHORT IS POSITIVE
      IMPLICIT REAL*8 (A-H,O-Z)
      REAL*8 X(40), XG(NG)
C SAVE GOOD DATA
      N=NG
      DO 1 I=1,N
    1 X(I)=XG(I)
      DO 2 I=1,NFOLD
      J=ISHORT
      IF (N .LE. 2) GOTO 4
      IF (ISHORT .LE. 0) GOTO 2
      NM3B2=(N-3)/2
      IF(J.GT.NM3B2) J=NM3B2
      N=N-J*2
      DO 3 K = 1, N
    3 X(K) = X(K+J)
    2 CALL FOLD(X,N)
    4 ML = (N+1)/2
```

```
      T = 0.5D0 * (X(ML)+X(N-ML+1))
      RETURN
      END




      SUBROUTINE GASTWT(ARRAY,NDIM,XGAST)
      IMPLICIT REAL*8(A-H,O-Z)
      DIMENSION ARRAY(NDIM)
      NM=(NDIM+1)/2
      NN=(NDIM+2)/2
      NL=(NDIM+2)/3
      NH=(NDIM-NL+1)
      IF(MOD(NDIM,3).EQ.0)GOTO 16
      XGAST=.3D0*(ARRAY(NL)+ARRAY(NH))+.2D0*(ARRAY(NM)+ARRAY(NN))
      RETURN
16    NL1=NL+1
      NH1=NH-1
      XGAST=.15D0*(ARRAY(NL)+ARRAY(NL1)+ARRAY(NH1)+ARRAY(NH))+.2D0*(AR
     *Y(NM)+ARRAY(NN))
      RETURN
      END




      SUBROUTINE GMA(X,V,DER)
      IMPLICIT REAL*8(A-H,O-Z)
      COMMON VAL
      SQRT2=1.414213562373095
      SQRTPI=.3989423
      PHI=DEXP(-.5*X**2)*SQRTPI
      CPHI=DERF(X/SQRT2)/2.
      V=(1.-VAL)*PHI-X*VAL*CPHI
      DER=-X*PHI-VAL*CPHI
      RETURN
      END




      SUBROUTINE HAMABC(X,N,A,B,C)
      IMPLICIT REAL*8(A-H,O-Z)
      REAL*8 X(N), D(40)
      ML = (N+1)/2
      T = 0.5D0 * (X(ML)+X(N-ML+1))
      DO 1 I=1,N
1     D(I)=DABS(X(I)-T)
      CALL SORT8(D,N)
      S = 0.5D0 * (D(ML)+D(N-ML+1))
      COUNT=0.0D0
      SUMBD=0.0D0
      DO 2 I=1,N
      IF (D(I) .LE. S) GOTO 2
      COUNT=COUNT+1.0D0
      SUMBD=SUMBD+1.0D0 /D(I)
2     CONTINUE
      IF(COUNT.LT.0.5D0) COUNT=0.00001
      TAILSZ=S*SUMBD/COUNT
      C = 8.0D0
      B = 4.5D0
```

```
      IF(TAILSZ.GT.0.60D0) GO TO 3
      IF(TAILSZ.LT.0.44D0 ) GOTO 4
      A=(75.0D0/8.0DC)*TAILSZ-25.CD0/8.CD0
      GO TO 5
   3  CONTINUE
      A=2.5D0
      GO TO 5
   4  CONTINUE
      A=1.0DC
   5  CONTINUE
      RETURN
      END




      SUBROUTINE HAMPEL(DATA,N,THETAN)
C         HAMPEL'S M-ESTIMATE OF 14 APRIL 1971
      IMPLICIT REAL*8(A-H,O-Z)
      REAL*8 DATA(N),DATAC(40)
C         LIMIT IS THE MAXIMUM NO. OF ITERATIONS ALLOWED
C         EPS IS THE RELATIVE ACCURACY REQUIRED
C         THETAN IS THE HAMPEL ESTIMATE OF LOCATION
      DATA LIMIT /20/, EPS /C.01D0/
      DATA XK1, XK2 /5.5D0, 2.D0/
      ML = (N+1)/2
      XMED = (DATA(ML)+DATA(N-ML+1))*0.5DC
      THETAN = XMED
  10  DO 30 I=1,N
  30  DATAC(I)=DABS(DATA(I)-XMED)
      CALL SORT8(DATAC,N)
      SIGMA = (DATAC(ML)+DATAC(N-ML+1)) * 0.5D0
      IF (SIGMA .EQ. 0.D0) RETURN
  50  THETA=XMED
      TMIN = -1.D75
      TMAX = 1.D75
      DO 70 J=1,LIMIT
      SPSI=0.
      SPSIP=0.
      DO 60 I=1,N
      STNX=(DATA(I)-THETA)/SIGMA
      SSTNX=DABS(STNX)
      IF (SSTNX .GT. XK1) GOTO 91
      IF (SSTNX .GT. XK2) GOTO 92
      PSIP = 1.D0
      PSI = STNX
      GOTO 93
  91  PSI = 0.DC
      PSIP = C.D0
      GOTO 93
  92  PSI = XK2 * DSIGN((SSTNX-XK1)/(XK1-XK2), STNX)
      PSIP = (XK2)/(XK2-XK1)
  93  SPSIP = SPSIP + PSIP
  60  SPSI=SPSI+PSI
      IF (SPSIP .LE. 0.D0) SPSIP = DFLOAT(N) / 2.DC
      THETAN=THETA+ SIGMA*SPSI/SPSIP
      IF (DABS(THETAN-THETA) .LT. EPS*DABS(THETAN)) RETURN
      IF (SPSI .GT. 0.D0) GOTO 71
      TMAX = THETA
      GOTO 72
```

```
   71 TMIN = THETA
   72 IF (THETAN .GE. TMAX .OR. THETAN .LE. TMIN) THETAN=(TMAX+TMIN)/
70    THETA=THETAN
      DO 73 I = 1, N
   73 DATAC(I) =( DATA(I)-THETAN)/SIGMA
      WRITE (6, 100) (DATAC(I), I=1, N)
  100 FORMAT ('0HAMPEL 2-PART DESCENDING M-ESTIMATE DOES NOT CONV.
     .   STANDARDIZED SAMPLE IS:'/(16F8.3))
      RETURN
      END




      SUBROUTINE HAM3MD (X, N, T, A, B, C)
      IMPLICIT REAL*8 (A-H,O-Z)
      REAL*8 X(N), X1(40)
C         RETURNS HAMPEL'S 3 PART DESCENDING M ESTIMATE.
C         A, B, AND C ARE ORDINATES SEPARATING 4 SECTIONS OF PSI CURVE
C         SLOPE IS 1 TO A, ZERO TO B, AND DESCENDS TO ZERO ABSCISSA TO
C         THE USER SHOULD ENSURE THAT A<B<C AND X IS SORTED REAL8.
C
C         METHOD OF SOLUTION IS 3 STROKES NEWTON FOLLOWED BY ONE
C         BINARY CHOP IN REPEATED CYCLES, WITH THE ANSWER INITIALLY
C         RESTRICTED TO 2.25 * SIGMA FROM THE SAMPLE MEDIAN.
      DATA ITLIM /20/
      ML = (N+1)/2
      XMED = (X(ML)+X(N-ML+1))*0.5D0
      DO 1 I = 1, N
    1 X1(I) = DABS(X(I)-XMED)
      CALL SORT8 (X1, N)
      SIGMA = (X1(ML)+X1(N-ML+1))*0.5D0
      BU = XMED + 2.25*SIGMA
      BL = XMED - 2.25*SIGMA
      T = XMED
      IT = 0
C         EVALUATE SUM OF PSI AND PSIP, ITS DERIVATIVE
    2 SPSI = 0.D0
      SPSIP = 0.D0
      IF (IT .GT. ITLIM) GOTO 30
      DO 12 I = 1, N
      XX = (X(I)-T)/SIGMA
      XD = DABS(XX)
      IF (XD .LE. A) GOTO 10
      IF (XD .LE. B) GOTO 11
      IF (XD .GT. C) GOTO 12
      PSIP = A/(B-C)
      PSI = DSIGN((XD-C)*PSIP,XX)
      GOTO 13
   10 PSI = XX
      PSIP = 1.D0
      GOTO 13
   11 PSI = DSIGN (A,XX)
      PSIP = 0.D0
   13 SPSI = SPSI + PSI
      SPSIP = SPSIP + PSIP
   12 CONTINUE
      IT = IT + 1
      IF (SPSIP .LE. 0.D0) SPSIP = 0.6D0*N
      SPSIP = SPSIP / SIGMA
```

```
      TOLD = T
      T = T + SPSI/SPSIP
      IF (DABS(T-TOLD) .LT. 0.001*SIGMA) RETURN
      IF (SPSI .GT. 0.D0) BL = TOLD
      IF (SPSI .LT. 0.D0) BU = TOLD
      IF (T .GT. BU .OR. T .LT. BL) GOTO 20
      IF (MOD(IT,4) .EQ. 3) GOTO 20
      GOTO 2
 20   T = (BU + BL) * 0.5D0
      IT = IT + 1
      GOTO 2
 30   DO 31 I = 1, N
 31   X1 (I) = (X(I)-T)/SIGMA
      IT = IT - 1
      WRITE (6, 2000) A, B, C, IT, (X1(J), J=1,N)
 000  FORMAT ('0HAMPEL 3 PART DESCENDING M ESTIMATE DOES NOT CONVERGE'/
     .  '      A, B, AND C ARE:', 3F10.4,'   AND THE TRIAL NUMBER WAS',
     .  I6,'   WITH STANDARDIZED SAMPLE:' / (16F8.3))
      RETURN
      END




      FUNCTION HDGSL1 (X, N)
         A VERSION OF THE HODGES LEHMANN ESTIMATE WHICH TAKES
         ADVANTAGE OF THE ASSOCIATION WITH WILCOXON'S TEST
      IMPLICIT REAL*8 (A-H,O-Z)
      REAL*8 X(N)
      DATA IMAX /25/
      ITER = 0
      T = 1.D21
      K = (N-1)*.707 + 1.999
      T1 = X(N-K+1)
      T2 = X(K)
      DT = T2-T1
      W1 = WILCXN (X, N, T1)
      W2 = WILCXN (X, N, T2)
 1    TOLD = T
      T = (T2-T1)/(W1-W2)*W1 + T1
      ITER = ITER + 1
      IF (ITER .GT. IMAX) GOTO 10
      IF (DABS(T-TOLD) .LE. 0.001D0*DT) GOTO 10
      W = WILCXN(X,N,T)
      IF (DABS(W) .LT. 0.5D0) GOTO 10
      IF (W .LT. 0.5D0) GOTO 2
      T1 = T
      W1 = W
      GOTO 1
 2    T2 = T
      W2 = W
      GOTO 1
 10   HDGSL1 = T
      RETURN
      END




      SUBROUTINE HH (X, N, T, K)
 HAMPELS HUBER
```

```fortran
C          HAMPEL'S MODIFICATION OF HUBER'S PROPOSAL 2 ESTIMATE
C          VECTOR X (LENGTH N) IS INPUT, LOCATION T IS OUTPUT.
C          ---> X IS ASSUMED TO BE SORTED <---
C          ASSOCIATED SCALE S IS COMPUTED INTERNALLY.
       IMPLICIT REAL*8 (A-H,O-Z)
       REAL*8 X(N)
       REAL K
       DIMENSION DD(100)
C
       DATA PI /3.14159265357828/
       DATA ISTOP/20/
C          VECTOR ITER STORES A FREQUENCY CHART OF THE PERFORMANCE.
       INTEGER ITER(20)
       DATA ITER /20*0/
C          **   COMPUTE INITIAL ESTIMATES S AND T   **
C          T IS THE MEDIAN AND S IS THE SYMMETRIZED INTERQUARTILE RANGE
       HUBERK = K
       ML = (N+1)/2
       T = (X(ML)+X(N-ML+1))*0.5D0
       ND4 = N/4
       DO 99 I99=1,N
       D=X(I99)-T
       DD(I99)=DABS(D)
   99  CONTINUE
       CALL SORT8 (DD, N)
       M1=(N+1)/2
       M2=(N+2)/2
       S=(DD(M1)+DD(M2))/(2.0D00*0.6754D00)
C          **   COMPUTE WINSORIZING COUNTS AND DECIDE ON STOPPING   **
       I = 0
       M1OLD = -1
       M3OLD = -1
   10  CONTINUE
       XL = T - S*HUBERK
       XU = T + S*HUBERK
       DO 11 J = 1, N
       IF (X(J) .GT. XL) GOTO 12
   11  CONTINUE
       J = N + 1
   12  M1 = J - 1
       DO 13 J = 1, N
       IF (X(N-J+1) .LT. XU) GOTO 14
   13  CONTINUE
       J = N + 1
   14  M3 = J - 1
       IF (M1 .EQ. M1OLD .AND. M3 .EQ. M3OLD) GOTO 50
   16  I = I + 1
       IF (I .GE. 20) GOTO 60
       FM1 = M1
       FM3 = M3
       FM2 = N - M3 - M1
       IF (FM2 .LT. 0.5D0) GOTO 60
       M1OLD = M1
       M3OLD = M3
C          **   UPDATE ESTIMATE SET   **
       XBAR = 0.D0
       M1P1 = M1 + 1
       NMM3 = N - M3
       DO 20 J = M1P1, NMM3
```

```
 20 XBAR = XBAR + X(J)
    XBAR = XBAR / FM2
    T = XBAR + HUBERK*S*(FM3-FM1)/FM2
    GOTO 10
 22 IF (M1 .GT. 0) M1 = M1 - 1
    IF (M3 .GT. 0) M3 = M3 - 1
    GOTO 16
 50 ITER(I) = ITER(I) + 1
    RETURN
 60 WRITE (6,1000) I, M1, M3, X
000 FORMAT ('0ABORTING HUBER ESTIMATE EVALUATION ON TRIAL',I4,
   .  '.  M1 AND M3 ARE', 2I5,', AND THE SAMPLE IS:' / (8D16.7))
    RETURN
    END




    SUBROUTINE HINGES (X, N, HL, HU)
    REAL X(N), HL, HU
    IL = (N+3)/4
    HL = X(IL)
    HU = X(N-IL+1)
    IF (MOD(N,4) .NE. 0) RETURN
    HL = 0.5 * (HL + X(IL+1))
    HU = 0.5 * (HU + X(N-IL))
    RETURN
    END




    SUBROUTINE HOGG1(ARRAY,NDIM,XHOGG)
       HOGG PROPOSAL IN LETTER TO JOHN TUKEY
    IMPLICIT REAL*8(A-H,O-Z)
    DIMENSION ARRAY(NDIM)
    CALL XMEAN(ARRAY,NDIM,XBAR)
    XKUR=0.D0
    XVAR=0.D0
    DO 7 I=1,NDIM
    X2=(ARRAY(I)-XBAR)**2
    XKUR=XKUR+X2**2
    XVAR=XVAR+X2
    XKUR=XKUR*DFLOAT(NDIM-1)**2/(XVAR**2*DFLOAT(NDIM))
    IF(XKUR.GT.4.5D0)GOTO 8
    IF(XKUR.LT.1.9D0)GOTO 9
    IF(XKUR.GT.3.1D0)GOTO 10
    XHOGG=XBAR
    RETURN
    CALL GASTWT(ARRAY,NDIM,XHOGG)
    RETURN
    CALL OUTMN(ARRAY,NDIM,XHOGG)
    RETURN
     CALL TRIMMD(ARRAY,NDIM,XHOGG)
    RETURN
    END




    SUBROUTINE HOGG2(X,N,HOGG)
PUBLISHED VERSION OF HOGG ESTIMATE
```

```
      IMPLICIT REAL*8(A-H,O-Z)
      DIMENSION X(N)
      CALL XMEAN(X,N,XBAR)
      XKUR=0.D0
      XVAR=0.D0
      DO 7 I=1,N
      X2=(X(I)-XBAR)**2
      XKUR=XKUR+X2**2
      XVAR=XVAR+X2
7     CONTINUE
      XKUR=XKUR*DFLOAT(N-1)**2/(XVAR**2*DFLOAT(N))
      IF(XKUR.LT.2.)GOTO 2
      IF(XKUR.LE.4.)GOTO 3
      IF(XKUR.LE.5.5)GOTO 4
      CALL XMEDIN(X,N,HOGG)
      RETURN
4     CALL TRIMMD(X,N,HOGG)
      RETURN
3     HOGG = XBAR
      RETURN
2     I=N/4
      IF(I.EQ.0)I=1
      J=N-I+1
       HOGG=0.
      DO 5 K=1,I
5     HOGG=HOGG+X(K)
      DO 6 K=J,N
6     HOGG=HOGG+X(K)
      HOGG=HOGG/(2.*DFLOAT(I))
      RETURN
      END




      SUBROUTINE HOOD (X,T,S,G,E)
      U=((X-T)**2)/S
      W=EXP(-0.5*U)
      G=(X-T)*W
      E=(U-1.)*W
      RETURN
      END




      SUBROUTINE H2 (X, N, T, HUBERK)
C     ***   COMPUTE HUBER PROPOSAL 2 ESTIMATOR    ***
C     VECTOR X (LENGTH N) IS INPUT, LOCATION T IS OUTPUT.
C     ---> X IS ASSUMED TO BE SORTED  <---
C     ---> N >= 4  <---
C     ASSOCIATED SCALE S IS COMPUTED INTERNALLY.
      IMPLICIT REAL*8 (A-H,O-Z)
      REAL*8 X(N)
      REAL HUBERK
C
      DATA PI /3.14159265357828/
      DATA ISTOP /20/
C     VECTOR ITER STORES A FREQUENCY CHART OF THE PERFORMANC
      INTEGER ITER(20)
      DATA ITER /20*0/
```

```
      PHI(Z) = 0.5D0 + DERF (Z/DSQRT(2.D0)) / 2.D0
C        **   COMPUTE SOME NECESSARY CONSTANTS    **
       PHIK = PHI (DBLE(HUBERK))
       BETA = 2.D0 * (HUBERK**2*(1.D0-PHIK) + PHIK-C.5D0
     .   - HUBERK/DSQRT(2.D0*PI)*DEXP(-C.5D0*HUBERK**2) )
C        **   COMPUTE INITIAL ESTIMATES S AND T    **
       ML = (N+1)/2
       T = (X(ML)+X(N-ML+1))*0.5D0
       ND4 = N/4
       F1 = DFLOAT (4*ND4 + 3 - N) / 4.D0
       RIQ = (1.D0-F1)*(X(N-ND4)-X(ND4+1)) + F1*(X(N-ND4+1)-X(ND4))
       S = RIQ / 1.34D0
C        **   COMPUTE WINSORIZING COUNTS AND DECIDE ON STOPPING    **
       I = 0
       M1OLD = -1
       M3OLD = -1
   10 CONTINUE
       XL = T - S*HUBERK
       XU = T + S*HUBERK
       DO 11 J = 1, N
       IF (X(J) .GT. XL) GOTO 12
   11 CONTINUE
       J = N + 1
   12 M1 = J - 1
       DO 13 J = 1, N
       IF (X(N-J+1) .LT. XU) GOTO 14
   13 CONTINUE
       J = N + 1
   14 M3 = J - 1
       IF (M1 .EQ. M1OLD .AND. M3 .EQ. M3OLD) GOTO 50
   16 I = I + 1
       IF (I .GE. 20) GOTO 60
       FM1 = M1
       FM3 = M3
       FM2 = N - M3 - M1
       IF (FM2 .LT. 0.5D0) GOTO 60
       M1OLD = M1
       M3OLD = M3
C        **   UPDATE ESTIMATE SET    **
       XBAR = 0.D0
       M1P1 = M1 + 1
       NMM3 = N - M3
       DO 20 J = M1P1, NMM3
   20 XBAR = XBAR + X(J)
       XBAR = XBAR / FM2
       S = 0.D0
       DO 21 J = M1P1, NMM3
   21 S = S + (X(J)-XBAR)**2
       SD1 = 0.D0
       IF (M1 .NE. M3) SD1 = (FM3-FM1)**2/FM2
       S = S / (DFLOAT(N)*BETA - (FM1+FM3+SD1)*HUBERK**2)
       IF (S .LE. 0.D0) GOTO 22
       S = DSQRT (S)
       T = XBAR + HUBERK*S*(FM3-FM1)/FM2
       GOTO 10
   22 IF (M1 .GT. 0) M1 = M1 - 1
       IF (M3 .GT. 0) M3 = M3 - 1
       GOTO 16
   50 ITER(I) = ITER(I) + 1
```

```
      RETURN
   60 WRITE (6,1000) I, M1, M3, X
 1000 FORMAT ('CABORTING HUBER ESTIMATE EVALUATION ON TRIAL',I4,
     .    '. M1 AND M3 ARE', 2I5,', AND THE SAMPLE IS:' / (8D16.7))
      RETURN
      ENTRY HFREQ
      WRITE (6,1001) ITER
 1001 FORMAT ('OTHE NUMBER OF ITERATIONS REQUIRED FOR THE HUBER',
     .    ' ESTIMATE HAS THE FOLLOWING FREQUENY DISTRIBUTION:' / (20
      RETURN
      END




      SUBROUTINE JAECKL (X, N, T)
C        JAECKEL ESTIMATE T OF X(1), ...,X(N)
      REAL*8 T
      REAL X(N)
      SIZE = N
      NH = (N+1)/2
      NB4 = (N+3) / 4
      SUM  =  0.0
      SSQ  =  0.0
      VMIN = 9999999.9
      DO 9 I = 1, NH
      NL = NH - I + 1
      NU = N - NL + 1
      END = X(NL) + X(NU)
      ESQ = X(NL)**2 + X(NU)**2
      SUM = SUM + END
      DN = NH - I
      W = SUM + END*DN
      DENOM = 2*I*(2*I-1)
      SSQ = SSQ + ESQ
      VAR = (SSQ + DN * ESQ - W**2 / SIZE) / DENOM
      IF (I .LT. NB4) GOTO 9
      IF (VAR .GT. VMIN) GOTO 9
      VMIN = VAR
      TWOI = 2 * I
      T = SUM / TWOI
    9 CONTINUE
      RETURN
      END




      SUBROUTINE JBT (X, N, T)
C        JAECKEL-BICKEL-TUKEY ESTIMATE OF 19 MAY 1971
      REAL*8 T
      REAL X(N)
      SIZE = N
      NH = (N+1)/2
      NB4 = (N+3) / 4
      SUM  =  0.0
      SSQ  =  0.0
      VMIN = 9999999.9
      DO 9 I = 1, NH
      NL = NH - I + 1
      NU = N - NL + 1
```

```
      END = X(NL) + X(NU)
      ESQ = X(NL)**2 + X(NU)**2
      SUM = SUM + END
      DN = NH - I
      W = SUM + END*DN
      DENOM = 2*I*(2*I-1)
      SSQ = SSQ + ESQ
      VAR = (SSQ + DN * ESQ - W**2 / SIZE) / DENOM
      IF (NH-I .NE. N/12 .AND. NH-I .NE. N/4) GOTO 9
      IF (VAR .GT. VMIN) GOTO 9
      VMIN = VAR
      TWOI = 2 * I
      T = SUM / TWOI
    9 CONTINUE
      RETURN
      END



      SUBROUTINE JKL(X,N,T)
C        JAECKEL-JOHNS ESTIMATE
      REAL X(N), U1(10), U2(10)
      IF (N .NE. 20) GOTO 425
      NU = 11
      ND = 10
      DO 5 I=1,5
      U1(I) = X(NU) - X(ND)
      U2(I) = U1(I)
      NU = NU + 1
    5 ND = ND - 1
      DO 9 I=6,9
      U1(I) = X(NU) - X(ND)
      U2(I) = U2(5)
      NU = NU + 1
    9 ND = ND - 1
      U1(10) = U1(9)
      U2(10) = U2(5)
      V11 = 0.
      V12 = 0.
      V22 = 0.
      DO 10 I=1,10
      V11 = V11 + U1(I)**2
      V12 = V12 + U1(I)*U2(I)
   10 V22 = V22 + U2(I)**2
      V11 = V11 * 25.
      V12 = V12 * 45.
      V22 = V22 * 81.
      T1 = 0.
      DO 19 I=2,19
   19 T1 = T1 + X(I)
      T2 = 0.
      DO 15 I=6,15
   15 T2 = T2 + X(I)
      D = V11 - 2.*V12 + V22
      TEST = 1000.*D/(V11+V22)
      IF(TEST.GE.1.0) GO TO 98
      C = 0.
      GO TO 99
   98 C = (V22 - V12)/D
```

```fortran
      IF(C.GT.1.5) C = 1.5
      IF(C.LT.-.5) C =  -.5
   99 T = C*T1/18. + (1.-C)*T2/10.
      RETURN
  425 ML = (N+1)/2
      T = (X(ML) + X(N-ML+1)) * 0.5D0
      END




      SUBROUTINE JOHNS (X, N, T)
C        VERNON JOHNS PROPOSAL OF 8 APRIL 1971 "SIMPLEST"
      IMPLICIT REAL*8 (A-H,O-Z)
      REAL X(N)
      REAL*8 N1, N2
      INTEGER R, RP1, S
      DATA R /1/
      IF (N .LT. 6) GOTO 3
      ND2 = N/2
      RP1 = R + 1
      S = (ND2 - R)/2
      IF (S .LT. 1) S = 1
      IRPS = R + S
      IRPSP1 = RP1 + S
      S1 = 0.D0
      S2 = 0.D0
      DO 1 J = RP1, IRPS
    1 S1 = X(J) + X(N+1-J) + S1
      MH = (N+1)/2
      DO 2 J = IRPSP1, MH
    2 S2 = X(J) + X(N+1-J) + S2
      D1 = 0.25D0 * (X(IRPS)-X(N+1-IRPS)+X(IRPSP1)-X(N-IRPS)
     .  +X(N+2-RP1)-X(RP1-1)+X(N+1-RP1)-X(RP1)   )
      D2 = 0.25D0*(X(N-IRPS)-X(IRPSP1)+X(N+1-IRPS)-X(IRPS))
      N1 = N - 2*R - 2*S
      RN1A = N + 2*R + 2*S
      N2 = N - 2*R
      C1A = DFLOAT(2*S)*RN1A/DFLOAT(2*R+S)/N2/D1**2 - 2.D0*N1/N2/D1
      C2A = 2.D0*N1/N2/D2**2 - 4.D0*DFLOAT(S)/N2/D1/D2
      SIGMAS = DFLOAT(N)/(2.D0*DFLOAT(S)*C1A + N1*C2A)
      T = (C1A*S1 + C2A*S2)/DFLOAT(N)*SIGMAS
      RETURN
    3 ML = (N+1)/2
      T = 0.5D0 * (X(ML)+X(N+1-ML))
      RETURN
      END




      SUBROUTINE JWTC2 (X, N, T)
C        IF SOME OUTSIDE CORNERS:  USE THE ONCE-C-SKIPPED MEDIAN
C        OUTSIDE SIDES BUT NOT CORNERS:  UNSKIPPED TRIMEAN
C        NOTHING OUTSIDE SIDES:  1-2-4-2-1 PENTAMEAN (EIGHTS-HINGES
C
C        JOHN TUKEY'S NOVEMBER 1970 PROPOSAL FOR A HOGG-TYPE
C        ESTIMATOR BASED ON SKIPPING INSTEAD OF KURTOSIS
      REAL X(N)
C
      ML = (N+1)/2
```

```
      XMED = 0.5D0 * (X(ML) + X(N-ML+1))
      CALL HINGES (X, N, XH1, XH2)
C        DECIDE ON TYPE OF ESTIMATE
      DH = XH2-XH1
      SM = AMAX1 (XH1-X(1), X(N)-XH2)
      IF (SM .GE. 2.*DH) GOTO 20
      IF (SM .GE. DH) GOTO 30
C        USE THE PENTAMEAN
      NB8=(N+7)/8
      NMNB8=N+1-NB8
      T = 4.0*XMED + 2.0*(XH1+XH2) + X(NB8)+X(NMNB8)
      T = T / 10.0
      RETURN
C        USE THE C-SKIPPED MEDIAN
   20 CALL OSKIP (X, N, 2.0, INIT, LEN)
      ML = (LEN+1)/2
      T = 0.5D0 * (X(INIT+ML-1)+X(INIT+LEN-ML))
      RETURN
   30 T = (XH1 + 2.*XMED + XH2) / 4.C
      RETURN
      END




      SUBROUTINE LFD (DATA, N, T)
C        LEAST FAVORABLE DISTRIBUTION
C        SUBROUTINES NEEDED
C          GMA(=PSIPRIME(K)/(PSIPRIME(K)+K*(PSI(K)-.5))-I/N)
C          RTNI(FINDS ZEROS OF GMA)
C          DAH2 (GIVES MODIFIED HUBER ESTIMATE)
C        USES COMMON FOR I/N OF GMA
      IMPLICIT REAL*8(A-H,O-Z)
      EXTERNAL GMA
      REAL*8 DATA(N)
      DIMENSION AK(40)
      COMMON V
      XST=1.
      EPS=.01
      IEND=20
      AN=N
      N1=N-1
      M=N1
      DO 11 I=1,N1
      V=I/AN
      CALL RTNI(X,F,DERF,GMA,XST,EPS,IEND,IER)
      AK(I)=X
      IF(X.GE..7D0) GOTO 11
      M=I
      GOTO 1
   11 CONTINUE
    1 I=1
    3 HUBERK=AK(I)
      CALL DAH2 (DATA, N, T, S, HUBERK, I)
      IQ=0
      DO 12 J=1,N
      X=DABS((DATA(J)-T)/S)
      IF(X.GT.AK(I)) IQ=IQ+1
   12 CONTINUE
      IF((IQ.LE.I).OR.(I.EQ.M )) GOTO 15
```

```
      I=MINO(IQ,M )
      GOTO 3
   15 CONTINUE
      RETURN
      END




      SUBROUTINE MELE (N,X,T)
C     MEAN LIKELIHOOD ESTIMATION OF LOCATION PARAMETER      R*4
      DIMENSION X(N)
      MAX=10
      EPS=5.E-4
      FN=N
      T=0.
      DO 210 I=1,N
  210 T=T+X(I)
      T=T/FN
      S=0.
      DO 220 I=1,N
  220 S=S+(X(I)-T)**2
      FN=N-1
      S=S/FN
      ITER=0
  250 ITER=ITER+1
      IF(ITER.GT.MAX) GO TO 700
      G=0.
      E=0.
      DO 260 I=1,N
      CALL HOOD (X(I),T,S,GG,EE)
      G=G+GG
  260 E=E+EE
      D=G/E
      T=T-D
      IF(ABS(D).GT.EPS) GO TO 250
      RETURN
  700 WRITE(6,13) MAX,D,EPS
      RETURN
   13 FORMAT (1H0,10X,'AFTER',I5,2X,'ITERATIONS, THE CHANGE IN THE
     1'LOCATION ESTIMATE,',E15.7,',',/1H0,10X,'IS STILL GREATER IN
     2'MAGNITUDE THAN',E15.7)
      END




      SUBROUTINE MI (A, N)
      IMPLICIT REAL*8 (A-H,O-Z)
      COMMON /DIM/ NROW, NCOL
      REAL*8 A(NROW, NCOL)
      DO 1 I = 1, N
    1 CALL SWP (A, N, I)
      RETURN
      END




      SUBROUTINE MIN1D(FUNCT,X,H,AMBDA,N,F,G,NUMF,IER,EPS,EST)
      IMPLICIT REAL*8(A-H,O-Z)
      REAL*8 H(1),X(1),G(1)
```

```
      LOGICAL CHECK
      IER=0
      DY=0.D0
      HNRM=0.D0
      GNRM=0.D0
      DO 10 J=1,N
      HNRM=HNRM+DABS(H(J))
      GNRM=GNRM+DABS(G(J))
10    DY=DY+H(J)*G(J)
      IF(DY)11,51,51
11    IF(HNRM/GNRM-EPS)51,51,12
12    FY=F
      ALFA=2.D0*(EST-F)/DY
      IF(X(N+1).GT.0.D0) ALFA=X(N+1)*ALFA/2.D0
      AMBDA=1.D0
      IF(ALFA)15,15,13
13    IF(ALFA-AMBDA)14,15,15
14    AMBDA=ALFA
15    ALFA=0.D0
16    FX=FY
      DX=DY
      DO 17 I=1,N
17    X(I)=X(I)+AMBDA*H(I)
171   IF(CHECK(X)) GOTO 50
      CALL FUNCT(X,F,G)
      NUMF=NUMF+1
      IF(F.LT.FX) RETURN
      FY=F
      DY=0.D0
      DO 18 I=1,N
18    DY=DY+G(I)*H(I)
      IF(DY)19,36,22
19    IF(FY-FX)20,22,22
20    AMBDA=AMBDA+ALFA
      ALFA=AMBDA
      IF(HNRM*AMBDA-1.D10)16,16,21
21    IER=2
      GOTO 51
22    T=0.D0
23    IF(AMBDA)24,36,24
24    Z=3.D0*(FX-FY)/AMBDA+DX+DY
      ALFA=DMAX1(DABS(Z),DABS(DX),DABS(DY))
      DALFA=Z/ALFA
      DALFA=DALFA**2-DX/ALFA*DY/ALFA
      IF(DALFA)51,25,25
25    W=ALFA*DSQRT(DALFA)
      ALFA=DY-DX+2.D0*W
      IF(ALFA)250,251,250
250   ALFA=(DY-Z+W)/ALFA
      GOTO 252
251   ALFA=(Z+DY-W)/(Z+DX+Z+DY)
252   ALFA=ALFA*AMBDA
      DO 26 I=1,N
26    X(I)=X(I)+(T-ALFA)*H(I)
      CALL FUNCT(X,F,G)
      NUMF=NUMF+1
      IF(F.LT.FX) GOTO 36
      IF(F.GT.FX) GOTO 28
      IF(F.LE.FY) GOTO 36
```

```
28      DALFA=0.D0
        DO 29 I=1,N
29      DALFA=DALFA+G(I)*H(I)
        IF(DALFA)30,33,33
30      IF(F-FX)32,31,33
31      IF(DX-DALFA)32,36,32
32      FX=F
        DX=DALFA
        T=ALFA
        AMBDA=T
        GOTO 23
33      IF(FY-F)35,34,35
34      IF(DY-DALFA)35,36,35
35      FY=F
        DY=DALFA
        AMBDA=AMBDA-ALFA
        GOTO 22
36      AMBDA=AMBDA-ALFA
        RETURN
50      AMBDA=.5*AMBDA
        DO 501 I=1,N
501     X(I)=X(I)-AMBDA*H(I)
        GOTO 171
51      CONTINUE
        IF(DY.GE.0.D0)IER=-2
        IF(HNRM/GNRM.LE.EPS)IER=-3
        IF(DALFA.LT.0.D0)IER=-1
        PRINT 54,IER
54      FORMAT(///1X,T10,'ERROR HAS OCCURED, IER=',I2,///)
        RETURN
        END




        SUBROUTINE MLE (X, N, T)
C           MAXIMUM LIKELIHOOD ESTIMATE
C           IT IS NECESSARY TO CHANGE THE CALL FOR EACH DISTRIBUTION R
        IMPLICIT REAL*8 (A-H,O-Z)
        REAL*8 E(4), X(N), LIK
        COMMON/BTFMLY/X1(40),N1
        EXTERNAL NORUNI
        DATA LIMIT, N2, EST /20, 2, 0.D0/
C           COME UP WITH SOME INITIAL VALUES TO TRY
        DO 1 I=1,N
1       X1(I)=X(I)
        N1=N
        ML = (N+1)/2
        E(1) = (X(ML)+X(N-ML+1))*0.5D0
        ND4 = N/4
        F1 = DFLOAT (4*ND4 + 3 - N) / 4.D0
        RIQ = (1.D0-F1)*(X(N-ND4)-X(ND4+1)) + F1*(X(N-ND4+1)-X(ND4))
        IF(RIQ.EQ.0.D0) RIQ=1.D-7
        E(2) = RIQ / 1.34D0
        CALL THETA(E,N2,NORUNI,EST,LIMIT,MODE,LIK)
        T=E(1)
        RETURN
        END
```

```
      SUBROUTINE MLTFMY(X,DATA,N2)
      IMPLICIT REAL*8(A-H,O-Z)
      REAL*8 DATA(N2), X(4), D(2)
      EXTERNAL TFMILY
      COMMON/BTFMLY/DAT(40),N
      COMMON/BAK/AK
      N=N2
      DO 1 I=1,N
1     DAT(I)=DATA(I)
      LIMIT=20
      N=N2
      ML=(N+1)/2
      XMED=(DATA(ML)+DATA(N-ML+1))*.5
      N4=N/4
      F1=(4.*N4+3.-N)/4.
      RIQ=(1.-F1)*(DATA(N-N4)-DATA(N4+1))+F1*(DATA(N-N4+1)-DATA(N4))
      S=RIQ/1.3496
      N=2
      MODE=1
      EST=0.
      X(1)=XMED
      X(2)=S
      DO 5 K=1,13
      IF(K.LE.7) AK=.5D0*(1+K)
      IF(K.GE.8) AK=K-3
      CALL TFMILY(X,FF,D)
      IF(K.GT.1) GOTO 3
      ST1=AK
      ST2=FF
      GOTO 5
3     IF(FF.GE.ST2) GOTO 6
      ST1=AK
      ST2=FF
5     CONTINUE
6     AK=ST1
      CALL THETA(X,N,TFMILY,EST,LIMIT,MODE,FF)
      RETURN
      END




      SUBROUTINE MSKIP (X, N, T, K, SKF)
C         MULTIPLIED SKIPPING (JWT 16 MAY 1971)
C         K IS THE MULTIPLIER
C         SKF IS THE SKIPPING FACTOR (2.0=C, FOR EXAMPLE)
C   BIG SKIP    SKIPS SOME AND THEN SOME MORE
      REAL*8 T
      REAL*4 X(N), K
      CALL OSKIP (X, N, SKF, INIT, LEN)
      NOUT=N-LEN
      NQ1=NOUT*K
      IF(NQ1.LT.1) NQ1=1
      NQ=(N-NOUT-1)/2
      IF(NQ.GT.NQ1) NQ=NQ1
      NBTTM=INIT+NQ
      NTOP=INIT+LEN-1-NQ
      C=0.0
      T=0.0
```

```
      DO 2 I=NBTTM,NTOP
      C=C+1.0
    2 T=T+X(I)
      T=T/C
      RETURN
      END




      SUBROUTINE MSKIPS (X, N, T, K, SKF)
C         MULTIPLIED SKIPPING (JWT 16 MAY 1971)
C         K IS THE MULTIPLIER
C         SKF IS THE SKIPPING FACTOR (2.C=C, FOR EXAMPLE)
C     BIG SKIP   SKIPS SOME AND THEN SOME MORE
      REAL*8 T
      REAL*4 X(N), K
      CALL OSKIP (X, N, SKF, INIT, LEN)
      NOUT=N-LEN
      NQ1=NOUT*K
      NQ=(N-NOUT-1)/2
      IF (NQ .GT. NQ1) NQ = NQ1
      NBTTM=INIT+NQ
      NTOP=INIT+LEN-1-NQ
      C=C.0
      ML = (NBTTM+NTOP)/2
      T = 0.25DO * (X(NBTTM)+X(ML)+X(NBTTM+NTOP-ML)+X(NTOP))
      RETURN
      END




      SUBROUTINE MSKIPO (X, N, T, K, SKF)
C         MULTIPLIED SKIPPING (JWT 16 MAY 1971)
C         K IS THE MULTIPLIER
C         SKF IS THE SKIPPING FACTOR (2.0=C, FOR EXAMPLE)
C     BIG SKIP   SKIPS SOME AND THEN SOME MORE
      REAL*8 T
      REAL*4 X(N), K
      CALL OSKIP (X, N, SKF, INIT, LEN)
      NOUT=N-LEN
      NQ1=NOUT*K
      NQ=(N-NOUT-1)/2
      IF (NQ .GT. NQ1) NQ = NQ1
      NBTTM=INIT+NQ
      NTOP=INIT+LEN-1-NQ
      C=C.0
      T=0.0
      DO 2 I=NBTTM,NTOP
      C=C+1.0
    2 T=T+X(I)
      T=T/C
      RETURN
      END




      SUBROUTINE MSKIP4 (X, N, T, K, SKF)
C         MULTIPLIED SKIPPING (JWT 18 MAY 1971)
C         K IS THE MULTIPLIER
```

```
      SKF IS THE SKIPPING FACTOR (2.0=C, FOR EXAMPLE)
BIG SKIP    SKIPS SOME AND THEN SOME MORE
      REAL*8 T
      REAL*4 X(N), K
      CALL OSKIP (X, N, SKF, INIT, LEN)
      NOUT=N-LEN
      NQ1=NOUT*K
      IF(NQ1.LT.1) NQ1=1
      NQ = (N-NOUT-2*N/5)/2
      IF(NQ.GT.NQ1) NQ=NQ1
      NBTTM=INIT+NQ
      NTOP=INIT+LEN-1-NQ
      C=0.0
      T=0.0
      DO 2 I=NBTTM,NTOP
      C=C+1.0
      T=T+X(I)
      T=T/C
      RETURN
      END


      SUBROUTINE NORUNI(X,V,D)
         GIVES - VALUE OF LN OF LIKELIHOOD OF NORMAL/UNIFORM AND DERIVA
      IMPLICIT REAL*8(A-H,O-Z)
      REAL*8 X(4),D(2)
      COMMON/BTFMLY/DATA(40),N
      THETA=X(1)
      SIGMA=X(2)
      LN5 = DLOG(0.5D0)
      RSIG2=1./SIGMA**2
      STORE1=0.
      STORE2=0.
      STORE3=0.
      M  = N
      DO 1 I=1,N
      CDATA=DATA(I)-X(1)
      IF(DABS(CDATA).GT.1.D-14) GOTO 2
      M=M-2
      STORE2=STORE2+1./SIGMA
      STORE3 = STORE3 - LN5
      GOTO 1
    2 CDATA2 = CDATA**2
      EXPR=DEXP(-.5*CDATA2*RSIG2)
      IF(EXPR.EQ.1.) GOTO 3
      TSTORE=CDATA*EXPR*RSIG2/(1.-EXPR)
      STORE1=STORE1+TSTORE-2./CDATA
      STORE2=STORE2+TSTORE*CDATA/SIGMA
      STORE3 = STORE3 - DLOG(1.-EXPR) + DLOG(CDATA2)
    1 CONTINUE
      V = -M * DLOG(SIGMA) + STORE3
      D(1)=STORE1
      D(2)=-N/SIGMA+STORE2
      RETURN
      END
```

```
      SUBROUTINE OLSHEN (X, N, T)
C         OLSHEN'S ESTIMATE AS OF 3/8/71
C         THOUGHT TO BE THE MLE FOR A T OF ABOUT 2.5 D.F.
C
C         IDEA IS TO DIVIDE EACH OBSERVATION BY AN ESTIMATE OF THE
C         SQUARE OF ITS VARIANCE (THE CONDITIONALLY SUFFICIENT STATI
C         IF WE KNEW THE VARIANCES OF THE INDIVIDUAL OBSERVATIONS
C         EXACTLY
C
      REAL X(N)
      ML = (N+1)/2
      T = (X(ML) + X(N-ML+1)) * 0.5
      IF (N .LT. 4) RETURN
      ND4 = N / 4
      F1 =  FLOAT (4*ND4 + 3 - N) / 4.D0
      RIQ = (1.0-F1) * (X(N-ND4) - X(ND4+1)) + F1*(X(N-ND4+1) - X(N
      HRIQSQ = RIQ**2 * 0.5
      I = 0
    1 I = I + 1
      TOLD = T
      SUM1 = 0.0
      SUM2 = 0.0
      DO 2 I = 1, N
      TERM = 1.0 / ((X(I)-T)**2 + HRIQSQ)
      SUM1 = SUM1 + X(I) * TERM
    2 SUM2 = SUM2 + TERM
      T = SUM1 / SUM2
      IF (I .GT. 10) RETURN
      IF (ABS(T-TOLD) .LE. 1.E-4*RIQ) RETURN
      GOTO 1
      END




      SUBROUTINE OSKIP (X, N, SKF, INIT, LEN)
C         ***   SEE INTRODUCTION TO EXPLORATORY DATA ANALYSIS   ***
C            X IS THE REAL*4 VECTOR OF OBSERVATIONS, LENGTH N.
C               --> X MUST BE SORTED IN ASCENDING ORDER <--
C            SKF IS THE SKIPPING FACTOR: C=2.0, T=1.5, S=1.0
C              S >= T >= C IN SEVERITY
C            SKIPPED SAMPLE IS X(INIT), LENGTH LEN (DEFINING A VEC
C            HL AND HU ARE THE LOWER AND UPPER HINGES
      REAL X(N)
      INIT = 1
      IEND = N
      IF (N-4) 60, 40, 10
   10 CALL HINGES (X, N, HL, HU)
   11 DIFF = (HU - HL) * SKF
      PL = HL - DIFF
      PU = HU + DIFF
   22 IF (X(INIT) .GE. PL) GOTO 25
      INIT = INIT + 1
      GOTO 22
   25 IF (X(IEND) .LE. PU) GOTO 29
      IEND = IEND - 1
      GOTO 25
   29 LEN = IEND - INIT + 1
      RETURN
C         FOR N = 4 CHECK TO SEE IF ONE END GAP IS TWICE THE RANGE
```

```
          OF THE OTHER THREE OBSERVATIONS
   40 IF (X(2)-X(1) .LE. 2.*(X(4)-X(2))) GOTO 45
      INIT = 2
      LEN = 3
      RETURN
   45 IF (X(4)-X(3) .LE. 2.*(X(3)-X(1))) RETURN
      LEN = 3
      RETURN
          FOR N = 3, CHECK TO SEE IF ONE GAP IS 10 TIMES THE OTHER
   6C IF (N .LE. 2) GOTO 29
      IF (X(2)-X(1) .LT. 10.*(X(3)-X(2))) GOTO 65
      INIT = 2
      LEN = 2
      RETURN
   65 IF (X(3)-X(2) .LT. 10.*(X(2)-X(1))) RETURN
      LEN = 2
      RETURN
      END




      SUBROUTINE OUTMN(ARRAY,NDIM,XOUTM)
      IMPLICIT REAL*8(A-H,O-Z)
      DIMENSION ARRAY(NDIM)
      DATA ALPHA/.5D0/
      X=DFLOAT(NDIM)*ALPHA*.5D0
      J=X
      XOUTM=0.D0
      IF(J.EQ.0)GOTO 14
      DO 21 I=1,J
   1  XOUTM=XOUTM+ARRAY(I)
   4  L=NDIM-J+1
      IF(L.GT.NDIM)GOTO 15
      DO 13 I=L,NDIM
   3  XOUTM=XOUTM+ARRAY(I)
   5  CONTINUE
      XOUTM=XOUTM+(X-DFLOAT(J))*(ARRAY(J+1)+ARRAY(L-1))
      XOUTM=XOUTM/(DFLOAT(NDIM)*(1.D0-ALPHA))
      RETURN
      END




      FUNCTION PB1 (X, N, START, K)
          RETURNS FIRST NEWTON APPROXIMATION TO HUBER ESTIMATE
          GIVEN VALUE "START", AS PROPOSED BY PETER BICKEL 8 OCT. 1970
          OR BEFORE.
      REAL X(N), START, K
      SUM = 0.0
      NCENT = 0
      DO 1 I = 1,N
      DIFF = X(I) - START
      IF (DIFF .GT. K) GOTO 2
      IF (DIFF .LT. -K) GOTO 3
      SUM = SUM + DIFF
      NCENT = NCENT + 1
      GOTO 1
    2 SUM = SUM + K
      GOTO 1
```

```
      3 SUM = SUM - K
      1 CONTINUE
        IF (NCENT .EQ. 0) GOTO 5
        PB1 = START + SUM / NCENT
        RETURN
      5 ML = (N+1)/2
        PB1 = (X(ML) + X(N-ML+1)) * 0.5
        RETURN
        END




        FUNCTION PB2 (X, N, START, K1)
C           RETURNS FIRST NEWTON APPROXIMATION TO HUBER ESTIMATE
C           GIVEN VALUE "START", AS PROPOSED BY PETER BICKEL 8 OCT. 1970
C           OR BEFORE.
        REAL X(N), START, K, K1
        ND4P1 = N/4+1
        RIQ = (X(N-ND4P1 + 1) - X(ND4P1))
        K = K1 * RIQ / 1.1805
        SUM = 0.0
        NCENT = 0
        DO 1 I = 1,N
        DIFF = X(I) - START
        IF (DIFF .GT. K) GOTO 2
        IF (DIFF .LT. -K) GOTO 3
        SUM = SUM + DIFF
        NCENT = NCENT  + 1
        GOTO 1
      2 SUM = SUM + K
        GOTO 1
      3 SUM = SUM - K
      1 CONTINUE
        IF (NCENT .EQ. 0) GOTO 5
        PB2 = START + SUM / NCENT
        RETURN
C           IF THE SLOPE IS ZERO, USE THE MEDIAN
      5 ML = (N+1)/2
        PB2 = (X(ML) + X(N-ML+1)) * 0.5
        RETURN
        END




        DOUBLE PRECISION FUNCTION PB3 (X, N, K1)
C           ONE STEP APPROXIMATION TO THE HUBER ESTIMATE
C           STARTING WITH THE MEDIAN, AND USING HAMPEL'S SCALE.
        REAL X(N), START, K, K1, X1(40)
        ML = (N+1)/2
        XMED = 0.5D0*(X(ML)+X(N-ML+1))
        DO 615 I = 1, N
    615 X1(I) = ABS (X(I) - XMED)
        CALL SORT (X1, N)
        START = XMED
        K = K1 * (X1(ML)+X1(N-ML+1)) / 1.3507
        SUM = 0.0
        NCENT = 0
        DO 1 I = 1,N
        DIFF = X(I) - START
```

```
      IF (DIFF .GT. K) GOTO 2
      IF (DIFF .LT. -K) GOTO 3
      SUM = SUM + DIFF
      NCENT = NCENT  + 1
      GOTO 1
    2 SUM = SUM + K
      GOTO 1
    3 SUM = SUM - K
    1 CONTINUE
      PB3 = START + SUM/NCENT
      RETURN
      END




      SUBROUTINE RTNI(X,F,DERF,FCT,XST,EPS,IEND,IER)
         FINDS ZEROS OF FCT
      IMPLICIT REAL*8(A-H,O-Z)
      IER=0
      X=XST
      TOL=X
      CALL FCT(TOL,F,DERF)
      TOLF=100.*EPS
      DO 6 I=1,IEND
      IF(F)1,7,1
      IF(DERF)2,8,2
      DX=F/DERF
      X=X-DX
      TOL=X
      CALL FCT(TOL,F,DERF)
      TOL=EPS
      A=DABS(X)
      IF(A-1.)4,4,3
      TOL=TOL*A
      IF(DABS(DX)-TOL)5,5,6
      IF(DABS(F)-TOLF)7,7,6
      CONTINUE
      IER=1
      RETURN
      IER=2
      RETURN
      END




      SUBROUTINE SH(X,N,T)
   THE ANDREWS(HAMPEL(JAECKEL(HUBER(FISHER)))) ESTIMATE
      COMPUTE M-ESTIMATE PSI FUNCTION IS SIN(X-T)/S
         VECTOR X (LENGTH N) IS INPUT, LOCATION T IS OUTPUT.
         ---> X IS ASSUMED TO BE SORTED  <---
         ---> N >= 4  <---
         ASSOCIATED SCALE S IS COMPUTED INTERNALLY.
      IMPLICIT REAL*8 (A-H,O-Z)
      REAL*8 X(N)
      DIMENSION DD(100)

      DATA PI /3.14159265357828/
      DATA ISTOP /20/
         VECTOR ITER STORES A FREQUENCY CHART OF THE PERFORMANCE.
```

```
          INTEGER ITER(20)
          DATA ITER /20*0/
C            **   COMPUTE INITIAL ESTIMATES S AND T    **
          HUBERK=PI
          ML = (N+1)/2
          T = (X(ML)+X(N-ML+1))*0.5D0
          ND4 = N/4
          III=0
  166     III=III+1
          DO 90 I99=1,N
          D=X(I99)-T
          DD(I99)=DABS(D)
  90      CONTINUE
          CALL SORT8(DD,N)
          M1=(N+1)/2
          M2=(N+2)/2
          S=(DD(M1)+DD(M2))/2.0D0
          S=S*2.1D0
C            **   COMPUTE WINSORIZING COUNTS AND DECIDE ON STOPPING    **
          I = 0
          M1OLD = -1
          M3OLD = -1
   10     CONTINUE
          XL = T - S*HUBERK
          XU = T + S*HUBERK
          DO 11 J = 1, N
          IF (X(J) .GT. XL) GOTO 12
   11     CONTINUE
          J = N + 1
   12     M1 = J - 1
          DO 13 J = 1, N
          IF (X(N-J+1) .LT. XU) GOTO 14
   13     CONTINUE
          J = N + 1
   14     M3 = J - 1
          IF (M1 .EQ. M1OLD .AND. M3 .EQ. M3OLD) GOTO 50
   16     I = I + 1
          IF (I .GE. 20) GOTO 60
          FM1 = M1
          FM3 = M3
          FM2 = N - M3 - M1
          IF (FM2 .LT. 0.5D0) GOTO 60
          M1OLD = M1
          M3OLD = M3
C            **   UPDATE ESTIMATE SET    **
          M1P1 = M1 + 1
          NMM3 = N - M3
          SS=0.D0
          SC=0.D0
          DO 20 J = M1P1, NMM3
          Z=(X(J)-T)/S
          SZ=DSIN(Z)
          CZ=DCOS(Z)
          SS=SS+SZ
          SC=SC+CZ
   20     CONTINUE
          T=T+S*DATAN(SS/SC)
          GOTO 10
   50     ITER(I) = ITER(I) + 1
```

```
      IF(III.LT.3) GO TO 166
      RETURN
   60 WRITE (6,1000) I, M1, M3, X
 1000 FORMAT ('0ABORTING ANDREWS-MODIFIED-HUBER ON TRIAL',I4,
     .   '. M1 AND M3 ARE', 2I5,', AND THE SAMPLE IS:' / (8D16.7))
      RETURN
      END



      SUBROUTINE SHORTH(ARRAY,NDIM,XCLOS)
      IMPLICIT REAL*8(A-H,O-Z)
      DIMENSION ARRAY(NDIM)
C  HALF IS ROUNDED UP IF SAMPLE SIZE IS ODD
      N=(NDIM+1)/2
      NDN=NDIM-N
      M=1
      L=N
      D=ARRAY(N)-ARRAY(1)
      DO 4 I=1,NDN
      D1=ARRAY(N+I)-ARRAY(I+1)
      IF(D1.GE.D)GOTO 4
      D=D1
      M=I+1
      L=N+I
    4 CONTINUE
      CALL XMEAN(ARRAY(M),N,XCLOS)
      RETURN
      END



      SUBROUTINE SKIP (X, N, SKF, INIT, LEN)
      REAL X(N)
      INIT = 1
      LEN = N
    1 LEN1 = LEN
      INIT1 = INIT
      CALL OSKIP (X(INIT), LEN1, SKF, INIT, LEN)
      INIT = INIT + INIT1 - 1
      IF (LEN1 .NE. LEN) GOTO 1
      RETURN
      END



      SUBROUTINE SYMJAE (X, N, T)
C          FOR COMPUTING VARIANCE, SYMMETRIZING IS AROUND MEDIAN
C          JAECKEL ESTIMATE T OF X(1), ...,X(N)
      IMPLICIT REAL*8 (A-H,O-Z)
      REAL*8 T, X(N)
      REAL XSYM(80)
      ML = (N+1)/2
      XMED = (X(ML)+X(N-ML+1))*0.5D0
      IL = 1
      IH = N
      D1 = XMED-X(IL)
      D2 = X(IH)-XMED
      DO 10 I = 1, N
```

```fortran
      IF (D2 .GT. D1) GOTO 11
      XSYM(I) = -D1
      IL = IL + 1
      D1 = XMED - X(IL)
      GOTO 10
11    XSYM(I) = -D2
      IH = IH - 1
      D2 = X(IH) - XMED
10    XSYM(2*N-I+1) = -XSYM(I)
      NT2 = N*2
      SIZE = NT2
      NH = N
      NB4 = (NT2 + 3)/4
      SSQ  =  0.0
      VMIN = 9999999.9
      DO 9 I = 1, NH
      NL = NH - I + 1
      NU = N - NL + 1
      ESQ = XSYM(NL)**2
      DN = NH - I
      DENOM = 2*I*(2*I-1)
      SSQ = SSQ + ESQ
      VAR = (SSQ + DN*ESQ)/DENOM
      IF (I .LT. NB4) GOTO 9
      IF (VAR .GT. VMIN) GOTO 9
      VMIN = VAR
      IMIN = I
9     CONTINUE
      ALPHA = (NH-IMIN)/DFLOAT(2*N)
      CALL TRMDMN (X, N, T, ALPHA)
      RETURN
      END




      SUBROUTINE SORT (V1,N)
      REAL V1(N)
C              SHELL SORT ALGORITHM CACM JULY 1964
      I = 1
1     I = I + I
      IF (I .LE. N) GOTO 1
      M = I - 1
2     M = M/2
      IF (M .EQ. 0) RETURN
      K = N - M
      DO 4 J = 1,K
      L = J
5     IF (L .LT. 1) GOTO 4
      IF (V1(L+M) .GE. V1(L)) GOTO 4
      X = V1(L+M)
      V1(L+M) = V1(L)
      V1(L) = X
      L = L - M
      GOTO 5
4     CONTINUE
      GOTO 2
      END
```

```
   SUBROUTINE SORT8 (V1, N)
   IMPLICIT REAL*8 (A-H,O-Z)
   REAL*8 V1(N)
             SHELL SORT ALGORITHM CACM JULY 1964
   I = 1
 1 I = I + I
   IF (I .LE. N) GOTO 1
   M = I - 1
 2 M = M/2
   IF (M .EQ. 0) RETURN
   K = N - M
   DO 4 J = 1,K
   L = J
 5 IF (L .LT. 1) GOTO 4
   IF (V1(L+M) .GE. V1(L)) GOTO 4
   X = V1(L+M)
   V1(L+M) = V1(L)
   V1(L) = X
   L = L - M
   GOTO 5
 4 CONTINUE
   GOTO 2
   END



   SUBROUTINE SWP (A, N, K)
   IMPLICIT REAL*8 (A-H,O-Z)
   COMMON /DIM/ NROW, NCOL
   REAL*8 A(NROW,NCOL)
   F = A(K,K)
   A(K,K) = 1.D0
   DO 4 I = 1, N
   IF (I .EQ. K) GOTO 4
   G = - A(I,K) / F
   A(I,K) = G
   DO 2 J = 1, N
   IF (J .EQ. K) GOTO 2
   A(I,J) = A(I,J) + A(K,J) * G
 2 CONTINUE
 4 CONTINUE
   DO 3 J = 1, N
 3 A(K,J) = A(K,J) / F
   RETURN
   END



   SUBROUTINE TFMILY(P,V,D)
      INPUT P(1)=THETA,P(2)=SIGMA,DATA(N),AK=DEG OF FREEDOM
      OUTPUT V=-VALUE OF LN OF LIKELIHOOD
        D(I)=DERIVATIVES
      NEEDS SUBROUTINE ALGMMA(=LN GAMMA)
   IMPLICIT REAL*8(A-H,O-Z)
   REAL*8 P(4),D(2),DATA(40)
   COMMON/BTFMLY/DATA,N/BAK/AK
   THETA=P(1)
   SIGMA=P(2)
```

```fortran
      SIGMAK=SIGMA**2*AK
      STORE1=0.D0
      STORE2=0.D0
      STORE3=0.D0
      DO 10 I=1,N
      DTTHTA=DATA(I)-THETA
      RATIO1=DTTHTA/(SIGMAK+DTTHTA**2)
      STORE1=STORE1+RATIO1
      RATIO2=DTTHTA*RATIO1
      STORE2=STORE2+RATIO2
10    STORE3=STORE3+DLOG(1.D0+(DTTHTA**2/SIGMAK))
      D(1)=-(AK+1.D0)*STORE1
      D(2)=1.D0/SIGMA*(N- (AK+1.D0)*STORE2)
      V=N*(-ALGMMA((AK+1.D0)/2.D0)+.5D0*DLOG(AK)+ALGMMA(.5D0*AK)
     *  +DLOG(SIGMA))+.5D0*(AK+1.D0)*STORE3
      RETURN
      END



      SUBROUTINE THETA(X,NDIM,FUNCT,EST,MAX,MODE,F2)
      IMPLICIT REAL*8(A-H,O-Z)
      REAL*8 X(1),X1(32),X2(32),G1(32),G2(32),ALFA(32),H(32),P(32
     * Y(32),PY(32),PE(32),ETA(4),BIGV(32)
      DATA ETA/1.D-2,1.D-4,1.D-16,1.D-16/
      IFLAG=0
      M=C
      N2=NDIM+1
      N1=NDIM+2
      X(N2)=2.
      X(N1)=0.
      NUMF=0
      IER=0
      DO 1 I=1,N1
1     X1(I)=X(I)
      CALL FUNCT(X1,F1,G1)
      NUMF=NUMF+1
      DO 2 I=1,NDIM
      X2(I)=X1(I)
      G2(I)=G1(I)
2     H(I)=-G1(I)
      F2=F1
      X2(N2)=X1(N2)
      X2(N1)=X1(N1)
6     CONTINUE
      KOUNT=0
      EPS=ETA(4)
      CALL MIN1D(FUNCT,X2,H,RO,NDIM,F2,G2,NUMF,IER,EPS,EST)
      IF(IER.NE.C)GOTO 100
      DO 3 I=1,N1
      BIGV(I)=X2(I)
3     ALFA(I)=X2(I)
      RO=-RO
      GG=0.D0
      DO 9 I=1,NDIM
9     GG=GG+G2(I)**2
      GG=DSQRT(GG)
4     DO 10 I=1,N1
      DO 5 J=1,N1
```

```
5       P(I,J)=0.D0
10      P(I,I)=1.D0
11      CONTINUE
        KOUNT=1
12      DO 15 I=1,NDIM
15      Y(I)=G2(I)
        Y(N2)=F2
        Y(N1)=ETA(1)
        V=0.D0
        DO 20 I=1,NDIM
20      V=V+X2(I)*G2(I)
        YA=0.D0
        DO 25 I=1,N1
25      YA=YA+Y(I)*ALFA(I)
        VYA=V-YA
        BIGV(KOUNT)=V
        DO 30 I=1,N1
        PY(I)=0.D0
        PE(I)=P(I,KOUNT)
        DO 30 J=1,N1
30      PY(I)=PY(I)+P(J,I)*Y(J)
        EPY=PY(KOUNT)
        IF(DABS(EPY).LT.ETA(3)) GOTO 130
        PY(KOUNT)=PY(KOUNT)-1.DC
        DO 50 I=1,N1
        DO 50 J=1,N1
5C      P(I,J)=P(I,J)-PE(I)*PY(J)/EPY
        DO 40 I=1,N1
        ALFA(I)=0.D0
        DO 40 J=1,N1
4C      ALFA(I)=ALFA(I)+P(I,J)*BIGV(J)
        DEL=0.D0
        DO 60 I=1,NDIM
60      DEL=DEL+G2(I)*(X2(I)-ALFA(I))
        IF(DABS(DEL).GT.ETA(4)) GOTO 65
        IF(IFLAG.EQ.1) GOTO 160
        IFLAG=1
        GOTO 130
65      IFLAG=0
        DO 61 I=1,N1
        H(I)=X2(I)-ALFA(I)
        IF(DEL.GT.0.D0) H(I)=-H(I)
61      CONTINUE
        DO 62 I=1,NDIM
        X1(I)=X2(I)
62      G1(I)=G2(I)
        F1=F2
        X1(N2)=X2(N2)
        X1(N1)=X2(N1)
        X2(N2)=ALFA(N2)
        X2(N1)=ALFA(N1)
        CALL MIN1D(FUNCT,X2,H,RO,NDIM,F2,G2,NUMF,IER,EPS,EST)
        IF(IER.NE.0) GOTO 100
        IF(DEL.GT.0.D0)RO=-RO
        GG=0.D0
        DO 86 I=1,NDIM
86      GG=GG+G2(I)**2
        GG=DSQRT(GG)
        KOUNT=KOUNT+1
```

```fortran
      M=M+1
      IF(((F1-F2).LE.ETA(2)).OR.(M.GT.MAX).OR.(GG.LT.ETA(1))) GOTO 160
      IF(KOUNT.LE.N1) GOTO 12
      GOTO 11
100   PRINT 110,IER
110   FORMAT(///,1X,T20,'THE PROGRAM HAS FAILED---IER= ',I2)
      RETURN
130   PRINT 140
140   FORMAT(///20X,'A RESTART HAS OCCURED'///)
      DO 150 I=1,NDIM
      X1(I)=X2(I)
      G1(I)=G2(I)
150   H(I)=-G1(I)
      F1=F2
      X1(N2)=X(N2)
      X1(N1)=X(N1)
      X2(N2)=X(N2)
      X2(N1)=X(N1)
      GOTO 6
160   DO 170 I=1,N1
170   X(I)=X2(I)
      RETURN
      END




      DOUBLE PRECISION FUNCTION TKUCHI (X, N)
C         TAKEUCHI'S ESTIMATE (JASA 1970)
C         WP SUPPLIES PROBABILITY Y(R) = X(J)
C         PJ SUPPLIES PROBABILITY Y(R) = X(I) AND Y(T) = X(J), T > R
      REAL*8 S(12,12),TR(12),S1(12,12),DSQRT
      REAL X(N), C(40)
      INTEGER R, T, R1, RP1, RM1, RM
      K = DSQRT (37.D0*N/20.D0)
      SP = 1.0
C         COMPUTE TR(1...K) AVERAGES OF SUBSAMPLE ORDER STATISTICS
C         CONDITIONAL ON THIS SAMPLE.  STRICTLY COMBINATORICS.
      DO 1 R = 1, K
      SP = SP * FLOAT(K-R+1) / FLOAT(N-R+1)
      TR(R) = 0.0
      NPRMK = N + R - K
      WP = SP
      DO 2 J = R, NPRMK
      TR(R) = TR(R) + WP*X(J)
      IF (J .EQ. NPRMK) GOTO 1
2     WP = WP * FLOAT(J*(N-J-K+R))/FLOAT((J-R+1)*(N-J))
1     CONTINUE
C
C         COMPUTE THE SUBSAMPLE VARIANCE-COVARIANCE MATRIX COND.
C         ON THIS SAMPLE.  ALSO STRICTLY COMBINATORIAL.
C
      PR = FLOAT(K) / FLOAT(N)
      KM1 = K - 1
      DO 3 R = 1 , K
      WP = PR
      NPRMK = N + R - K
      IF (R .EQ. K) GOTO 10
      PR = PR * FLOAT(K-R)/FLOAT(N-R)
      PT = PR
```

```
      RP1 = R + 1
      DO 4 T = RP1, K
      NPTMK = N + T - K
      PI = PT
      S(R,T) = 0.D0
      DO 6 I = R, NPRMK
      IPTMR = I + T - R
      PJ = PI
      DO 7 J = IPTMR, NPTMK
      S(R,T) = S(R,T) + PJ*X(I)*X(J)
      IF (J .EQ. NPTMK) GOTO 9
    7 PJ = PJ * FLOAT ((J-I)*(N-J-K+T))/ FLOAT ((J-I-T+R+1)*(N-J))
    9 IF (I .EQ. NPRMK) GOTO 8
    6 PI = PI * FLOAT((N-K+R-I)*I)/FLOAT((I-R+1)*(N-T+R-I))
    8 S(R,T) = S(R,T) - TR(R)*TR(T)
      S(T,R) = S(R,T)
    4 PT = PT * FLOAT(K-T)/FLOAT(N-T)
   10 S(R,R) = -TR(R)**2
      DO 5 J = R, NPRMK
      S(R,R) = S(R,R) + WP*X(J)**2
      IF (J .EQ. NPRMK) GOTO 3
    5 WP = WP * FLOAT(J*(N-J-K+R))/FLOAT((J-R+1)*(N-J))
    3 CONTINUE
C     SYMMETRIZE S
      DO 13 R = 1, K
      DO 13 T = 1, K
   13 S1(R,T) = (S(R,T)+S(K-R+1,K-T+1))/2.D0
      CALL DIMEN1(S1,12,12)
      CALL MI (S1, K)
      SUM2 = 0.D0
      TKUCHI = 0.D0
      DO 12 R = 1, K
      SUM1 = 0.D0
      DO 11 T = 1, K
   11 SUM1 = SUM1 + S1(T,R)
      SUM2 = SUM2 + SUM1
   12 TKUCHI = TKUCHI + SUM1 * TR(R)
      TKUCHI = TKUCHI / SUM2
      RETURN
      END



      DOUBLE PRECISION FUNCTION TRIMEN (X, N)
      REAL X(N)
      IF (N-3) 2, 3, 4
    4 IF (N .EQ. 4) GOTO 42
      M1 = (N+1) / 2
      CALL HINGES (X, N, H1, H2)
      TRIMEN = (H1 + H2 + X(M1) + X(N-M1 + 1)) / 4.D0
      RETURN
    3 TRIMEN = (X(1)+X(3)+X(2)+X(2))/4.D0
      RETURN
    2 TRIMEN = (X(1)+X(N))/2.D0
      RETURN
   42 TRIMEN = (X(1)+X(4)+3.D0*(X(2)+X(3))) / 8.D0
      RETURN
      END
```

```
      SUBROUTINE TRIMMD(ARRAY,NDIM,XTRIMD)
      IMPLICIT REAL*8(A-H,O-Z)
      DIMENSION ARRAY(NDIM)
      XTRIM D=0.DO
      O=DFLOAT(NDIM)*.25DO
      NO=0
      OW=1.DO-(O-DFLOAT(NO))
      NO=NO+2
      MO=NDIM-NO+1
      DO 1 I=NO,MO
1     XTRIM D=XTRIM D+ARRAY(I)
      XTRIM D=XTRIM D+OW*(ARRAY(NO-1)+ARRAY(MO+1))
      XTRIM D=XTRIM D/(DFLOAT(NDIM)-2.DC*O)
      RETURN
      END




      SUBROUTINE TRMDMN (X, N, XLOC, ALPHA)
      IMPLICIT REAL*8 (A-H,O-Z)
      REAL*8 X(N)
      RN1A = DFLOAT(N) * ALPHA
      N1A = IDINT(RN1A) + 1
      N1 = N1A + 1
      N2 = N - N1A
      XLOC = C.DO
      DO 1 I = N1, N2
1     XLOC = XLOC + X(I)
      IF (N1A .LE. 0) GOTO 2
      REM = DFLOAT (N1A) - RN1A
      XLOC = XLOC + REM * (X(N1A) + X(N2+1))
2     XLOC = XLOC / (DFLOAT(N)-RN1A*2.DC)
      RETURN
      END




      FUNCTION WILCXN (X, N, T)
C         WILCOXON SIGNED-RANK TEST
C         SUM OF POS RANKS - SUM OF NEG RANKS
C         A NEGATIVE ANSWER MEANS T IS TCO HIGH
      IMPLICIT REAL*8 (A-H,O-Z)
      REAL*8 X(N)
      LOGICAL SKIP
      IW = 0
      NU = N
      NL = 1
      D1 = T - X(NL)
      D2 = X(NU) - T
      SKIP = .FALSE.
      DO 1 I = 1, N
      IF (SKIP) GOTO 20
      IF (D1 .LT. D2) GOTO 1C
      IF (D1 .EQ. D2) GOTO 20
      IW = IW - N + I - 1
      NL = NL + 1
      D1 = T - X(NL)
```

```
      GOTO 1
 2C NL = NL + 1
      D1 = T - X(NL)
      NU = NU - 1
      D2 = X(NU) - T
      SKIP = .TRUE.
      GOTO 1
 30 SKIP = .FALSE.
      GOTO 1
 10 IW = IW + N - I + 1
      NU = NU - 1
      D2 = X(NU) - T
  1 CONTINUE
      WILCXN = IW
      RETURN
      END


      SUBROUTINE XMEAN (ARRAY, NDIM, XBAR)
      IMPLICIT REAL*8 (A-H,O-Z)
      DIMENSION ARRAY(NDIM)
      XBAR=0.DO
      DO 1 I=1,NDIM
      XBAR=XBAR+ARRAY(I)
      XBAR=XBAR/DFLOAT(NDIM)
      RETURN
      END


      SUBROUTINE XMEDIN(ARRAY,NDIM,XMEDN)
      IMPLICIT REAL*8(A-H,O-Z)
      DIMENSION ARRAY(NDIM)
      M=(NDIM+1)/2
      L=(NDIM+2)/2
      XMEDN=.5DO*(ARRAY(M)+ARRAY(L))
      RETURN
       END
```

12A. Random Number Computer Programs

```
GLOBAL PROCEDURE RANDOM(R14);
  BEGIN COMMENT RANDOM NUMBER GENERATOR FROM IBM SYSTEMS JOURNAL
  VOL. 8, NUMBER 2, 1969, P. 136.   ARTICLE BY LEWIS, GOODMAN, MILLER.;
  COMMENT PERMUTATION OF 500 RANDOM NUMBERS ACCORDING TO MOSES AND
OAKFORD, P. 153.  COMPUTER LANGUAGE IS IBM 360 DEPENDENT:  PL360
IS DESCRIBED IN JACM, JAN. 1968, BY WIRTH.  COMPILER MAY BE OBTAINED
FROM STANFORD UNIVERSITY.;

    GLOBAL DATA RNDATAXX BASE R1;
    INTEGER STARTV = 724287, MULT = 16807;
    LOGICAL CHAR = #40000000, P = #7FFFFFFF;
    INTEGER NPERMUTE = 500, NTAKE = 501;
    ARRAY 500 SHORT INTEGER ORDER = (
    190, 175, 133, 81, 260, 269, 164, 318, 107, 240, 415, 22, 105, 46
    71, 163, 116, 227, 446, 384, 242, 67, 332, 48, 128, 21, 451, 154,
    381, 359, 386, 365, 300, 231, 176, 135, 497, 167, 120, 168, 217,
    193, 108, 344, 87, 99, 351, 214, 134, 209, 399, 447, 408, 483, 19
    322, 98, 339, 445, 27, 206, 255, 226, 188, 371, 41, 368, 257, 313
    328, 342, 20, 64, 45, 285, 224, 82, 264, 153, 162, 3, 363, 319,
    436, 88, 455, 189, 83, 349, 335, 177, 284, 38, 91, 184, 245, 157,
    262, 94, 183, 304, 372, 458, 465, 76, 46, 369, 228, 145, 201, 170
    464, 406, 466, 321, 210, 60, 92, 410, 413, 459, 44, 4, 472, 63,
    85, 348, 477, 78, 494, 419, 113, 343, 33, 171, 150, 9, 290, 460,
    484, 439, 253, 258, 19, 361, 31, 457, 402, 306, 496, 43, 334, 416
    112, 93, 411, 283, 74, 425, 169, 15, 57, 1, 109, 441, 100, 90, 45
    489, 129, 12, 315, 137, 492, 277, 131, 378, 141, 267, 291, 197,
    204, 256, 346, 65, 422, 136, 358, 298, 219, 95, 401, 272, 409, 10
    252,
    36, 7, 50, 195, 355, 96, 463, 347, 102, 336, 340, 288, 199, 111,
    396, 360, 152, 147, 495, 158, 400, 323, 427, 191, 397, 62, 297,
    498, 200, 35, 443, 24, 5, 39, 296, 341, 26, 292, 265, 377,
    311, 114, 234, 407, 470, 388, 289, 440, 149, 279, 215, 438, 301,
    302, 196, 230, 139, 420, 121, 69, 119, 203, 144, 221, 79, 186, 16
    155, 18, 449, 305, 316, 450, 444, 433, 448, 462, 124, 293, 391,
    123, 490, 480, 357, 353, 182, 32, 222, 434, 432, 248, 392, 366,
    212, 110, 263, 160, 286, 320, 247, 148, 337,
    179, 13, 479, 428, 159, 10,
    241, 376, 324, 181, 362, 488, 58, 330, 161, 2, 243, 356, 405, 17,
    56, 233, 115, 310, 398, 299,
    23, 225, 281, 308, 418, 295, 106, 146, 140, 72,
```

```
       312, 275, 394, 364, 216, 246, 125, 166, 6, 303, 482,
       461, 11, 456, 412, 103, 68, 417, 220, 70, 273, 173, 229,
       276, 211, 345, 424, 205, 156, 395,
       485, 238, 185, 49, 294, 379, 331, 404, 500, 430, 499, 47, 53, 37,
       374, 138, 329, 54, 474, 143, 382, 178, 165, 122, 239, 453, 73, 367,
       370, 59, 352, 34, 478, 25, 237, 84, 387, 389, 259, 213, 435, 326,
       40, 236, 207, 354, 476, 268, 86, 266, 317, 442, 414, 118, 223, 491,
       194, 101, 325, 172, 174, 249, 218, 473, 475, 52, 151, 380, 375,
       198, 309, 282, 421, 8, 180, 127, 307, 486, 29, 30, 251, 244, 469,
       97, 350, 14, 51, 117, 187, 429, 274, 423, 254, 208, 232, 270, 28,
       338, 132, 493, 385, 333, 66, 250, 61, 373, 80, 471, 280, 202, 89,
       77, 481, 487, 42, 278, 55, 327, 271, 390, 130, 75, 452, 403, 126,
       431, 314, 142, 393, 437, 467, 261, 235, 426, 287, 383             );
     ARRAY 500 REAL NUMBERS;
     ARRAY 500 INTEGER INTNUMBER SYN NUMBERS;
     REAL THISRANDOM SYN B1;

     R0 := NTAKE + 1;  IF R0 >= NPERMUTE
       THEN BEGIN
         STM (R2,R3,B13(28));
         FOR R0 := 1 STEP 1 UNTIL NPERMUTE DO
           BEGIN COMMENT GENERATE ALL RANDOM NUMBERS FOR THIS SET;
           R3 := STARTV * MULT / P;  STARTV := R2;
           R2 := R2 SHRL 7 + CHAR;  COMMENT FLOATING REPRESENTATION;
           R3 := R0 - 1 SHLL 1;  R3 := ORDER(R3) - 1 SHLL 2;
           INTNUMBER(R3) := R2;  COMMENT STORE NUMBER ACCORDING TO PERM;
           END;
         R0 := 0;  LM (R2, R3, B13(28));
       END;
     NTAKE := R0;  R0 := R0 SHLL 2;  R1 := @NUMBERS + R0;
     F01 := 0L;  F0 := F0 + THISRANDOM;
   END.
GLOBAL PROCEDURE SNDY1(R14);
   BEGIN COMMENT SET STARTING VALUE OF RANDOM GENERATOR;
     R1 := B1;  R0 := B1;  COMMENT LOAD ARGUMENT;
     BEGIN
       EXTERNAL DATA RNDATAXX BASE R1;
       INTEGER STARTV;
       R0 := R0 + STARTV AND #7FFFFFFF;  STARTV := R0;
     END;
   END.



     FUNCTION RANORM(X)
     IMPLICIT REAL*8 (A-H,O-Z)
     LOGICAL GCASE
     DATA PI /3.14159265358979/
```

```
      DATA GCASE /.TRUE./
C           DRAW FROM THE GAUSSIAN DISTRIBUTION
C        COMPUTES GAUSSIAN RANDOM BY BOX-MULLER-KNUTH TECHNIQUE
      IF (GCASE) GOTO 1
      RANORM = S
      GCASE = .TRUE.
      RETURN
    1 U1 = 2.DO * RANDOM(DUMMY) - 1.0DO
      U2 = 2.DO * RANDOM(DUMMY) - 1.0DO
      S = U1*U1 + U2*U2
      IF (S .GT. 1.0DO) GOTO 1
      S = DSQRT(-2.DO * DLOG(S)/S)
      RANORM = U1 * S
      S = U2 * S
      GCASE = .FALSE.
      RETURN
      END
```

12B. Random Sample Generation Procedures

Situation	Nsamp	Alg	SNDY1 Start	First sample: Lowest	Highest
ALL NORM 5	640	1	6088260	-1.031	0.323
ALL NORM 10	1000	1	4581962	-1.091	1.410
ALL NORM 20	1000	1	8306491	-2.433	1.089
ALL NORM 40	640	1	3937975	-2.592	2.037
ALL CAUC 5	640	3	152819	-0.399	2.833
ALL CAUC 10A	1000	3	1614952	-27.50	3.985
ALL CAUC 20	1000	3	8437741	-4.756	6.483
ALL CAUC 40	640	3	4382523	-99.20	226.5
5% 3N 20	640	2	1217821	-2.020	6.595
10% 3N 20	640	2	1501784	-1.791	2.545
15% 3N 20	640	2	8599742	-2.465	4.786
25% 3N 20	640	2	753676	-3.840	2.067
50% 3N 20	1000	2	8546839	-3.530	6.088
20% 3N 10	640	2	1837759	-0.947	3.349
5% 10N 20	640	2	4613218	-7.393	1.457
ALL 1/U 10	640	3	2021246	-9.441	12.78
ALL 1/U 20	640	3	118897	-19.74	52.95
25% 1/U 20	640	4	1256385	-2.227	1.568
ALL D-EX 20	640	5	5387000	-1.361	4.207

Algorithms:

1. Draw consecutively from RANORM.

2. Draw consecutively from RANORM. If there are k contaminants, multiply the first k Normals drawn by the contaminating scale.

3. Draw first from RANORM to get a numerator. Then draw from RANORM or RANDOM to get a divisor. Repeat until sample is completed.

4. Draw numerator from RANORM. Draw t from RANDOM. If t < .25 draw denominator from RANDOM, otherwise the denominator is 1. Repeat for each observation.

5. Draw numerator from RANORM. Draw r from RANDOM. The denominator is sqrt(1/(-2*ln(r))).

    Other situations were not generated by the method above. If you are interested in reproducing them, please contact W. H. Rogers at Stanford University.

13. DUAL-CRITERION PROBLEMS IN ESTIMATION

As we have seen in Chapter 6 there are advantages to assessing estimate quality in terms of looking at two criteria at once. The simplest case, which we shall treat first, is one in which both criteria are -- or are simple linear codings of -- variances for two different cases (distributions or contamponents). Having one criterion a linear combination of variances for diverse cases is a simple generalization. Other cases, of at least as great interest, arise when the variances are disguised by gentle nonlinearities -- as when variances are converted into efficiencies or deficiencies -- and when variances (or simple functions thereof) for diverse cases are combined in a nonlinear way (e.g., the maximum). Treating such cases requires an inner iterative loop, but does not seem likely to make manipulation too much more difficult.

13A VARIANCE CRITERIA: SETS OF ESTIMATES

If our criteria are the variances of estimates under cases A and B, it is convenient to use "Avar" and "Bvar" for the variances, and "Acov" and "Bcov" for the covariances. If we have a set,

$$T_1, T_2, \ldots, T_k,$$

of estimates, it is very natural to consider all linear combinations

$$a_1 T_1 + a_2 T_2 + \cdots + a_k T_k$$

that satisfy

$$a_1 + a_2 + \cdots + a_k = 1$$

as a natural and convenient class of estimates.
Clearly

$$Avar\{T\} = \Sigma a_i^2 \cdot Avar \, T_i \quad + \quad \underset{i \neq j}{\Sigma \Sigma} a_i a_j \cdot Acov\{T_i, T_j\}$$

$$Bvar\{T\} = \Sigma a_i^2 \cdot Bvar \, T_i \quad + \quad \underset{i \neq j}{\Sigma \Sigma} a_i a_j \cdot Bcov\{T_i, T_j\}$$

310

By varying the a's, the points

$$(\text{Avar}\{T\}, \ \text{Bvar}\{T\})$$

will cover a certain accessible set in the (Avar, Bvar) plane. We typically want to make <u>both</u> coordinates small. (In geographic terms we want to move both West and South.)
    Because we deal with variance criteria, the shape of the SW margin of the accessible set has certain properties.

13A1 Linear Combinations

Consider any unbiased linear combination

$$Z = \psi T + (1-\psi)U = U + \psi(T - U) = T + (1-\psi)(U - T).$$

Then Avar and Bvar are of the form

$$\text{Avar}\{U\} + 2\psi \, \text{Acov}\{U, T - U\} + \psi^2 \cdot \text{Avar}\{T - U\}$$

and

$$\text{Bvar}\{T\} + 2(1-\psi)\text{Bcov}\{T, U - T\} + (1-\psi)^2 \text{Bvar}\{U - T\}$$

which takes the forms

$$a + 2e\psi + c\psi^2 = \left(a - \frac{e^2}{c}\right) + \left(c \ \psi + \frac{e}{c}\right)^2$$

$$b + 2f(1-\psi) + d(1-\psi)^2 = \left(b - \frac{f^2}{d}\right) + d\left(1 - \psi + \frac{f}{d}\right)^2$$

for suitable a,b,c,d,e,f. If we put

$$\theta = \frac{\psi + \dfrac{e}{c}}{\left(1 - \psi + \dfrac{f}{d}\right) + \left(\psi + \dfrac{e}{c}\right)} = \frac{\psi + \dfrac{e}{c}}{1 + \dfrac{f}{d} + \dfrac{e}{c}}$$

we have

$$1 - \theta = \frac{1 - \psi + \dfrac{f}{d}}{1 + \dfrac{f}{d} + \dfrac{e}{c}}$$

so that the Avar and Bvar take the forms

$$\left(a - \frac{e^2}{c}\right) + c\left(1 + \frac{f}{d} + \frac{e}{c}\right)^2 \theta^2 = a^* + c^*\theta^2$$

$$\left(b - \frac{f^2}{d}\right) + d\left(1 + \frac{f}{d} + \frac{e}{c}\right)(1 - \theta)^2 = b^* + d^*(1 - \theta)^2 \ .$$

If we now rescale and displace the two axes appropriately, so that the (Avar, Bvar) plane becomes the (x,y) plane, where

$$x = \frac{1}{c^*} \, \text{Avar}\{Z\} - \frac{a^*}{c^*}$$

$$y = \frac{1}{d^*} \, \text{Bvar}\{Z\} - \frac{b^*}{d^*}$$

then the linear combinations

$$Z = \psi T + (1-\psi)U$$

are plotted along the curve

$$x = \theta^2$$

$$y = (1-\theta)^2$$

of which the SW quadrant is of most interest to us.

Exhibit 13-1 shows a part of this curve, with its SW quadrant solid and other portion dashed.

Since the curve is convex and open to the NE, the included area lies to the SW of any line segment joining two of its points that runs SE-NW. (Three instances are illustrated in Exhibit 13-1.) As a consequence, the SW margin of any accessible set is convex.

* The principal combinations *

If we now set

$$T^* = \left(- \frac{e}{c}\right)T + \left(1 + \frac{e}{c}\right)U$$

for which

Exhibit 13-1

THE WORKING CURVE THROUGH THE PRINCIPAL COMBINATIONS

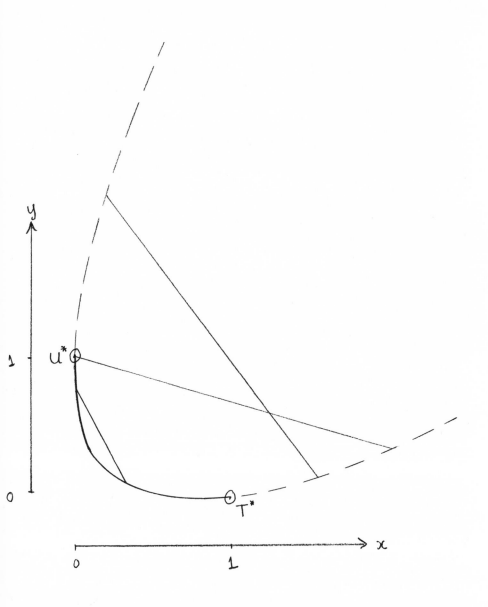

$$\psi = -\frac{e}{c} \quad \text{and} \quad \theta = 0$$

and

$$U^* = \left(1 + \frac{f}{d}\right)T + \left(-\frac{f}{d}\right)U$$

for which

$$1 - \psi = -\frac{f}{d} \quad \text{and} \quad 1 - \theta = 0$$

so that the (x,y) coordinates of T* and U* are (1,0) and (0,1) respectively. Thus T* and U* (also shown in Exhibit 13-1, correspond to the ends of the SW quadrant.

Given any two estimates, T and U, their unbiased linear combinations are equally well described in terms of the principal combinations T* and U*, which are such that:
- the desirably extreme combinations are just the POSITIVE combinations of T* and U*;
- the shape of the extreme curve is just that of the solid curve in Exhibit 13-1 (subject to different rescaling on both axes).

More explicitly, we have

$$1 + \frac{f}{d} = 1 + \frac{\text{Acov}\{U,T-U\}}{\text{Avar}\{T-U\}} = \frac{\text{Acov}\{T,T-U\}}{\text{Avar}\{T-U\}}$$

$$1 + \frac{e}{c} = 1 + \frac{\text{Bcov}\{T,U-T\}}{\text{Bvar}\{U-T\}} = \frac{\text{Bcov}\{U,T-U\}}{\text{Bvar}\{T-U\}}$$

so that

$$T^* = \frac{\text{Acov}\{T,T-U\}}{\text{Avar}\{T-U\}} T - \frac{\text{Acov}\{U,T-U\}}{\text{Avar}\{T-U\}} U$$

$$U^* = -\frac{\text{Bcov}\{T,U-T\}}{\text{Bvar}\{U-T\}} T + \frac{\text{Bcov}\{U,U-T\}}{\text{Bvar}\{U-T\}} U \ .$$

Thus it is not difficult to find the principal combinations of any two estimates and hence the corresponding (Avar,Bvar) points.

It is interesting to note that T* and U* can be characterized either by:
- T* minimizes Bvar and U* minimizes Avar
or by

$$- \text{Acov}\{U^*, T^*-U^*\} = \text{Bcov}\{T^*, U^*-T^*\} = 0.$$

There is often some interest in the point V^* corresponding to $\theta = 1/2$, whose Avar and Bvar are $a^* + c^*/4$ and $b^* + d^*/4$. Since we have

$$\text{Avar}\{U^*\} = a^*$$

$$\text{Avar}\{T^*\} = a^* + c^*$$

$$\text{Bvar}\{U^*\} = b^* + d^*$$

$$\text{Bvar}\{T^*\} = b^*$$

it follows that

$$\text{Avar}\{V^*\} = \frac{3}{4}\text{Avar}\{U^*\} + \frac{1}{4}\text{Avar}\{T^*\}$$

$$\text{Bvar}\{V^*\} = \frac{3}{4}\text{Bvar}\{T^*\} + \frac{1}{4}\text{Bvar}\{U^*\}$$

As either Exhibit 13-1 or Exhibit 13-2 shows, the SW-quadrant is nearly circular. The center of the circle passing through T^*, U^* and V^* lies at C^*, whose coordinates are:

$$\frac{7}{8}\text{Avar}\{T^*\} + \frac{1}{8}\text{Avar}\{U^*\}$$

and

$$\frac{1}{8}\text{Bvar}\{T^*\} + \frac{7}{8}\text{Bvar}\{U^*\} \ .$$

As Exhibit 13-2 further shows, the SW-quadrant lies between .99 and 1.00 times the common distance of T^*, U^*, and V^* from C^*.

Rescaling the two axes differently will, of course, change this near circularity into near ellipticity.

Exhibit 13-2

THE NEAR CIRCULARITY OF THE SW QUADRANT

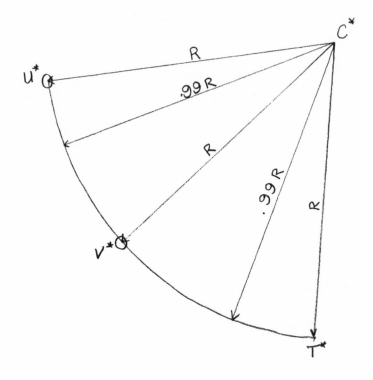

13B VARIANCE CRITERIA: BAYESIAN BLUES

Since we desire to go as far West AND South as we can, and since these uttermost points lie on a convex curve, it is natural to seek these points in terms of the direction of the tangent to the extreme curve. This direction need not be thought of as anything meaningful in itself. It is sufficient to know that, as we pass through all directions, we will catch all points on the extreme curve.

A convenient way to get all directions in the (U,V) plane is to look at all

$$U \cdot \sin^2\phi + V \cdot \cos^2\phi$$

for all ϕ in, say, $(0°,90°)$. (Notice that ϕ , though a convenient parameter, is NOT the slope.) This amounts to looking at

$$\cos^2\phi \cdot \text{Avar}\{T\} + \sin^2\phi \cdot \text{Bvar}\{T\}$$

for some ϕ and asking for the T in our family that minimizes this. If we call the resulting estimate T_ϕ , we know that

$$\{T_\phi \mid 0° \le \phi \le 90°\}$$

covers the extreme curve at the SW margin of the accessible set.

Though there is no logical necessity, statisticians seem likely to feel more familiar with the situation if they interpret $\cos^2\phi$ and $\sin^2\phi$ as probabilities associated with the cases concerned, after which

$$\text{var}_\phi = \cos^2\phi \cdot \text{Avar} + \sin^2\phi \cdot \text{Bvar}$$

can be interpreted as the variance of a mixed (Bayesian) situation. We may then interpret minimizing

$$\text{var}_\phi\{T\} = \text{var}_\phi\{a_1 T_1 + a_2 T_2 + \cdots + a_k T_k\}$$

as finding the

Bayesian BLUE

(where, as usual, BLUE = Best Linear Unbiased Estimate)

corresponding to the estimates $T_1, T_2, \ldots, T_k$.

To find our extreme curve, we have only to calculate (a reasonable number of points on) the curve of Bayesian BLUEs.

Thus, if we are given two variance criteria, each set of estimates (finite or infinite) generates (with some labor) a corresponding extreme curve. If we know the k variances and $\frac{1}{2}k(k-1)$ covariances for each case, we are prepared

- to locate the curve,
- to specify which linear combinations of the T_i are needed to attain the located point.

This process allows us, then, to construct attainable curves based on any set of estimates we find interesting.

13C VARIANCE CRITERIA; ESTIMATING THE BOUNDING CURVE

A generalization of the approach of David Hoaglin (Princeton thesis, 1971) allows us to find and delineate the extreme curve for ALL invariant estimates. If $\vec{y}$ is the sample (from a distribution) or polysample (from a contamponent), we can collect $\vec{y}$'s into configurations or orbits, $\vec{c}$, i.e., equivalence classes under location and scale transformations, such that all $\vec{y}$'s in $\vec{c}$ differ only by location and scale. If we are to have an estimate invariant under location and scale transformations, we can make only one choice per configuration, since once we know the estimate for one $\vec{y}$ in $\vec{c}$, we know it for all other $\vec{y}$'s in $\vec{c}$.

Hoaglin has developed practical (though not easy) computing techniques for evaluating, for any assumed distribution:

(1)   The total likelihood of $\vec{c}$.
(2)   The best choice to make.
(3)   The variance, conditional on the configuration for this best choice.
(4)   The rate at which the conditional variance increases when the choice is changed.

If we use a simple notation, we have, for distributions A and B,

- likelihoods: $L_A(\vec{c})$, $L_B(\vec{c})$ (which need to

be known only up to one multiplicative constant for

A-likelihoods and one multiplicative constant for B-likelihood).

- variance formulas:

$$H_A(\vec{c}) + J_A^2(\vec{c})(t - t_{oA}(\vec{c}))^2$$

$$H_B(\vec{c}) + J_B^2(\vec{c})(t - t_{oB}(\vec{c}))^2$$

(where t is the choice).

If we change our choice at one configuration $\vec{c}$, from, say, t_o to $t_o+\delta$, the conditional variance changes are

$$2J_A(\vec{c})(t_o - t_{oA}(\vec{c}))\delta$$

and

$$2J_B(\vec{c})(t_o - t_{oB}(\vec{c}))\delta$$

which are to be weighted by

$$L_A(\vec{c})d\vec{c}$$

and

$$L_B(\vec{c})d\vec{c}$$

respectively. The ratio of the changes in Avar and Bvar for a change in choice at this $\vec{c}$ is thus

$$\frac{L_A(\vec{c})J_A(\vec{c})(t_o - t_{oA}(\vec{c}))}{L_B(\vec{c})J_B(\vec{c})(t_o - t_{oB}(\vec{c}))} .$$

If this ratio is not essentially constant, we can improve one or both variances by making partially compensating changes at $\vec{c}$'s where the ratio differs. (Moreover, the ratio will be negative, else a change at one $\vec{c}$ could improve both variances.) Thus we may parametrize the extreme situations by setting this ratio equal to

$$- \frac{\sin^2\phi}{\cos^2\phi}$$

319

for any fixed ϕ and all $\vec{c}$.

The result of such a choice is

$$t_o = \frac{\cos^2\phi\, L_A(\vec{c})J_A(\vec{c})t_{oA}(\vec{c}) + \sin^2\phi\, L_B(\vec{c})J_B(\vec{c})t_{oB}(\vec{c})}{\cos^2\phi\, L_A(\vec{c})J_A(\vec{c}) + \sin^2\phi\, L_B(\vec{c})J_B(\vec{c})}$$

$$= \theta\, t_{oA}(\vec{c}) + (1-\theta)t_{oB}(\vec{c})$$

where

$$\theta = \frac{\sin^2\phi \cdot L_B(\vec{c})J_B(\vec{c})}{\cos^2\phi\, L_A(\vec{c})J_A(\vec{c}) + \sin^2\phi\, L_B(\vec{c})J_B(\vec{c})}$$

for which

$$\mathrm{Avar}\{t_o | \vec{c}\} = H_A(\vec{c}) + J_A(\vec{c})\theta^2(t_{oB}(\vec{c}) - T_{oA}(\vec{c}))^2$$

$$\mathrm{Bvar}\{t_o | \vec{c}\} = H_B(\vec{c}) + J_B(\vec{c})(1-\theta)^2\left(t_{oB}(\vec{c}) - t_{oA}(\vec{c})\right)^2 .$$

If we have a "sample" of $\vec{c}$'s, drawn from some known case, so that we know the two sets of weights $W_A(\vec{c})$ and $W_B(\vec{c})$ appropriate for treating these $\vec{c}$'s as samples of configurations from either case, we can estimate the Avar and Bvar of t_o, which we now write t_ϕ, as

$$\mathrm{Avar}\{t_\phi\} = \frac{1}{\Sigma W_A}\, \Sigma\, W_A\left[H_A + J_A\theta^2\left(t_{oA} - t_{oB}\right)^2\right]$$

$$\mathrm{Bvar}\{t_\phi\} = \frac{1}{\Sigma W_B}\, \Sigma\, W_B\left[H_B + J_B(1-\theta)^2\left(t_{oA} - t_{oB}\right)^2\right]$$

where we have suppressed the dependence of every quantity (including θ) on $\vec{c}$.

Except for very substantial computational labor, all is clear.

13D TWO VARIANCES: COMBINED PICTURES

How then might we wish to look at a simple two-variance situation, say Gaussian variance and a 25% contaminated variance? We now naturally give some special consideration to the following sets of estimates:
- all trims (where we need consider only integer trims, as the others are linear combinations of these);
- all hubers (where it will probably suffice to approximate by taking about a half dozen well-spread-out instances);
- all moderately high performance estimates in the study;
- all estimates in the study;
- all invariant estimates, whether or not in the study.

To these five sets of estimates there will correspond 5 curves in the (Avar,Bvar) plane. The first two can be calculated, without substantial effort, from the variances and covariances already at hand. The third can be calculated similarly. However, the increase in dimensionality -- from 10 or 6 to perhaps 25 or 30 -- will bring in needs for care, both in computational technique and in allowance for serendipitious fitting. All this will be worse for the fourth. To calculate the fifth curve we must work hard.

Given these five curves on a sheet of graph paper, what can we learn? One important question, answered by the closeness or distance of any of the first four curves to the fifth curve, is this:
- To what extent do the estimates of the set explore the aspects of our samples that must be (linearly) available in order to do well for the two cases together?

If we are not doing well, we should stop and try to ask where we are failing. Is the gap toward one end of the SW quadrant or toward the other, or toward both, or everywhere?

If we have values of t_ϕ for a set of configurations (as we will need to have in calculating t_ϕ's Avar and Bvar), we can regress t_ϕ on our set of estimates covering the configurations, weighted for

$$(\cos^2\phi)(\cos A) + (\sin^2\phi)(\cos B)$$

and look hard to see what we are missing.

It is hard to believe that if we have all these curves that we can avoid
- either learning that we can do well enough with what we have,
- or learning something about what is missing from our armamentarium.

13E   FURTHER MODIFICATIONS

13E1   Individual Efficiencies and Deficiencies

If we work with efficiencies or deficiencies, defined as

$$\text{efficiency} = \frac{\text{standard variance}}{\text{variance}}$$

$$\text{deficiency} = 1 - \frac{\text{standard variance}}{\text{variance}} ,$$

we are working with monotone functions of variances. Our pictures will be slightly distorted (perhaps usefully), but the same estimates will correspond to points on each kind of extreme curve.
Computationally. we will probably work with vari-
ances, re-expressing the final results. Nothing of great novelty is to be expected.

13E2   Mixtures of Variances (Weighted Linear Mixtures)

If one or both criteria involve weighted linear combinations of variances, all goes as before, except for more detailed work. The BLUEs are a little more Bayesian, but their calculation is not seriously different; it is as easy to mix covariances as to mix vari-
ances. The invariant extreme calculation is a little longer. The calculation of the H's, J's, and t_o's
involves integration over more distributions, but the results are to be combined into the same number of summaries as before.

13E3 Monotone Nonlinear Combination of Variances

Some of the criteria that we wish to give most attention to are monotone nonlinear combinations of variances. Two examples are the Root Mean Positive Square and the maximum deficiencies.

These two examples illustrate two main classes which seem to deserve different approaches.

In the first class, illustrated by RMPS deficiencies and RMPS modified discrepancies, the dependence of the criterion on the variances is differentiable (with minor exceptions) and relatively smoothly so.

Replacing, in an inner iterative loop,

$$h(v_1, v_2, \ldots, v_k) = h(\vec{v})$$

by the tangent-plane approximation,

$$h(\vec{v}_o) + \Sigma \left(\frac{\partial h}{\partial v_i}\right)_o v_i - \Sigma \left(\frac{\partial h}{\partial v_i}\right)_o v_{oi}$$

where the "o" subscripts indicate evaluation at $\vec{v}_o$ will ordinarily be satisfactory as a means of obtaining the extreme curve, whose points correspond to estimates that are now NOT Bayesian BLUEs (for any preassigned mix of distributions).

The case where one criterion is

$$\max(v_1, v_2, \ldots, v_k)$$

requires rather different treatment, either some sort of alternating algorithm or a constrained procedure in which relations $v_i = v_j$, once attained, become constraints.

323

13F BOUNDS FOR VARIANCES - TWO SITUATION CASES - CRAMER RAO METHODS - BOUNDS DEPEND ON N

The classical characteristic of the Cramer-Rao inequality is that it gives a bound for

(variance)(sample size)

that is independent of n. Since the sharp bound decreases with n (as illustrated in Hoaglin 1971, for example), the single situation bound is thoroughly asymptotic.

13F1 Similar Methods, Other Results

As we shall soon see, methods which lead to a simple argument for the Cramer-Rao bound for a single situation can be applied, in dual situations, to produce a less asymptotic sort of bound.

13F2 A Dual Formulation

Our concern is with two slippage specifications for random samples of size n, based, respectively, on the scalar specifications

$$dF_\theta(x) = dF(x-\theta) = f(x-\theta)dx$$

$$dG_\theta(x) = dG(x-\theta) = g(x-\theta)dx$$

Near some θ_0, then

- we may as well put $\theta_0 = 0$,
- if we may differentiate under the integral sign, we have, using,

$$\frac{\partial}{\partial\theta} \Pi dF(x_i) = -f^*(\vec{x})\Pi dF(x_i)$$

$$\frac{\partial}{\partial\theta} \Pi dG(x_i) = -g^*(\vec{x})\Pi dG(x_i)$$

where

$$f^*(\vec{x}) = \Sigma \frac{f'(x_i)}{f(x_i)} = \frac{\partial}{\partial\theta} \log \Pi f_\theta(x_i)$$

$$g^*(\vec{x}) = \Sigma \frac{g'(x_i)}{g(x_i)} = \frac{\partial}{\partial\theta} \log \Pi g_\theta(x_i)$$

the result that

$$\frac{\partial}{\partial\theta} \{ave\{T|F_\theta\}\}|_{\theta=0} = -\int T(\vec{x})f*(\vec{x}) \ dF(\vec{x})$$

$$\frac{\partial}{\partial\theta} \{ave\{T|G_\theta\}\}|_{\theta=0} = -\int T(\vec{x})g*(\vec{x}) \ dG(\vec{x})$$

- in particular, if $T \equiv 1$,

$$-\int f*(\vec{x}) \ dF(\vec{x}) = 0 = -\int g*(\vec{x}) \ dG(\vec{x})$$

- note further that the two values of Fisher information are

$$I(F) = \int [f*(\vec{x})]^2 dF(\vec{x})$$

$$I(G) = \int [g*(\vec{x})] \ dG(\vec{x}) \ .$$

13F3 Local Invariance

The condition that a function $T(\vec{x})$ should behave locally like an invariant estimate near $\theta = 0$ are, for location estimation, namely

$$0 = ave\{T(\vec{x})|F\} = \int T(\vec{x}) \ \Pi dF(x_i)$$

$$0 = ave\{T(\vec{x})|G\} = \int T(\vec{x}) \ \Pi dG(x_i)$$

and local invariance, namely

$$1 = \frac{\partial}{\partial\theta} ave\{T(\vec{x})|F_\theta\}|_{\theta=0} = -\int T(\vec{x})f*(\vec{x}) \ \Pi dF(x_i)$$

and

$$1 = -\int T(\vec{x})g*(\vec{x}) \ \Pi dG(x_i) \ .$$

Any (finite-variance) invariant estimate of location for such a specification can be made unbiased by subtracting a suitable constant. It will then be a locally unbiased, locally invariant function of either sample, and will satisfy the conditions above. Thus a sharp bound for variances (two specifications!) of locally unbiased,

locally invariant functions will, of necessity, be a
bound -- perhaps blunt -- for locally invariant estimates.

13F4 Melding the Variances

We will get all extremal pairs of variances (one for each
specification) if we find the minimum of

$$\cos^2\phi \cdot \text{var}\{T(\vec{x})|F\} + \sin^2\phi \cdot \text{var}\{T(\vec{x})|G\}$$

for all ϕ, $0 \leq \phi \leq \pi/2$. This melded variance can be
written, using

$$\text{ave}\{T(\vec{x})|F\} = 0 = \text{ave}\{T(\vec{x})|G\}$$

which we plan to assume, as

$$\cos^2\phi \cdot \int T^2(\vec{x}) \, \Pi dF(x_i) + \sin^2\phi \cdot \int T^2(\vec{x}) \, \Pi dF(x_i)$$

$$= \int T^2(\vec{x}) \, dH(\vec{x})$$

where

$$dH(\vec{x}) = \cos^2\phi \cdot \Pi dF(x_i) + \sin^2\phi \cdot \Pi dG(x_i) \ .$$

Our desire is to minimize this for each fixed ϕ.

Since $dH(\vec{x})$, $\Pi dF(x_i)$ and $\Pi dG(x_i)$ are all non-
negative and smooth, there are functions $j(\vec{x})$ and $k(\vec{x})$
such that

$$\Pi dF(x_i) = j(\vec{x}) \, dH(\vec{x})$$

$$\Pi dG(x_i) = k(\vec{x}) \, dH(\vec{x})$$

which satisfy

$$\cos^2\phi \cdot j(\vec{x}) + \sin^2\phi \cdot k(\vec{x}) = 1$$

except where $dH(\vec{x})$ is identically zero. Furthermore,

$$j(\vec{x}) = \frac{\Pi dF(x_i)}{\cos^2\phi \cdot \Pi dF(x_i) + \sin^2\phi \cdot \Pi dG(x_i)}$$

$$k(\vec{x}) = \frac{\Pi dG(x_i)}{\cos^2\phi \cdot \Pi dF(x_i) + \sin^2\phi \cdot \Pi dG(x_i)}$$

(except where dH is identically zero, and the values of $j(\vec{x})$ and $k(\vec{x})$ do not matter).

We now write I_{abcd}, with any number subscripts for

$$\int a(\vec{x})b(\vec{x})c(\vec{x})d(\vec{x}) \ dH(\vec{x}) \ .$$

As an example, consider

$$I(F) = \int [f*(\vec{x})]^2 \ \Pi dF(x_i) = \int [f*(\vec{x})]^2 j(\vec{x}) \ dH(\vec{x}) = I_{f*f*j}$$

similarly

$$I(G) = I_{g*g*k} \ .$$

The local invariance conditions become

$$1 = -I_{Tf*j} \quad \text{and} \quad 1 = -I_{Tg*k} \ .$$

13F5 Lagrange Minimization

Leaving aside the trivial question of an additive constant and the local estimation conditions, the minimization problem, with Lagrange multipliers, involves

$$\int [T^2(\vec{x}) + 2\alpha T(\vec{x})f*(\vec{x})j(\vec{x}) + 2\beta T(\vec{x})g*(\vec{x})k(\vec{x})] \ dH$$

$$= I_{TT} + 2\alpha \ I_{Tf*j} + 2\beta I_{Tg*k} \ .$$

The integrand is minimized, for given α and β when

$$T(\vec{x}) = -\alpha f*(\vec{x})j(\vec{x}) - \beta \ g*(\vec{x})k(\vec{x}) \ .$$

Inserting this expression in the local invariance conditions gives

$$1 = \alpha \ I_{f*f*jj} + \beta I_{f*g*jk} \ ,$$

and

$$1 = \alpha \ I_{f*g*jk} + \beta I_{g*g*kk}$$

whence

$$\alpha = \frac{I_{g*g*kk} - I_{f*g*jk}}{I_{f*f*jj} I_{g*g*kk} - I^2_{f*g*jk}} \quad ,$$

and

$$\beta = \frac{I_{f*f*jj} - I_{f*g*jk}}{I_{f*f*jj} I_{g*g*kk} - I^2_{f*g*jk}} \quad .$$

13F6   The Minimum

Substituting these values, we find

$$\text{var}\{T(\vec{x})|H\} = I_{TT}$$

$$= \alpha^2 I_{f*f*jj} + 2\alpha\beta I_{f*g*jk} + \beta^2 I_{g*g*kk}$$

which reduces to

$$\frac{I_{f*f*jj} - 2I_{f*g*jk} + I_{g*g*kk}}{I_{f*f*jj} \cdot I_{g*g*kk} - I^2_{f*g*jk}} \quad .$$

13F7   When $n \to \infty$

When $n \to \infty$, $\Pi dF(x_i)$ and $\Pi dG(x_i)$ will be, if
$F \neq G$, concentrated in different places. As a result,
approximately,

$$j(\vec{x})k(\vec{x}) \sim 0$$

$$j(\vec{x})j(\vec{x}) \sim j(\vec{x})/\cos^2\phi$$

$$k(\vec{x})k(\vec{x}) \sim k(\vec{x})/\sin^2\phi$$

and the var$\{T|H\}$, for $n \to \infty$, is nearly

$$\frac{\dfrac{1}{\cos^2\phi} I_{f^*f^*j} + \dfrac{1}{\sin^2\phi} I_{g^*g^*k}}{\dfrac{1}{\cos^2\phi \sin^2\phi} I_{f^*f^*j} \cdot I_{g^*g^*k}} = \frac{\sin^2\phi}{I_{f^*f^*j}} + \frac{\cos^2\phi}{I_{g^*g^*k}}$$

$$= \frac{\sin^2\phi}{I(F)} + \frac{\cos^2\phi}{I(G)}$$

which is what we would expect if (as we should be able to do as $n \to \infty$) we attained the Cramer-Rao bound for each situation separately.

13F8 But When n is Finite

When n is finite, the product $j(\vec{x})k(\vec{x})$ is not zero (usually nowhere zero) and we do not have such a reduction. Since $dH = \cos^2\phi \cdot \Pi dF(x_i) + \sin^2\phi \cdot \Pi dG(x_i)$ it is sufficient, in order to approximate the variance, to evaluate $j(\vec{x})$, $k(\vec{x})$, $f^*(\vec{x})$ and $g(\vec{x})$, both for $\vec{x}$'s simulating a sample from F and for $\vec{x}$'s simulating a sample from G. (Note that $j(\vec{x})$ and $k(\vec{x})$ depend only on ϕ and $\Pi dF(x_i)/\Pi dG(x_i)$ while $f^*(\vec{x})$ and $g^*(\vec{x})$ do not depend on ϕ.

Thus calculation of this bound should be relatively easy.

13G BOUNDS FOR VARIANCES-TWO SITUATION CASES-OTHER METHODS

We know lower bounds for unbiased estimates in single situations. The Cramer Rao inequality is one of these. For robust investigations, we would like to know how small a variance can be attained simultaneously over two or more situations. In this section, several bounds are presented. These bounds may be readily calculated by Monte Carlo.

Many formulations may be studied. The one chosen here is a linear combination of variances Avar, Bvar corresponding to two sampling situations. We find bounds for

$$\text{Var}\{T,\alpha\} = \alpha \, \text{Avar}\{T\} + \beta \, \text{Bvar}\{T\} \ .$$

The special case $\beta = 1 - \alpha$ may be interpreted in a Bayesian model in which the entire sample comes from one of two populations. Note this does not correspond to the case of each observation coming from a mixture of two distributions.

13G1 A Lower Bound Based on Minimum Variance Estimates

Consider two distributions F_A, F_B and let T_A, T_B be minimum variance unbiased estimates for F_A, F_B respectively. Let Avar, Bvar denote variances with respect to F_A, F_B. Let T be an unbiased estimate. We establish an inequality for $\alpha \, \text{Avar}\{T\} + (1-\alpha)\text{Bvar}\{T\}$. Assume without loss of generality that $E_A\{T_A\} = E_B\{T_B\} = 0$.*
Rewriting $T = T_A + T - T_A$ yields

$$\alpha\text{Avar}\{T\} + (1-\alpha)\text{Bvar}\{T\} = E_A\{T^2\} + (1-\alpha)E_B\{T^2\} \ .$$

The first term may be rewritten

$$\alpha E_A\{T^2\} = \alpha E_A\{T_A^2 + (T-T_A)^2 + 2T_A(T-T_A)\} \ .$$

Since by assumption T_A is minimum variance,
$E_A\{T_A(T-T_A)\} = 0$ and

$$\alpha E_A\{T^2\} = \alpha E_A\{T_A^2\} + \alpha E_A\{T-T_A\}^2$$

$$= \alpha\text{Avar}\{T_A\} + \alpha E_A\{T-T_A\}^2 \ .$$

*in the section ave is denoted by E.

Hence

$$\alpha A var\{T\} + (1-\alpha)B var\{T\}$$

$$= \alpha A var\{T_A\} + (1-\alpha)B var\{T_B\} + \alpha E_A\{T-T_A\}^2 + (1-\alpha)E_B\{T-T_B\}^2 .$$

Now the last two terms may be written as integrals

$$\int (T-T_A)^2 \alpha f_A(x) \ dx + \int (T-T_B)^2 (1-\alpha)f_B(x) \ dx .$$

The combined integrand is minimized, for each x, by T_{MIN} where

$$2\alpha(T_{MIN}-T_A)f_A(x) + 2(1-\alpha)(T_{MIN}-T_B)f_B(x) = 0$$

That is, when

$$T_{MIN} = \frac{\alpha f_A(x)T_A + (1-\alpha)f_B(x)T_B}{\alpha f_A(x) + (1-\alpha)f_B(x)}$$

Thus, for each x and all α with $0 \le \alpha \le 1$, the value of T_{MIN} lies between the values of T_A and T_B. The sum of the integral is at least as large as their value when $T = T_{MIN}$, namely,

$$\int (T_B-T_A)^2 \left[\frac{(1-\alpha)^2\alpha f_B^2 f_A + (1-\alpha)\alpha^2 f_A^2 f_B}{(\alpha f_A(x) + (1-\alpha)f_B(x))^2} \right] \ dx$$

$$= \int (T_B-T_A)^2 \frac{\alpha(1-\alpha)f_A f_B}{\alpha f_A(x) + (1-\alpha)f_B(x)} \ dx$$

$$= \int (T_B-T_A)^2 \frac{1}{\frac{1}{1-\alpha}\frac{1}{f_B} + \frac{1}{\alpha f_A}} \ dx$$

$$= \int (T_B-T_A)^2 \frac{\alpha f_A(x)}{\frac{\alpha}{1-\alpha}\frac{f_A}{f_B} + 1} \ dx$$

$$= \int (T_B-T_A)^2 \left(\frac{(1-\alpha)f_B}{1 + \frac{1-\alpha}{\alpha}\frac{f_B}{f_A}} \right) \ dx .$$

Combining these results yields the following inequality:

$$\alpha A\text{var}\{T\} + (1-\alpha)B\text{var}\{T\} > \alpha A\text{var}\{T_A\} + (1-\alpha)B\text{var}\{T_B\}$$

$$+ \int (T_B - T_A)^2 \; \frac{1}{\dfrac{1}{\alpha f_A(x)} + \dfrac{1}{(1-\alpha)f_B(x)}} \; dx \; .$$

(A cruder inequality replaces this last term with

$$\int \frac{(T_B - T_A)^2}{2} \; \min\{\alpha f_A(x), (1-\alpha)f_B(x)\} \; dx \; .)$$

This bound may be evaluated for a range of α by Monte Carlo.

13G2   A Variance Bound Based on the Normal/Independent Form

If a variable $z = x/y$ where x is $N(0,1)$ and y is independently distributed, $G(y)$, the expectation of any function of z may be written $E_z\{\cdot\} = E_y E_{x/y}\{\cdot\}$. Frequently a bound can be found for this inner expectation. In particular, since the conditional distribution of x_i/y_i is $N(0, 1/y_i^2)$, the (conditional) minimum variance unbiased estimate of location has variance $1/\Sigma y_i^2$. Thus if T is an unbiased estimate of location

$$\text{var}\{T\} = E_z\{T^2\} \geq E_y\{1/\Sigma y_i^2\} \; .$$

Since the Cauchy distribution may be expressed as the ratio of two independent normal variables, it follows that, for this distribution,

$$\text{Var}\{T\} \geq E\{1/\chi_n^2\} = 1/(n-2) \; .$$

Alternatively the contaminated normal distribution

$$F(x) = (1-\alpha)\Phi(x) + \alpha\Phi(kx)$$

may be expressed as x/y where

$$y = \begin{cases} 1 & \text{with probability } 1-\alpha \\ 1/k & \text{with probability } \alpha \end{cases} \quad .$$

Hence

$$E\{1/\Sigma\, y_i^2\} = \Sigma\, b(n,r,\alpha)[1/(n-r+r/k)]$$

Any mixture of distributions may be derived by mixing the distributions of the y's. This result makes it possible to readily obtain bounds for variances in situations where one or more distributions are mixed.

NOTE: The results of this appendix could give numerical results concerning pairs of variances and depending on n. They have not yet been so used.

14. INTEGRATION FORMULAS, MORE OR LESS REPLACEMENT FOR MONTE CARLO

14A FUNDAMENTALS

14A1 The Cost of Weighting

A Gauss integration or averaging formula takes the form

$$\Sigma \, w_i \, \phi(u_i)$$

when we deal with a uniformly distributed u. The weights w_i are all positive, larger near the ends of the interval. Everyone concerned with squeezing down variability for constant computing effort is likely to point to the inequality of the w_i as a source of unnecessary variance when equal effort is applied to the assessment of each u_i

To this there are two answers:
- In most cases one can allocate effort to the evaluation of the different u_i in view of the weight to be applied (and, more importantly, in view of the effort × variance product for that i).
- Allocation to allow for varying w is unlikely to be worth the effort.

The simplest support for the second answer is the increase in variance associated with the difference in weights, which is the excess over 1 of the expression

$$\frac{\Sigma w_i^2 \Sigma 1}{\Sigma w_i \Sigma w_i}$$

which, when evaluated for Gauss averaging formulas, gives the results in Exhibit 14-1.

14A2 What Is to Be Averaged?

The naive idea is that in reducing the scope of a Monte Carlo by using an integration formula to average over one coordinate we need to integrate quite irregular functions. As we shall now see, this is usually not so.

Exhibit 14-1

LOSS (IN VARIANCE TERMS) DUE TO THE
USE OF CONSTANT SAMPLING EFFORT AT EACH POINT OF
AN n-POINT GAUSSIAN AVERAGING (INTEGRATION) SCHEME

n	loss
3	6%
5	12%
6	11%
10	16%
20	20%
40	22%

Note: The loss can of course be avoided by sampling with
intensity proportional to w_i^2.

* The independent case *

Consider (u,v) and $\phi(u,v)$ with u and v independent and easily sampled and ave$\{\phi\}$ desired.
Write

$$\phi(u,v) = \alpha_u + \beta_v + \gamma_{uv}$$

with

$$\text{ave}\{\gamma_{uv}|u\} = 0 = \text{ave}\{\gamma_{uv}|v\}$$

and hence

$$\text{cov } \alpha_u \text{ with } \gamma_{uv} = 0 = \text{cov } \beta_v \text{ with } \gamma_{uv}$$

(Note that cov α_u with β_v = 0 by independence.).
Suppose that we consider an integration formula $\Sigma_i w_i \phi(u_i)$ for averaging over u. We know (§15D1) that the variance of ave$\{\phi\}$ will be minimized by assigning different v's to each u. But let us start with the less efficient two-way case.
One way to say things is to take v_1 to v_J, to be denoted by v_j, and calculate

$$\frac{1}{J} \sum_j \sum_i w_i \phi(u_i, v_j) = \frac{1}{J} \sum_i \sum_j w_i \phi(u_i, v_j) \ .$$

If we think of this as in the left-hand form, we are integrating (by an approximate numerical formula) J times, once for each j. The functions we integrate are $\phi(u_i, v_j)$ as a function of u, whose behavior may be quite detailed.
If we think of it as on the right, however, we can write what we calculate

$$\sum_i w_i\left(\frac{1}{J} \sum_j \phi(u_i, v_j)\right) = \sum_i w_i\left(\alpha_{u_i} + \frac{1}{J} \sum_j \beta_{v_j} + \frac{1}{J} \sum \gamma_{u_i v_j}\right)$$

whose average value over the selection of the v_j is

$$\sum w_i(\alpha_{u_i} + \beta_0 + 0) = \beta_0 + \sum w_i \alpha_{u_i} \ .$$

So far as bias is concerned, we are integrating $\chi(u) = \alpha_u$
$= \text{ave}\{\phi(u,v)\}$ which is often much more nicely behaved
$\quad\;\; v$
than the individual $\phi(u,v_j)$ for v_j fixed.

If we free the v's, so that we draw J_i v_{ij}'s for
each i, we calculate

$$\sum_i w_i \alpha_{u_i} + \frac{1}{J_i} \sum \beta_{v_{ij}} + \frac{1}{J_i} \sum \gamma_{u_i v_{ij}}$$

whose average is still

$$\sum_i w_i \alpha_{u_i} + \beta_0 + 0$$

* Yet more *

Suppose u, v are not independent. Put

$$\alpha_u = \text{ave}\{\phi(u,v)\}$$
$$\quad\;\; v$$

$$\delta_{uv} = \phi(u,v) - \alpha_u \; .$$

Then

$$\text{ave}\{\delta_{uv}|u\} = 0$$

$$\phi(u,v) = \alpha_u + \delta_{uv}$$

and

$$\sum w_i \left(\frac{1}{J_i} \sum \delta_{u_i v_{ij}} \right) = \sum w_i \left(\alpha_{u_i} + \frac{1}{J_i} \delta_{u_i v_{ij}} \right)$$

whose average is

$$\sum w_i \alpha_{u_i} \; .$$

In general, then, we need our integration formulas to
work well for the functions found by averaging over the
sampled variable(s).

337

* Allocation *

So much for bias; what of variance? Let

$$\sigma_u^2 = \text{var}\{\delta_{uv}|u\} \ .$$

Let there be J_i v's for u_i, as above, drawn independently. Then

$$\text{var} \ \frac{1}{J_i} \ \Sigma \ \phi(u_i,v_{ij}) = \frac{\sigma_{u_i}^2}{J_i}$$

so that

$$\text{var} \ w_i\left(\frac{1}{J_i} \ \Sigma \phi(u_i,v_{ij})\right) = \frac{w_i^2\sigma_{u_i}^2}{J_i}$$

and

$$\text{var} \ \Sigma w_i\left(\frac{1}{J_i} \ \Sigma \phi(u_i,v_{ij})\right) = \Sigma \frac{w_i^2\sigma_{u_i}^2}{J_i} \ .$$

The usual maximization gives

$$J_i \propto w_i J_{u_i}$$

and

$$N\cdot\text{var}\left(\Sigma w_i \ \frac{1}{J_i} \ \Sigma\phi(u_i,v_{ij})\right) = \left(\Sigma \ w_i\sigma_{u_i}\right)^2 \sim (\text{ave } \sigma_u)^2 \ .$$

To do a good job of allocating the J_i we need to know something about how σ_u^2 varies with u. To do this calls for either deep insight or a preliminary experiment with at least partial two-way structure.

* Essentials *

We now see the consequences of the

$$\phi(u,v) = \alpha_u + \delta_{uv}$$

split. The bias problems of numerical integration depend only on the α's; the variance problems only on the δ's.

* Choice *

When will it pay to use numerical integration? General and precise answers clearly need more knowledge than we are likely to have, but writing out the formal comparisons may enlighten us.

The bias is a function of the contortions of α_u, and may as well be called just

$$\text{bias}\{\alpha_u\}$$

with { } to emphasize it is a functional. There is also the functional

$$\sigma_u^2 = \underset{u}{\text{var}}\{\alpha_u\}$$

Simple sampling ought to be weighted proportionally to σ_u^2 and, so far as δ_{uv} goes, ought also to yield

$$N \cdot \text{var(sample mean)} \sim (\text{ave } \sigma_u)^2 .$$

Overall it yields

$$N \cdot \text{var(sample mean)} \sim \sigma_\alpha^2 + (\text{ave } \sigma_u)^2$$

which to be compared with

$$N \cdot \text{mse(integro-sample mean)} \sim N(\text{bias}\{\alpha_u\})^2 + (\text{ave } \sigma_u)^2.$$

Thus the choice depends upon

$$N \cdot (\text{bias}\{\alpha_u\})^2 \gtrless \sigma_\alpha^2$$

339

and it pays to use the integration formula up to a cer-
sample size, namely about

$$N = \frac{\sigma_\alpha^2}{(bias\{\alpha_u\})^2} \, .$$

* More realistically *

In practice, given $\phi(u,v)$ with σ_α^2 large, we
would hope to avoid simple sampling of u. Rather we
would use stratification, antithesis, overt regression,
etc. to do a better job.
Almost any such procedure can be thought of as
writing

$$\alpha_u = \alpha_{eff,u} + \alpha_{ineff,u}$$

where the sterilized part of α, $\alpha_{ineff,u}$, has no effect
on the variability of the sampling result (though it
still contributes to the bias if we use an integration
formula). We have now to replace

$$\sigma_u^2 = var\{\alpha_u\}$$

by

$$\sigma_{eff,\alpha}^2 = var\{\alpha_{eff,u}\}$$

which can be much smaller.
The crossover value of N is, of course, equivalently reduced.
Judgment as whether to integrate or not boils, then,
to making judgments about

$$(bias\{\alpha_u\})^2$$

and

$$\sigma_{eff,\alpha}^2 \, .$$

No other losses are involved, since we do not need to
repeat v's.

If the splitting $\alpha_u = \alpha_{eff,u} + \alpha_{ineff,u}$ is by regression on well-behaved variables, we can replace

$$(bias\{\alpha_u\})^2$$

by

$$(bias\{\alpha_{eff,u}\})^2$$

when we wish.

14B EXAMPLES FOR A SINGLE COORDINATE

Essentially all Monte Carlo calculations are reasonably taken as calculations of appropriate averages. (Calculations of probabilities are averages of the corresponding indicator functions.) If the averaging is over $(u, \vec{v})$, where given u, it is easy to sample from $\vec{v}$ (in particular when u and $\vec{v}$ are independent), then we can consider replacing sampling over u (with its variance) by approximate integration over u (with its bias -- to numerical analysts, its truncation error). To this end we can use any integration formula that seems appropriate. To make good choices, we need to understand the behavior of the integration formula in question from as many points of view as possible.

In 14A2 we noticed that the integration formula is really applied to

$$\psi(u) = \underset{\vec{v}}{ave} \; \phi(u, \vec{v})$$

when averaging u. The importance of this is the usually greater smoothness of $\psi(u)$, as compared to that of $\phi(u, \vec{v}_0)$ for individual $\vec{v}_0$.

We give special attention to so-called Gauss integration formulas, both because they seem interesting and because they seem less familiar to statisticians.

14B1   Monomials and Chebyshev Polynomials

Polynomials play a large role in our thinking about a variety of questions of numerical analysis. Ancient history has had much to do with our inefficient habits of thinking of polynomials as sums of monomials

$$1 + x - \frac{1}{2}x^2 + \frac{1}{8}x^3$$

rather than as sums of Chebyshev polynomials (abbreviatable "Chebyrials")

$$\frac{3}{4}T_0(x) + \frac{35}{32}T_1(x) - \frac{1}{4}T_2(x) + \frac{1}{32}T_3(x) \ .$$

Here

$$T_0(x) \equiv 1$$

$$T_1(x) \equiv x$$

$$T_2(x) \equiv 2x^2 - 1$$

$$T_3(x) \equiv 4x^3 - 3x$$

$$\text{etc.}$$

(Note how much faster the coefficients of $T_n(x)$ tend to decrease in this example. This tends to happen for most smooth functions.) This decrease is more relevant since the bounds

$$|x^n| \leq 1, \text{ for } -1 \leq x \leq 1$$

$$|T_n(x)| \leq 1, \text{ for } -1 \leq x \leq 1$$

are both best possible.
Other useful relations are

$$T_n(x) = \cos[n(\cos^{-1}x)] \quad \text{for} \quad -1 \leq x \leq 1$$

(The companion relation is $T_n(x) = \cosh[n(\cosh^{-1}x)]$ for $|x| > 1$.) and

$$1 = T_0(x)$$

$$x = T_1(x)$$

$$x^2 = \frac{1}{2}[T_2(x) + T_0(x)]$$

$$x^3 = \frac{1}{4}[T_3(x) + 3T_1(x)]$$

$$x^4 = \frac{1}{8}[T_4(x) + 4T_2(x) + 3T_0(x)]$$

$$x^5 = \frac{1}{16}[T_5(x) + 5T_3(x) + 10T_1(x)]$$

$$x^6 = \frac{1}{32}[T_6(x) + 6T_4(x) + 15T_2(x) + 10T_0(x)] .$$

(Note that the coefficients are halves of binomial distributions with $p = 1/2$.).

We will find it helpful to do our polynomial thinking in terms of Chebyrials.

We will also find it helpful to consider $T_n(x)$ with n not necessarily integral. We speak then simply of Chebyshev functions. All Chebyshev functions ripple back and forth between $+1$ and -1, rippling faster as we approach the ends of the interval. Among them Chebyrials are distinguished by having their value of $+1$ or -1 at each end of the interval.

14B2 Gauss Quadrature

If we wish to average (or integrate) functions $\phi(u)$ on some interval, conveniently taken as $-1 \le u \le +1$ (when the average becomes half the integral) we have a variety of choices. One is to pick n points, u_1 to u_n, and n weights, w_1 to w_n, such that

$$\Sigma w_i \phi_J(u_i)$$

gives the exact average for many functions $\phi_J(u)$. Since we have n u_i and n w_i, it is plausible to try to do this for $2n$ different functions. If we succeed, the resulting formula will be exact for all linear combinations, $\Sigma c_J \phi_J(u)$, of the chosen functions.

The points and weights that do this for all polynomials of degree $\leq 2n-1$ make up what are called Gauss quadrature formulas (when adjusted to give the integrals) or Gauss averaging formulas (when adjusted to give the averages). We report next on the behavior of these formulas for $n = 10$ and $n = 20$. In actual Monte Carlo, we are likely to use more than 20 points, so if there is only a single u the numbers we look at are conservative. If there are two u's, which are likely to interact, so that 20 points for each leads to a total of 400 points, the numbers are perhaps typical. If we want to deal with three probably-interacting u's, so that 10 points for each become 1000, we may be a little pressed to do as well as the numbers that follow suggest.

We shall measure the error of averaging as a fraction of the largest value of the function averaged, taking note of where this fraction rises to (10^{-10}), 10^{-5}, and $10^{-2.5}$ -- these are mean square errors of (10^{-20}), 10^{-10} and 10^{-5} respectively. Note that simple sampling of a quantity uniformly distributed on $-1 \leq \phi \leq +1$ requires the following sample sizes to reach these particular mean square errors: (8×10^{18}), 8×10^8, and 8333. If evaluation is expensive we may not afford reaching even the last of these. Simple sampling alone is never going to reach the middle one. The first one is so precise as only to be of interest to someone not engaged in Monte Carlo.

14B3 Standard Examples

* Polynomials *

If we look at polynomials, emphasizing the Chebyrials, $T_n(u)$, we find:

1) All polynomials of degree ≤ 19 (for $n = 10$) ≤ 39 (for $n = 20$) are averaged exactly (to the precision of our arithmetic).

2) All odd polynomials (only odd powers) of any degree are averaged equally exactly.

3) For $n = 10$, $T_{20}(u)$, $T_{22}(u)$, $T_{42}(u)$, $T_{62}(u)$, $T_{64}(u)$, $T_{82}(u)$, $T_{84}(u)$, $T_{86}(u)$, and presumably so on, are averaged very poorly.

4) For $n = 20$, $T_{40}(u)$, $T_{72}(u)$, $T_{82}(u)$ and presumably so on are equally poorly averaged.

5) Intermediate $T_j(u)$, at least through $j = 100$,

are averaged rather better, though rarely as well as $10^{-2.5}$.

In judging the implications of these results, we need to remember that the coefficient of $T_{20}(u)$ is usually very much smaller than that of u^{20}, often by a factor approaching 2^{19}, and that the corresponding factor for $T_{40}(u)$ and u^{40} is 2^{39}. If polynomials approximate the function at all reasonably, then, the errors of Gauss averaging are likely to be quite small.

It is natural to try to compare the errors of Gauss averaging formulas with those of simple equally-spaced averaging formulas. For case (A) the error for any even Chebyrial $T_{2k}(u)$ is greater than $10^{-2.5}$, beginning with $k = 1$, for both 10-point and 20-point formulas.

* Chebyshev functions *

The Chebyshev functions $T_n(x)$, for n not necessarily an integer, look very much alike, waving back and forth between -1 and $+1$. The Chebyrials arising when n is an integer are only distinctive in having $T_n(\pm 1) = \pm 1$. Why should we confine our attention to the isolated polynomials?

20 equally spaced points do about $10^{2.5}$ times as bad as 20 Gauss points. The former do not stay below $10^{-2.5}$ for even $n < 1$. The latter do that well for $n < 27$.

It is probable that a statistician contemplating Monte Carlo should give much more attention to

$$\left| \frac{\text{error}}{\text{function half-width}} \right| < 10^{-2.5} \text{ for } T_s(x) \text{ with } 0 \le s \le 27$$

than to

error $\equiv 0$ for $T_n(x)$ with integer n from 0 to 39.

Looking at the Chebyshev functions leaves us with a much sounder ground than looking at Chebyrials alone.

* Exponential functions *

If $\phi(u)$ is an exponential function of maximum value 1, namely $e^{-c(u+1)}$ or $e^{+c(u-1)}$, the error of Gauss averaging rises to our thresholds at the following values of c_j and of e^{2c}, the latter being the ratio of the highest to lowest value for the interval concerned.

(relative error)	10^{-10}	10^{-5}	$10^{-2.5}$
n = 10:			
value of c:	8.0	> 8	??
value of e^{2c}:	9,000,000		
n = 20:			
value of c:	78	??	??

We see, adopting the Monte Carlo plausible value of $10^{-2.5}$, that exponentials are reasonably satisfactorily handled.

Equally spaced formulas give errors less than $10^{-2.5}$ of e^c (the maximum value of e^{-cu}) only for $c \leq$ about 1.6 (10-point) or $c \leq$ about 3.5 (20-point).

* Centered Gaussian densities *

If $\phi(u)$ has the shape of a Gaussian probability density centered at zero (we will continue to keep the largest value of $|\phi(u)|$ equal to 1), then the averaging is of high quality so long as the density is appreciably truncated. (Since the Gauss (quadrature) u_i tend to hug ± 1 and shun 0, we cannot expect to do too well for narrow functions centered at zero.) The narrownesses at which we reach the thresholds, in terms of σ for the distribution and of interval length in σ's, are:

	(10^{-10})	10^{-5}	$10^{-2.5}$
n = 10:			
value of σ:	(0.52)	0.27	0.17
length:	(3.9σ)	7.4σ	11.8σ
n = 20:			
value of σ:	(0.19)	0.125	< .10σ
length:	(10.5σ)	16σ	> 20σ

For interval lengths less than 4σ or 5σ, the equal-spaced (type A) formulas do worse -- often much worse -- than the corresponding Gauss formula. For greater interval lengths the situation is reversed. Only infrequently would we strongly prefer the equal spacing.

* Fourials *

If $\phi(u) = \sin ku = \sin c\pi u$, the Gauss average, like most quadrature formulas will be exact (zero) by symmetry. We have thus only to look at the case $\phi(u) = \cos ku = \cos c\pi u$, where integer values of c will appear in a Fourier series for the whole interval. (We shall abbreviate these special cosines, often called Fourier functions, as "Fourials".)

Contrary to more familiar quadrature formulas, the error of Gauss averaging applied to cosines is of constant sign until it becomes quite large. The threshold values of k and the numbers of Fourials below threshold as follows:

Error threshold	(10^{-10})	10^{-5}	$10^{-2.5}$
n = 10:			
value of k:	(5.2)	9.2	13
Fourials:	(1 clear)	3 clear	4 clear
n = 20:			
value of k:	(18.2)	25.5	31.0
Fourials:	(5 clear)	8 clear	9 clear

Here we see only moderately satisfactory performance for Gauss averaging, at least if we want to count Fourials.

This is particularly so since equally spaced formulas are exact for many Fourials.

All-in-all, we have to conclude that, both relatively and absolutely:

If we are to deal with relatively smooth functions, Gauss averaging is likely to do very well indeed.

14B4 Uniform Grids

The naive competitors of Gauss averaging are formulas using equally spaced values, either

(A) $\quad \dfrac{1}{n}\left[\dfrac{1}{2}\phi(-1) + \phi\left(-1 + \dfrac{2}{n}\right) + \cdots + \phi\left(1 - \dfrac{2}{n}\right) + \dfrac{1}{2}\phi(+1)\right]$

or

(B) $\quad \frac{1}{n}\left[\phi\left(-1+\frac{1}{n}\right) + \phi\left(-1+\frac{3}{n}\right) + \cdots + \phi\left(1-\frac{3}{n}\right) + \phi\left(1-\frac{1}{n}\right)\right]$

These can, if we wish, be thought of as repeated uses of (a) the "trapezoidal rule", and (b) a simple midpoint formula. It is helpful to ask how well they do for the examples already considered, and to contrast the results with those for Gauss averaging.

* Simpson's rule *

Rather better results can, of course, be obtained by repeated use of Simpson's rule. We leave details to the reader.

14B5 Heuristics

How can we conveniently think about the results we have seen? Two main remarks seem likely to lead to good judgment:

- Polynomials -- and functions with rapidly converging Taylor series -- almost inevitably wiggle more violently toward the ends of an interval. Gauss formulas, designed to be exact for many polynomials also do well with other functions that wiggle more toward the end of the interval.

- Only sines and cosines wiggle uniformly over intervals. Some functions wiggle most in the interior of the interval of interest: for example, a Gaussian density that is nearly zero anywhere near either end of the interval. For such functions, equally spaced points or, usually better, iterated Simpson's rule do better than Gauss formulas.

The authors' response to this is to expect:

- to ordinarily use Gauss formulas rather than equally-spaced or Simpson formulas;

- to be quite free in moving Monte Carlo off a single coordinate -- or a few coordinates -- by using Gauss formulas whenever this is easy.

NOTE: These techniques were not called upon in the present study.

15. MONTE CARLO FOR CONTAMINATED GAUSSIANS

15A OUTLINE

If we are to assess some average or combination of averages -- as when we assess the variance of an estimate -- over all samples from a contaminated Gaussian distribution, we are almost inevitably driven to Monte Carlo. The question is how to do the Monte Carlo effectively. This account tries to assemble the results and insights of three accompanying sections (15B, 15C, 15D) and lay out as complex a proposal as seems worthy of possible consideration.

The first step should undoubtedly be to move from samples from contaminated Gaussians to dual samples of fixed numbers from each of two fixed Gaussians. (Any sample from a contaminated Gaussian is a simple binomial mixture of such dual samples.)

One important consequence of such a step needs to be emphasized. We automatically give up configurations (naturally called "outer configurations") for the whole sample and replace them by pairs of configurations, one for each part of the dual sample (naturally called "inner configurations").

* Coordinates, two systems *

Given n_1 and n_2 with $n_1 + n_2 = n$, we can think of our data as two Gaussian samples. From each we can calculate a mean m, a sample variance s^2, and a configuration $\vec{c} = (\vec{x} - m)/s$. Hence

$$m_1, \ s_1^2, \ \vec{c}_1, \ m_2, \ s_2^2, \ \vec{c}_2$$

are six independent components with well-known distributions.

We can combine m_1 and m_2 into m_B and s_1^2 and s_2^2 into s_B^2, such that m_B and s_B^2 have the usual distributions and respond to location and scale. A suitable difference m_Δ can be made independent of m_B and $b = s_1^2/s_B^2$ is independent of s_B^2. If we choose m_B and s_B^2 properly, and take

$$m_B, \; s_B^2, \; m_\Delta, \; b, \; \vec{c}_1, \; \vec{c}_2$$

as coordinates, m_Δ, b, $\vec{c}_1$, $\vec{c}_2$ will specify a location-and-scale configuration $\vec{c}$. Since the distribution of m_B, s_B^2 is simple, the value of

$$\text{ave}\{y^2 | \vec{c}\} = A(\vec{c}) = A(m_\Delta, \; b, \; \vec{c}_1, \; \vec{c}_2)$$

where y is covariant, is easily found from the value of y for one exemplar, say

$$\{m_B, \; s_B^2, \; \vec{c}\} = \{0, \; 1, \; \vec{c}\} \; .$$

* Reasonable choices *

The main Monte Carlo problem of finding $\text{ave}\{y^2\}$ is then finding

$$\underset{m_\Delta, b}{\text{ave}} \; \underset{c_1}{\text{ave}} \; \underset{c_2}{\text{ave}} \; A(m_\Delta, \; b, \; \vec{c}_1, \; \vec{c}_2) \; .$$

On general grounds (see Section 15D) this is to be done by no-way sampling, but there is reason to separate the averaging into three independent factors because:

1) Averaging over either m_Δ or b or both can be done by a numerical integration formula if desired.

2) Averaging over a $\vec{c}_i$ of small dimension (1, 2, or 3) can be done (see 15B) with special effectiveness. Averaging over dimensions 4 to 6 requires somewhat more thought.

3) Averaging over a $\vec{c}_i$ of larger dimension (7 to 20) seems most likely to be helped by regression or polynomials (one or two well-selected coordinates). (See 15C.)

In each case, efficient calculation will be enhanced by optimum allocation of effort. Whether we can do better by averaging over $\vec{c}_1$ and $\vec{c}_2$ together, using the technique mentioned in (3) requires careful thought.

15B MORE OR LESS ANTITHETIC SAMPLING OF GAUSSIAN LOCATION AND SCALE CONFIGURATIONS IN LOW AND MODERATELY LOW DIMENSIONS: A First Exploration

Particularly when we deal with a sample from a Gaussian distribution -- either as our final sample or as an aid to generating a final sample -- we are likely to be concerned with separating samples into three parts

location,   scale,   configuration

and trying to treat these parts separately.

In the Gaussian case, integration over location and scale are often easily accomplished and the configuration is uniformly distributed over the surface of a hypersphere. We explore the problem of sampling over such a surface.

15B1 A Direct Approach

In general we face restrictions

$$\Sigma c_i = 0 \qquad \Sigma c_i^2 = C \qquad \text{(say } C{=}1 \text{ for convenience).}$$

A natural geometry to look at is the equiangular hyperplane, where $\Sigma c_i = 0$ is automatically fulfilled. This cuts the unit sphere in a hypercircle, on which the configurations are spread isotropically.

We now turn to special values of n.

$$* \ n = 1 \ *$$

The second condition falls away, and $c_1 = 0$ always.

$$* \ n = 2 \ *$$

The hypercircle consists of two points; the two solutions $\frac{+1}{\sqrt{2}}, \frac{-1}{\sqrt{2}}$ and $\frac{-1}{\sqrt{2}}, \frac{+1}{\sqrt{2}}$ are equivalent.

* n = 3 *

The hyperplane is an ordinary plane and the hyper-circle an ordinary circle. If we insist, as we may without loss of generality, that the c_i are ordered, then we have a uniform distribution on one-sixth of a circle extending between two points where the c_i appear like
```
   x                    x
x     x    and    x        x.
```
This may be parameter-ized as

$$\cos\left(\frac{\pi}{3}\,u\right),\ \cos\left(\frac{\pi}{3}\,u + \frac{2\pi}{3}\right),\ \cos\left(\frac{\pi}{3}\,u + \frac{4\pi}{3}\right)$$

with u uniform on $(0,1)$.
    Antithetic sampling for u with paired intervals is now quite feasible.

* n = 4 *

The hypercircle is now an ordinary sphere and the hyperplane is a 3-space. Adding the requirement that the c's be ordered now confines us to one-twenty-fourth of the sphere, to a spherical triangle with corners where the c_i appear like
```
          x
          x    x
          x
```
and
```
          x    x
          x    x
```
and
```
               x
          x    x    ,
               x
```
namely

$$(-a,-a,-a,3a)\quad \text{with}\quad 12a^2 = 1$$

$$(-b,-b,+b,+b)\quad \text{with}\quad 4b^2 = 1$$

$$(-3a,a,a,a)\quad \text{with}\quad 12a^2 = 1\ .$$

We could use

$$\sin \theta \; d\theta \; d\phi = d(\cos \theta) \; d\phi$$

as a surface element with any convenient pole and trans-
fer a curvilinear plane triangle into a selected twenty-
fourth, but it is not easy to know how to play antitheses
in a plane curvilinear triangle. It seems best to change
our approach.

For quite different reasons, of course, one might
like to plot, say, the difference between the conditional
average squares of two estimates over a twenty-fourth in
order to see what is going on.

15B2 A Sorting Approach

Let us begin by considering more or less antithetic
sampling in a plane triangle. Let us approach this by
drawing a large sample from the distribution in question.
We can then select 7 points in the triangle as follows
(vertices and midpoints):

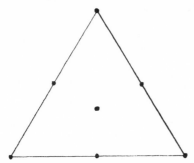

and divide the sample points into 7 piles according as
to which of the 7 points they are closest.

Two branches are now open to us:

1) We can move our boundaries so as to find equal
numbers of sample points in each of the 7 piles. Sets
of 7 points, one from each pile at random are then sen-
sibly stratified. A little matching could then make them
plausibly somewhat antithetic.

2) We can stick to our boundary and split all or a
randomly selected some of the points in each small pile
to get equal numbers -- alternatively we can delete

points in large piles to do the same, and then use sets of 7 points with weights, the weights being proportional to the reciprocal of the (average) splitting or of the fraction retained.

Further work may suggest modifications and elaborations.

* Configurations with n = 4 *

Using a similar process on the configuration sphere -- namely on the curvilinear triangle that is our selected twenty-fourth seems likely to be quite effective. The overfrequenting of the corners makes some allowance for a tendency -- that may or may not exist -- for the conditional surface to be more wiggly there.

All that we have to do is to draw samples, force them onto the sphere (by location and scale) and then onto the chosen twenty-fourth (by ordering) and find their inner products with the 10 selected unit vectors, assigning them to the pile corresponding to the largest inner product.

* Configurations with n = 5? *

Now the corresponding region is a curvifacial tetrahedron on the (hyper)spherical surface in 4-space. It seems plausible to select
    1) the four corners
    2) the centers of the four faces
    3) the center of gravity
as 9 sorting points. Whether the 6 centers of edges - or some further inner points should be added is in doubt.

* n = 6? *

One might try an analog:
    1) the five corners
    2) the ten CG's for three corners associated in a face
    3) the CG of all 5.

15C TAKING ADVANTAGE OF TWO COORDINATES
IN UNSIMPLIFIED MONTE CARLO

Consider averaging $\phi(u,v,\vec{w})$ where u and v are scalars and it is not easy to sample with u and v fixed. Suppose, however, that it is much easier to generate samples of $(u,v,\vec{w})$ than to evaluate ϕ .

If u and v are well-chosen scalar coordinates, we may be able to gain considerable precision by taking proper advantage of them. Under the assumptions, the use of antithesis or of Gauss averaging seems less than attractive. Presumably, then, we should consider stratification (of regression type) or regression, with a view to deciding by trial which to use.

15C1 The Well-behaved Case

Let us begin by proceeding as if the joint distribution of u and v as initially expressed, has neither singularities nor near singularities, leaving to later sections some suggestions for the treatment of such difficulties. We shall use specific sample sizes, like 3000, 2000, 200, etc. to fix the general ideas. Actual practice may well use rather similar values, but need not.

We draw a first sample of (u,v) -- presumably, though not necessarily, as the first coordinates of a sample of $(u,v,\vec{w})$. Let this be a sample of 3000. Order the sample on u. For any contiguous collection of u's, there is a batch of v's from which medians, quartiles or 8ths may be calculated. Here we calculate "moving" 8ths on v and smooth then by a suitable process. The result is seven curves

$$V_1(u), \ V_2(u), \ \ldots, \ V_7(u)$$

such that the 8 sets of inequalities

$$v < V_1(u)$$

$$V_1(u) \le v < V_2(u)$$

$$\text{---------}$$

$$V_6(u) \le v < V_7(u)$$

$$V_7(u) \le v$$

have crudely equal probability for (u,v) given u. We can also find the empirical 8ths of u, say U_1 to U_7. Together, the 7 "verticals", $u = U_j$, and the 7 curves, $V_i(u)$, divide the (u,v) plane into $64 = 8 \times 8$ regions, crudely of equal probability.

We can now draw a second sample of, say, 2000 $(u,v,\vec{w})$ and sort them into the 64 regions according to their values of u and v. (Since we expect about 30 per region, the chance of getting at least 10 in all cells is quite large.) We now select a stratified sample of 200 -- the working sample -- as follows:

1) From each of the 36 inner regions, take 2 (u,v)'s at random from those falling in region from the second sample.

2) From each of the 24 edge regions, take 4 (u,v)'s at random (as in (1)).

3) From each of the 4 corner regions, take 8 (u,v)'s at random.

(The total selected is $2(36) + 4(24) + 8(4) = 72 + 96 + 32 = 200$.) The weight associated with any one of the 200 $(u,v,\vec{w})$'s is proportional to the ratio

$$\frac{\text{\# entering region in the new sample}}{\text{\# selected for actual use}} \ .$$

We shall call this weight q, label the regions with running index k, and label the (u,v)'s drawn for actual use with two indices, k and h. (Clearly h goes from 1 to n_k, where n_k is 2, 4 or 8 according to the type of region.)

We now evalutate $\phi(u,v,\vec{w})$ for each of the 200 $(u,v,\vec{w})$'s in the working sample. The simple estimate

$$\frac{\Sigma\Sigma \, q_k \phi(u_{kh}, v_{kh}, \vec{w}_{kh})}{\Sigma\Sigma \, q_k}$$

where the double sums are over the 200 combinations of k and h, is then an unbiased stratified estimate of one ϕ. In view of the stratification, and the randomness of selection (both of the 2000, and of the 2, 4 or 8 from each region = each stratum), the variance of this estimate is equal to the average value of

$$\frac{\Sigma\Sigma \ q_k^2 \cdot s_k^2}{(\Sigma\Sigma \ q_k)^2}$$

where, for any k, we have

$$s_k^2 = \frac{1}{n_k-1} \ \Sigma_h \ \left[\phi(u_{kh}, v_{kh}, \vec{w}_{kh}) - \overline{\phi}_k\right]^2$$

and

$$\overline{\phi}_k = \frac{1}{n_k} \ \Sigma_h \ \phi(u_{kh}, v_{kh}, \vec{w}_{kh}) \ .$$

This stratified estimate -- and this estimate of its variance -- is a base line upon which we can try to improve.

15C2 Some Regression Alternatives

It may well be that we can do better by relatively simple regression than by stratification. What would be reasonable carriers to try?

If u and v are well (i.e., usefully) expressed, polynomials in u and v may be a reasonable choice. Rather than separating terms into main effects and inter-actions (which might make sense for very specially chosen (u,v) -- and which, regretfully, we must assess as unlikely to happen), the natural first step is to sort out the polynomials by degree. How far is it reasonable to go?

Although we have a total of 200 points, the guarantee of a firm structure rests on the 8 × 8 pattern of regions. Accordingly, it would not seem reasonable to go beyond the 7th (or perhaps the 6th) degree in u. (It may not be reasonable to go quite so far in v.) With only 64 guaranteed "distinct" observations, it ought not to be wise to fit more than 30 or 40 coefficients, at the very most. The situation is thus as in Exhibit 15-1.

We can consider going through the 6th (or possibly 7th) degree if the empirical fit seems to call for it.

To use regression, our natural approach is the following:

1) Calculate polynomial values (perhaps, after a pilot trial, for a roughly orthogonal basis) for each of the 200 (u,v)'s.

2) Calculate successive regressions of increasing degree and evaluate the estimated variance of the

Exhibit 15-1

DEGREES OF FREEDOM ABSORBED IN POLYNOMIAL REGRESSION

Degree	Degrees of Freedom	
	Added	Total
Constant	1	1
Linear	2	3
Quadratic	3	6
Cubic	4	10
4th	5	15
5th	6	21
6th	7	28
7th	8	36

weighted mean residual. Choose a regression that seems
highly effective. Take the corresponding weighted mean
residual as an estimate of the average residual. Record
also the estimated variance of this estimate.

    3) Draw a third sample of, say, 2000 (u,v)'s.

    4) Evaluate the chosen regression (or regressions)
for each of these 2000 (u,v)'s, taking their mean as an
estimate of the regression's average and a suitable multiple of their s^2 as an estimate of the variance of
this estimate.

    5) Combine the estimates and estimated variances
from (2) and (4), in view of

$$\phi \equiv \text{regression} + \text{residual}$$

and the independent estimation of the ave(regression) and
ave(residual) in (4) and, implicitly as zero, in (2).

    6) Compare with previous candidates, including
stratification.

15C3 Re-expression

    We may believe, or suspect, that u and v are not
well expressed as bases for point polynomials. It is
reasonable to try alternatives (in reasonable number) and
seize on the alternative that seems to do best.

    In some cases we will understand

$$\underset{\vec{w}}{\text{ave}} \ (u, v, \vec{w})$$

well enough to introduce square-roots, logarithms or
other standard functions as likely ways of re-expressing
u and v, either separately or together.

    Failing insight, or perhaps to supplement it, we
can try some generally hopeful re-expressions. One of
the simplest of these follows a logit (or flog) pattern
as follows:

    1) Order the 2000 values of u from the 3rd
sample.

    2) Define u' by

$$u' = \log \frac{i}{n+1-i} \quad \text{when u takes its ith value}$$

whenever there is no tie at rank i. If there is a tie,

then either (i) use the mean i for the tie or (ii) use the lowest values of each of i (numerator) and n+1-i (denominator) for the tie.

To evaluate u' for the working sample, we use linear interpolation among the paired values for the third sample.

Fixing up v can be another matter. (The technique just discussed for u → u' may be satisfactory. It may not.) A more delicate approach may be worth trying. Let g run from 1 to 7, and set x_1 = log(1/7), x_2 = log(1/3), x_3 = log(3/4), x_4 = 0, x_5 = log(4/3), x_6 = log 3, x_7 = log 7. For a suitable collection of u's, let

$$a_u + b_u \cdot V_g(u) + c_u \cdot V_g^2(u)$$

be the regression (over g) of x_g on $V_g(u)$, subject to the constraint that the regression is monotone over the range where (u,v) pairs occur. (If necessary, reduce or annul c_u.)

Now smooth $\{a_u\}$, $\{b_u\}$, and $\{c_u\}$, and then fit a reasonable expression (long polynomial in u, perhaps?) to each of $\{a_u\}$, $\{b_u\}$, and $\{c_u\}$. Let the approxima- be a(u), b(u), and c(u).

Now let

$$V" = a(u) + b(u) \cdot V + c(u) \cdot V^2 .$$

(Crudely, but only crudely, V" will have an approximately logistic distribution conditional on u.)

Other schemes can be -- and no doubt will be -- invented.

15C4 How Dangerous Is the Short Cut?

The procedure just outlined uses three large samples of either (u,v) or (u,v,$\vec{w}$). By doing so everything is kept clean. It is quite plausible to use a single large sample for all three. This introduces biases into both the estimate of ave{ϕ} and the estimate of this esti- mate's variance. (It could be worth-while to consider the covariance of estimate for the two averages: regression and residual.)

While some trials may be illuminating, it seems likely to be relatively safe to short-cut in this way, especially if the covariance is allowed for.

15C5 Dealing with Singularities -- Example

A plausible choice of two scalar coordinates for Gaussian configurations offers us a chance to illustrate one technique for dealing with singularities in the distribution of (u,v). Suppose $\{c_i\}$ is the result of drawing a sample $\{y_i\}$ from a unit Gaussian and then coding it linearly to attain $\Sigma c_i = 0$, $\Sigma c_i^2 = 1$.

Two coordinates plausibly relevant to such configurations are, for configurations of roughly 20 values each:

1) $u = \Sigma c_i^4$

2) a measure of central spread, for example (assuming the c's ordered)

$$v = (c_{13} - c_8) + (c_{12} - c_9) + (c_{11} - c_{10}) \, .$$

Both u and v have finite maximum and minimum values, the minimum of v being zero and the other three positive. It will help our progress to ask what patterns of c_i give these extremes. Since the analysis is easier for even numbers, we shall treat only the even case here.

The configuration

for which $c_i = 0$ and $c_i^2 = 2na^2$, whence $a = 1/\sqrt{2n}$, has all c_i^2 equal. Hence it minimizes $u = \Sigma c_i^4 = \Sigma (c_i^2)^2$ given $\Sigma c_i^2 = 1$. At the same time it makes $v = 3(2a) = 6a$.

361

It is relatively easy to see that this is the maximum
that v can attain, since it is only when the other 7
and 7 oberservations coincide with the central 3 and 3
observations that the 3 differences can have so large a
sum.

Consider next the configuration

$$2n-1 \left\{ \begin{array}{c} \vdots \\ \vdots \\ \vdots \\ \vdots \end{array} \right. \qquad \cdot$$

$$-b \qquad (2n-1)b$$

for which $\Sigma c_i = 0$ and $\Sigma c_i^2 = \{(2n-1)^2 + (2n-1)\}b^2$ so

that $b^2 = 1/(2n-1)(2n)$ and

$$u = \Sigma c_i^4 = \{(2n-1) + (2n-1)^4\}b^4$$

is close to unity. This is the unique maximum of u,
and, since v = 0, a minimum of v. This minimum is non-
unique, since we can fix, for 2n = 20, for example,
$c_8 = c_9 = c_{10} = c_{11} = c_{12} = c_{13}$ and then fix the other

14 to satisfy $\Sigma c_i = 0$, $\Sigma c_i^2 = 0$. Roughly, then, v will
attain its minimum, 0, on a 14 - 2 = 12 dimensional sub-
space.

The region of the (u,v) plane realized thus takes
the form shown in Exhibit 15-2.

We find two "singularities", at (u_{min}, v_{max}) and
$(u_{max}, 0)$, respectively. We would like to "spread out"
v at each of these points. A natural spreading should
generate level curves all of which run into both singularities. The level curves, it would seem, should also
have different shapes as they enter either singularity.

A simple way to attain this is to consider as a new
coordinate a sum of two slopes

$$v^* = \frac{v-0}{u_{max}-u} + \frac{v - v_{max}}{u - u_{min}} \quad .$$

The level curve $v^* = A$ corresponds to

Exhibit 15-2

THE REGION REALIZED IN (u,v) SPACE

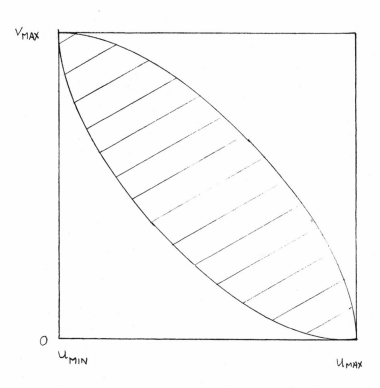

$$(v-0)(u-u_{min}) + (v-v_{max})(u_{max}-u) = A(u_{max}-u)(u-u_{min})$$

which reduces to

$$-u_{min} \cdot v + v \cdot u_{max} + v_{max} \cdot u - v_{max} u_{max} = A(u_{max}-u)(u-u_{min})$$

which shows that each level curve can be given as v equal to a quadratic in u.

(Recoding to $(0,1)$ for both variables reduces the equation to

$$v + u - 1 = A(1-u)u$$

or

$$v = +1 - u + A(1-u)u$$

which may be easier to consider.)

Having aquired a new v^* -- and a new u^* by a similar process with u and v interchanged-- it is now open to us to work with
1) polynomials in u and v (as always)
2) polynomials in u^* and v
3) polynomials in u and v^*
4) polynomials in u^* and v^* .

It may well be reasonable to try all four of these, and then use the one that seems to do the most for us in lowering the estimated variance of estimate.

15D MISCELLANEOUS NOTES ON MONTE CARLO

The sections that follow are brief explorations into a variety of relatively simple questions in the use of Monte Carlo. They were stimulated by the problem of Monte Carlo for contaminated normal distributions, but are not at all confined to that case.

15D1 On Taking Disadvantage of Independent Components in Simple Experimental Sampling

Suppose $x = (u,v)$ is so distributed that u and v are independent, and that we are to assess $\text{ave}\{\phi(x)\}$ by sampling. Suppose further that we can sample u separately, and simply, v separately and simply, or u, v together and simply. Which is it desirable to do?

* Comparison of variability *

Let

$$\phi(x) = \alpha_u + \beta_v + \gamma_{uv}$$

where

$$\text{ave}\{\gamma_{uv}\} = 0, \ \text{ave}\{\gamma_{uv}|u\} = 0 = \text{ave}\{\gamma_{uv}|v\}$$

$$\text{cov}\{\alpha_u \ \text{with} \ \gamma_{uv}\} = 0$$

$$\text{cov}\{\beta_v \ \text{with} \ \gamma_{uv}\} = 0$$

and put $\sigma_R^2 = \text{var}\{\alpha_u\}$, $\sigma_C^2 = \text{var}\{\beta_v\}$, $\sigma_I^2 = \text{var}\{\gamma_{uv}\}$ then, for a total of rc values:
- no-way sampling (u,v sampled independently, rc pairs)

$$\text{var} = \frac{1}{r}\left(\sigma_R^2 + \sigma_C^2 + \sigma_I^2\right)$$

- one-way sampling (r values of u; c values of v for each u)

$$\text{var} = \frac{1}{r}\sigma_R^2 + \frac{1}{rc}\left(\sigma_C^2 + \sigma_I^2\right)$$

or, interchanging rows and columns,

$$\text{var} = \frac{1}{c}\sigma_R^2 + \frac{1}{rc}\left(\sigma_R^2 + \sigma_I^2\right)$$

- two-way sampling (r values of u; c values of v; all rc combinations)

$$\text{var} = \frac{1}{r}\sigma_R^2 + \frac{1}{c}\sigma_C^2 + \frac{1}{rc}\sigma_I^2 .$$

Clearly "no-way sampling", that is sampling (u,v) together and simply is the best of the four.
If we consider the two-way optimized for rc = constant, we easily find

$$\frac{1}{r} = \frac{\sigma_C^2}{\lambda - \sigma_I^2} \quad \text{and} \quad \frac{1}{c} = \frac{\sigma_R^2}{\lambda - \sigma_I^2}$$

whence

$$\lambda - \sigma_I^2 = \sqrt{rc \ \sigma_C^2 \sigma_R^2}$$

$$\frac{1}{r} = \frac{1}{\sqrt{rc}} \ \sqrt{\sigma_C^2/\sigma_R^2} \quad \text{and} \quad \frac{1}{c} = \frac{1}{\sqrt{rc}} \ \sqrt{\sigma_R^2/\sigma_C^2}$$

so that

$$\text{var} = \frac{2}{\sqrt{rc}} \ \sqrt{\sigma_R^2 \sigma_C^2} + \frac{1}{rc} \sigma_I^2$$

which now decreases as $1/\sqrt{rc}$.

* Structured sampling? *

There can be advantages to two-way sampling, but if they are in variance terms they must come from an ability to do NONsimple analysis for one or both of u or v, for example, stratification; antithetic or postantithetic sampling on both u and v. (Regression, including post-stratification, for separate u or v can always be used for (u,v) together. Stratification, antithetic sampling, or postantithetic sampling on either u or v alone can always be done for (u,v) together.) It is probably necessary to gain by doing something special for both before structured sampling has to be preferred to no-way sampling.

* An important comment *

If the average dependence on u -- and on v -- is reasonable smooth and simple enough to approximate well, the use of structured samples need not be as disadvantageous as we might think. Starting with structured samples and guiding ourselves to good forms of regression means that, as we take more points, the effective sizes of σ_R^2 and σ_C^2, which will refer mainly to residuals from an ideal regression, will decrease. As a result the term

$$\frac{2}{\sqrt{rc}} \ \sqrt{\sigma_R^2 \sigma_C^2} = \frac{2}{\sqrt{n}} \ \sqrt{\sigma_R^2 \sigma_C^2}$$

will decrease faster than $1/\sqrt{n}$. Whether it will decrease like $1/n$, at least for reasonable n, or not,
is unclear.

15D2 Possibilities and Difficulties in Postregression (the Scalar Case)

Suppose that we have a rather good "black box" to fit
a smooth $q(\lambda)$ to pairs of values (λ, ϕ). Suppose that
we wish to assess $ave\{\phi(u)\}$ where u is cheap but $\phi(u)$
is expensive and where $\lambda(u)$ is a plausible carrier.
Then we could consider the path to be described. (Numbers
like 10, 100, 1000 and $100/10 = 10$ are purely illustra-
tive.)
Let us begin with 10 sets each of 10 u's, so drawn
that each can serve as a subsample. Calculate $\phi(u)$ for
all 10. Calculate 1000 or 10,000 u's and the corresponding $\lambda(u)$'s for reference. Then repeat the following all
10 possible ways.
Leave out one set of 10 u's. Apply the black box to
the $90(\lambda(u), \phi(u))$'s to find a corresponding $q(\lambda)$. Use
the 1000 or 10,000 $\lambda(u)$'s to estimate $ave\{q(\lambda(u))\}$, use
this estimate in a regression estimate of $ave\{\phi(u)\}$
based on the 10 set aside (λ, ϕ) pairs. (Quite likely
to use coefficient 1 in this regression is good enough
and probably better.)
After repeating 10 times (with a total of 10 different $q(\lambda)$'s) the results can then be combined into a single
estimate of $ave\{\phi(u)\}$.
The quality of the result clearly depends on the
quality of $q(\lambda)$, particularly on its extrapolability.
This may mean that weighted sampling, with low weights
for extreme $q(\lambda(u))$, may be well worth while.
It also means that good black boxes are especially
valuable.
If we use weighted sampling, hopefully with low
weights at the extremes of $q(\lambda)$, we should neglect the
weights in fitting $q(\lambda)$, but use them, as we must, in
estimating averages.

15D3 Some Techniques in Postantithetic Sampling

Suppose (1) that it is cheap to sample u's (at
unspecified complexity), (2) that it is felt that antithetic sampling according to $\lambda(u)$ would be helpful, and
(3) that it is not easy to do this directly. What might
we do?

* Postantithetic grouping *

If we wish, we can calculate a batch of u's, order
them by λ(u), and then select from them antithetically.
We can then use a random one or more of the selected
groups for our (presumably noncheap) calculation.

* Doing something smoothish *

Since we are working with ordered values of a sample
of λ(u), our gaps will be exponential. As a result of
their nonconstancy, our antitheses will not be as good as
we might desire. Calculating still more u's, thus making
the ratio of total size to group size larger is some help.
An alternate approach may be of interest.

Let us draw a quite large sample of u's, and order
by λ(u). Suppose the total number, N, to be divisible
by 10. Do the following:

1) Pick an integer from 1 to 10, uniformly at
random.

2) Retain the lowest 10 and the highest 10 of the
original N, each with weight 1/10.

3) Retain every 10th one of the remaining N-20,
starting with the kth where k was picked in step 1.
(These all have unit weight.)

4) Group antithetically from those that remain.

5) Select one or more groups equiprobably at random.

A possible virtue of this scheme over simple antithetic selection is that we stabilize spacings without
being forced to use close together values in some possible
groups.

Another possible virtue is its preservation of end
values with low weights.

NOTE: These techniques, developed in connection with the
present study, were not used in it.

REFERENCES

Sections

Abramowitz, M. and Stegun, J. A. (1964).
Handbook of Mathematical Functions. N.B.S.,
Washington.

2C3d   Barnard, G. A., Jenkins, G. M. & Winston, C. B.
(1962).   Likelihood inference and time series.
J.R.Statist. Soc. A 125, 321-352.

1, 3B1   Bickel, P. J. (1965).   On some robust estimates
of location.   Ann. Math. Stat. 36, 847-858.

2C2   Bickel, P. J. (1971).   A note on approximate
(M) estimates.   Submitted for publication.

2E2, 3E5   Bickel, P. J. and Hodges, J. L. (1967).   The
asymptotic theory of Galton's test and a
related simple estimate of location.
Ann. Math. Stat. 38, 73-89.

Birnbaum, A. and Laska, E. (1967).   Optimal
robustness: a general method with applications
to linear estimates of location.   J. Am. Stat.
Assoc. 62, 1230-1240.

4G   Box, G. E. P. and Muller, M. A. (1959).   A note
on the generation of random normal deviates.
Ann. Math. Stat. 29, 610-613.

7A7   Chambers, J. (1971).   Algorithm 410, Partial
Sorting [M1].   Comm. ACM 14, 357-8.

3E2   Chernoff, H. (1964).   Estimation of the mode.
Ann. Inst. Stat. Math. 16, 31-41.

2B   Chernoff, H., Gastwirth, J., and Johns, M. V.
(1967).   Asymptotic distributions of linear
combinations of functions of order statistics
with application to estimation.   Ann. Math.
Stat. 38, 52-72.

369

Sections

        Crow, E. and Siddiqui, M. (1967). Robust estimation of location. J. Am. Stat. Assoc. 62, 353-389.

4C   Dixon, W. and Tukey, J. (1968). Approximate behavior of the distribution of Winsorized t (Trimming/Winsorization 2) Technometrics 10, 83-98.

2E3   Ferguson, T. S. (1967). Mathematical Statistics - a Decision Theoretic Approach. Academic Press.

3A   Filippova, A. A. (1962). Mises' theorem on the asymptotic behavior of functionals of empirical distribution functions and its statistical applications. Theory of Probability and its Applications 7, 24-57.

1   Filliben, J. J. (1969). Simple and robust linear estimation of the location parameter of a symmetric distribution. Ph.D. Thesis, Princeton University.

4C   Fraser, D. A. S. (1968). The Structure of Inference, John Wiley and Sons, New York.

1, 2B2, Gastwirth, J. (1966). On robust procedures.
3B2   J. Am. Stat. Assoc. 61, 929-948.

1, 5B   Gastwirth, J. and Cohen, M. (1970). Small sample behavior of some robust linear estimates of location. J. Am. Stat. Assoc. 65, 946-73.

1, 3B1, Hampel, F. (1968). Contributions to the theory
5E, 5F of robust estimation. Ph.D. Thesis, Berkeley.

3A, 5F Hampel, F. (1971). A qualitative definition of robustness. To appear in Ann. Math. Stat.

Sections

6B5   Harris, T. E. and Tukey, J. W. (1949). Development of large-sample measures of location and scale which are relatively insensitive to contamination. Memorandum Report 31, Statistical Research Group, Princeton University.

4E   Hoaglin, D. (1971). Optimal invariant estimation of location for three distributions and the invariant efficiency of some other estimators. Ph.D. Thesis, Princeton University.

2E2   Hodges, J. L. (1967). Efficiency in normal samples and tolerance of extreme values for some estimates of location. Proc. 5th Berkeley Symp. Math. Stat. and Prob. 1, 163-186.

2E2   Hodges, J. L. and Lehmann, E. L. (1963). Estimates of location based on rank tests. Ann. Math. Stat. 34, 598-611.

2E1, 3E3   Hogg, R. (1967). Some observations on robust estimation. J. Am. Stat. Assoc. 62, 1179-1186.

3E3   Hogg, R. (1969). Personal communication.

1, 2C1   Huber, P. J. (1964). Robust estimation of a location parameter. Ann. Math. Stat. 35, 73-101.

2B3   Jaeckel, L. B. (1971). Some flexible estimates of location. To appear in Ann. Math. Stat.

6   Johnson, N. L., Nixon, E., Amos, D. E. and Pearson, E. S. (1963). Table of percentage points of Pearson curves for given $\sqrt{\beta_1}$ and β_2 expressed in standard measure. Biometrika 50, 459-495.

4G   Knuth, D. E. (1969). Fundamental Algorithms. Addison-Wesley, Reading, Massachusetts.

REFERENCES (cont.)

Sections

2C2 LeCam, L. (1956). On the asymptotic theory of estimation and testing hypotheses. <u>Proc. IVth Berkeley Symp. Math. and Prob.</u> 129-157.

1 Leone, F., Jayachanchan, T. and Eisenstat, S. (1967). A study of robust estimators. <u>Tech.</u> 9, 652-660.

4G Lewis, P. A. W., Goodman, A. S. and Miller, J. M. (1969). <u>IBM Systems Journal</u> 8, 136-146.

4G Marsaglia, G. (1968). Random numbers fall mainly in the planes. <u>Proc. of the National Academy of Sciences</u> 61, 25-28.

 Merrington, M. and Pearson, E. S. (1958). An approximation to the distribution of non-central t. <u>Biometrika</u> 45, 484-491.

4G Moses, L. E. and Oakford, R. F. (1963). <u>Tables of Random Permutations</u>. Allen & Unwin, London.

2B2 Mosteller, F. (1947). On some useful "inefficient" statistics. <u>Ann. Math. Stat.</u> 17, 377-408.

2C2 Neyman, J. (1950). Contributions to the theory of the X^2 test. <u>Proc. Ist. Berkeley Symp. Math. Stat. Prob.</u>

 Pearson, E. S. (1963). Some problems arising in approximating to probability distributions using moments. <u>Biometrika</u> 50, 95-111.

 Pearson, E. S. (1954-1966). <u>Biometrika Tables for Statisticians</u>, Vol. I, Cambridge, The University Press (1st ed. 1954, 3rd ed. 1966).

Sections

Pearson, E. S. and Tukey, J. W. (1965). Approximate means and standard deviations based on distances between percentage points of frequency curves. Biometrika 52, 533-546.

4E   Pitman, E. J. G. (1939). The estimation of the location and scale parameters of a continuous population of any given form. Biometrika 30, 391-421.

2B   Sarhan, A. and Greenberg, B. (1962). Contributions to Order Statistics. John Wiley and Sons, New York.

1, 3E4   Takeuchi, K. (1969). A uniformly asymptotically efficient robust estimator of location. NYU, Courant Institute Technical Report IMM 375.

Tukey, J. W. (1970). Exploratory Data Analysis. (Limited Preliminary Edition) Addison-Wesley, Reading, Massachusetts.

3E2   Venter, J. H. (1967). On estimation of the mode. Ann. Math. Stat. 38, 1446-1455.

3A   Von Mises, R. (1947). On the asymptotic distribution of differentiable statistical functions. Ann. Math. Stat. 18, 309-348.